Electric Energy

An Introduction

Second Edition

POWER ELECTRONICS AND APPLICATIONS SERIES

Muhammad H. Rashid, Series Editor
University of West Florida

PUBLISHED TITLES

Advanced DC/DC Converters
Fang Lin Luo and Hong Ye

Alternative Energy Systems: Design and Analysis with Induction Generators, Second Edition
M. Godoy Simões and Felix A. Farret

Complex Behavior of Switching Power Converters
Chi Kong Tse

DSP-Based Electromechanical Motion Control
Hamid A.Toliyat and Steven Campbell

Electric Energy: An Introduction, Second Edition
Mohamed A. El-Sharkawi

Electrical Machine Analysis Using Finite Elements
Nicola Bianchi

Modern Electric, Hybrid Electric, and Fuel Cell Vehicles: Fundamentals, Theory, and Design
Mehrdad Eshani, Yimin Gao, Sebastien E. Gay, and Ali Emadi

Uninterruptible Power Supplies and Active Filters
Ali Emadi, Abdolhosein Nasiri, and Stoyan B. Bekiarov

Electric Energy

An Introduction

Second Edition

Mohamed A. El-Sharkawi

CRC Press
Taylor & Francis Group
Boca Raton   London   New York

CRC Press is an imprint of the
Taylor & Francis Group, an **informa** business

CRC Press
Taylor & Francis Group
6000 Broken Sound Parkway NW, Suite 300
Boca Raton, FL 33487-2742

© 2009 by Taylor & Francis Group, LLC
CRC Press is an imprint of Taylor & Francis Group, an Informa business

Library of Congress Cataloging-in-Publication Data

El-Sharkawi, Mohamed A.
    Electric energy : an introduction / by Mohamed A. El-Sharkawi. -- 2nd ed.
        p. cm. -- (Power electronics and applications ; no. 9)
    Includes index.
    ISBN-13: 978-1-4200-6219-9
    ISBN-10: 1-4200-6219-0
    1. Power electronics. I. Title. II. Series.

TK7881.15.E525 2008
621.3--dc22   2008008269

Visit the Taylor & Francis Web site at
http://www.taylorandfrancis.com

and the CRC Press Web site at
http://www.crcpress.com

Dedication

*This book is dedicated to my wife Fatma
and my sons Adam and Tamer.
The book is also dedicated to all engineers,
without whom we would still be living in caves.
A special dedication goes to the
fathers of our electric power systems:
Nikola Tesla and Thomas Edison.*

Contents

Preface

The first course on electric energy in engineering schools is traditionally taught as an energy conversion course. This was justified during most of the last century, as power was the main topic of electrical engineering. Nowadays, the field includes a number of new specializations such as digital systems, computer engineering, communications, imaging, and networks. With the field being so widespread, energy conversion turned into a topic for students specializing only in power. Consequently, a large number of schools have decided to move their energy conversion classes from the core curricula to the elective curricula. Other schools with limited resources have dropped the energy conversion courses altogether.

In recent years, renewed interest in electric energy has emerged due to several reasons. Among them are the ongoing search for renewable energy, the societal impact of blackouts, the environmental impact of generating electricity, and the lack of knowledge by most electrical engineers in fundamental subjects such as electric safety and generation of electricity. In addition, the new ABET criteria in the United States encourages the development of curricula that underline broad education in engineering, contemporary engineering, and the impact of engineering solutions in a global and societal context. All these requirements can be met by restructuring the introductory electric energy course and making it relevant to all electrical engineering and most mechanical engineering students. This is the main objective of the book.

The book is authored to assist schools that wish to a course with a wider view of the key aspects of electric energy while maintaining high levels of modeling and analysis. Most of the topics in this book are related to issues encountered daily and, therefore, should be of great interest to all engineering students. Most of the chapters in the book are structured to be standalone topics, so instructors can pick and choose the chapters they want to teach. The instructors can also select the sequence in which they prefer to teach the chapters. Most of the examples in the book are from real systems, with real data to make the course relevant to all students.

Among the topics covered in this second edition are energy resources, renewable energy, power plants and their environmental impacts, electric safety, power quality, power market, blackouts, and future power systems. In addition, the topics of electromechanical conversion, transformers, power electronics, power quality, and three-phase systems are included in the book to address the needs of some school for teaching these important topics. However, these traditional topics are written to be more relevant to all students. For example, the sections on electric motors include linear and levitated motors as well as stepper motors.

Chapter 1 is dedicated to the history of developing the electric energy system. The innovative milestones started with the Greek philosopher Thales of Miletus around 600 BC, passing through William Gilbert (1544–1603), Alessandro Guiseppe Antonio Anastasio Volta (1745–1827), André-Marie Ampère (1775–1836), Georg Simon Ohm (1789–1854), Michael Faraday (1791–1867), Hippolyte Pixii (1808–1835), Antonio Pacinotti (1841–1912), Thomas Alva Edison (1847–1931), and Nikola Tesla (1856–1943). This second edition includes key inventors in power electronics such as John Ambrose Fleming (1849–1945) and Julius Edgar Lilienfeld (1881–1963). The chapter addresses the battle between the direct current (dc) and alternate current (ac) systems that were created by Thomas Edison and Nikola Tesla. It analyzes the merits and demerits of each system.

In Chapter 2, the modern power network is described by three distinct systems: generation, transmission, and distribution. Each one is identified and discussed in terms of its overall operation. At the consumer voltage levels, the chapter addresses the frequency and voltage standards worldwide and gives reasons for the differences among the various standards. This chapter is expanded in this second edition and it includes information on control centers.

Chapter 3 deals with energy resources, which are often divided into three categories: fossil fuel, nuclear fuel, and renewable resources. Fossil fuels include oil, coal, and natural gas. The renewable energy resources include hydroelectric, wind, solar, hydrogen, biomass, tidal, and geothermal. All these resources are also classified as primary and secondary resources. The primary resources include fossil fuel, nuclear fuel, and hydroelectric energy. The secondary resources include all renewable energy minus hydroelectricity. The chapter discusses the known reserves as well as the world consumptions of the primary resources. In this second edition, all data are updated with 2006 statistics.

In Chapter 4, primary resource power plants are studied. This includes the description of the various designs, the main components of the power plants, the theory of operation, and their modeling and analyses. The chapter has several examples of key calculations for the hydroelectric, fossil fuel, and nuclear plants.

Chapter 5 deals with the environmental impact of generating electricity. Each of the primary resources is associated with some air, water, and land pollutions. Although there is no pollution-free method for generating electricity, the primary resources are often accused of being responsible for most of the negative environmental impacts. In this chapter, the various pollution problems associated with the generation of electricity are presented, discussed, and put into perspective.

In Chapter 6, the renewable energy methods are presented. Renewable energy is a loosely used phrase to describe energy generated in any manner from resources other than fossil or nuclear fuels. They include hydroelectric, wind, solar, wave and tidal, geothermal, and hydrogen. The chapter describes each of these technologies, the main components of each system, the merits and demerits of each technology, the mathematical modeling, and the energy calculations of key systems. In this second edition, several enhancements are made. These include detailed models for the photovoltaic (PV) cells and modules, descriptions of how various fuel cells operate, hydrogen economy, modeling of fuel cells, and modeling of tidal energy systems.

Chapter 7 is a review of the ac circuit analysis that is required for subsequent chapters. Usually, the topics in this chapter are covered in earlier courses. However, for completion, they are also included in this book.

Chapter 8 discusses the three-phase systems. In additional to mathematical analysis, the chapter discusses the reasons for using the three-phase systems and the various connections in power systems.

Chapter 9 deals with electric safety, a topic often overlooked. Students will learn what constitutes harmful level of currents and their biological effects. They will also learn about primary and secondary shocks and how they are affected by various variables such as currents, voltages, frequencies, and pathways. Body and ground resistance calculations are studied as well as their effects on the step and touch potentials. The impact of ground resistance on the level of electric shock is explained by various common safety scenarios. Various world safety standards for neutral and ground are discussed and justifications for three-prong plugs and outlets are made. Furthermore, safety equipment for household applications such as the ground fault interrupter is presented. In this edition, stray voltage is explained and its hazard is analyzed. The world's practices with regard to neutral and ground wires are explained and their safety is analyzed. Also, magnetic field and its biological effects are discussed.

Chapter 10 deals with various power electronic devices and circuits. Because of the power electronic revolution, the boundary between dc and ac is becoming invisible. Besides the main solid-state devices such as the silicon-controlled rectifier (SCR), insulated gate bipolar transistor (IGBT), and metal oxide semiconductor field effect transistor (MOSFET), the chapter analyzes the four types of converters: ac/dc, dc/dc, dc/ac, and ac/ac. For each of these converters, at least one circuit is discussed and analyzed in detail using waveforms and mathematical models. In this edition, the discussion of the ac/ac converter is expanded.

Chapter 11 is an introduction to transformers, and includes single- and three-phase transformers as well as autotranformers and transformer banks. Detailed analyses for ideal and actual transformers are given.

Chapter 12 discusses the essential electric machines such as the induction motor, synchronous machines, and the stepper motor. All these machines are addressed in the context of their applications to make the machine theory more relevant to all students. For example, linear and magnetically levitated trains are used for applications of the induction machine, and the hard disk drive and printer are associated with stepper motors. Besides the basic machine modeling, students can analyze application scenarios such as a train powered by a linear induction motor traveling under certain road and wind conditions. In this edition, we have added sections on dc motors and single-phase motors.

Chapter 13 is a new chapter in the second edition. It deals with the imperfection of the ac waveform. It includes analyses of voltage problems such as flicker and sag, and frequency problems. Besides the analyses, the chapter identifies the impact of imperfect waveforms on other system components as well as methods to alleviate the problems associated with power quality.

Chapter 14 deals with the operation of the power system. The chapter describes secure and insecure operations as well as stable and unstable conditions of the power system. The difference between radial and network structures as related to power system reliability is analyzed. The effect of a power imbalance on the power system stability is studied. The concepts of power angle, system capacity, and transmission limits are introduced. Energy demand and the energy trade between utilities are addressed as an economical necessity as well as a reliability obligation. The concept of spinning reserve and its impact on system reliability is given. The various conditions that could lead to outages and blackouts are addressed, and three major blackout scenarios in the United States are described.

Chapter 15, which is new for this edition, deals with the future directions in power system research. It includes several topics such as the smart grid and space power systems.

Acronyms and Abbreviations

ac	Alternating current
BJT	Bipolar junction transistor
BTU	British thermal unit
BWR	Boiling water reactor
CGS	Centimeter–gram–second
CO_x	Carbon oxide
Cp	Coefficient of performance
dc	Direct current
Diac	Bilateral diode
DMFC	Direct methanol fuel cell
DT	Darlington transistor
EF	Electric field
EMF	Electromagnetic field
EPA	Environmental Protection Agency
FACTS	Flexible alternating current transmission system
FC	Fuel cell
FET	Field effect transistor
GFCI	Ground fault circuit interrupter
GPR	Ground potential rise
GTO	Gate turn-off SCR
HSM	Hybrid stepper motor
IEEE	Institute of Electrical and Electronics Engineers
IGBT	Insulated gate bipolar transistor
IHD	Individual harmonic distortion
IM	Induction motor
ISCCS	Integrated Solar Combined Cycle System
KE	Kinetic energy
LIM	Linear induction motor
Maglev	Magnetically levitated linear induction motor
MCFC	Molten carbonate fuel cell
MF	Magnetic field
MOSFET	Metal oxide semiconductor field effect transistor
MPE	Maximum permissible exposure
NAAQS	National ambient air quality standards
NASA	National Aeronautics and Space Administration
NERC	North American Electric Reliability Council
NO_x	Nitrogen oxide
OHGW	Overhead ground wire
OSHA	Occupational Safety and Health Administration
P	Real power
PE	Potential energy
PEM	Proton exchange membrane
pH	Potential of hydrogen
PMSM	Permanent magnet stepper motor
Pu	Plutonium

PV	Photovoltaic
Q	Reactive power (in ac circuits)/Thermal Energy (in power plants)
REM	Radiation equivalent man
rms	Root mean square
rpm	Revolutions per minute
S	Complex power, apparent power
SCR	Silicon-controlled rectifier
SIDAC	Silicon diode for alternating current
SO_x	Sulfur oxide
SOFC	Solid oxide fuel cell
TEC	Thermal energy constant
THD	Total harmonic distortion
TL	Transmission line
TSR	Tip speed ratio
U	Uranium
UCTE	Union for the Co-ordination of Transmission of Electricity
UPS	Uninterruptible power supply
VF	Voltage fluctuation
VRSM	Variable reluctance stepper motor
WHO	World Health Organization
WLIM	Wheeled linear induction motor
xfm	Transformer

1 History of Power Systems

A large number of great scientists created wonderful innovations that led to the electric power systems as we know it today. Although electricity was discovered around 600 BC, it was not until the end of the nineteenth century that we could have electric energy on demand by flipping a switch at every home, school, office, or factory. Today, dependence on electric power is so entrenched in our societies that we cannot imagine our life without electricity. We indeed take electricity for granted, so when we experience an outage, we realize how our life is dependent on electricity. This dependency creates a formidable challenge to engineers to make the power system the most reliable and efficient complex system ever built by man.

The road to creating an electric power system began when the Greek philosopher Thales of Miletus discussed electric charge around 600 BC. The Greeks observed that when rubbing fur on amber, electric charge would build up on the amber. The charge would allow the amber to attract light objects such as hair. However, the first scientific study of the electric and magnetic phenomena was done by the English physician William Gilbert (1544–1603). He was the first to use the term electric, which is a derivation from the Greek word for amber ($\eta\lambda\varepsilon\kappa\tau\rho o\nu$). The word amber itself was derived from the Arabic word *Anbar*. Indeed, several volumes are needed to justifiably credit the geniuses behind the creation of our marvelous power system. Unfortunately, because of the lack of space, we shall only highlight the milestone developments.

A good starting spot in history would be the middle of the eighteenth century when the Italian scientist Alessandro Guiseppe Antonio Anastasio Volta (1745–1827) showed that galvanism occurred whenever a moist substance was placed between two different metals. This discovery eventually led to the first battery in 1800. Figure 1.1 shows a model of Volta's battery that is displayed in the U.S. Smithsonian Museum in Washington, DC. The battery was the source of energy used in subsequent developments to create magnetic fields and electric currents. Today, we use the unit "volt" for electric potential in honor of this great Italian inventor.

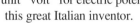

When the Danish Professor Hans Christian Oersted (1777–1851) was working on an experiment that involved the use of battery, he noticed that a compass needle has deflected from its normal heading of the magnetic north. This was the first reported discovery of the electromagnetic force created by the electric current. This discovery is the basis for the design of electromechanical devices such as motors and actuators. This important relationship between electricity and magnetism was not interpreted by Oersted, but was later explained by André-Marie Ampère. In honor

FIGURE 1.1   Model of Volta's battery. (Image courtesy of the Smithsonian Museum.)

of Oersted, his name is used as a unit for the magnetic field intensity in the centimeter–gram–second (CGS) system.

The French mathematician and physicist André-Marie Ampère (1775–1836) was the first to explain the ambiguous link between magnetism and electric currents that Oersted could not rationalize. His work is known as Ampere's law that relates the magnetic field to the electric current. Ampere's law eventually led to the development of electro-magnetic devices such as motors, generators, and transformers. Today, we use ampere as the unit of electric current in honor of this French scientist.

The German scientist Georg Simon Ohm (1789–1854) was the first to discover the electrical resistance. He noticed that the current passing through a wire increased when the cross section of the wire increased, and was reduced when the length of the wire increased. By this discovery, Ohm explained the fundamental relationship between the electric current, the electromotive force, and the resistance. His theory is known as *Ohm's law*. Although simple, the theory opened the door for all electrical circuit analysis and designs. We use ohm as the unit for resistance (or impedance) in his honor.

The next genius is the English chemist and physicist Michael Faraday (1791–1867). He was a superb matriculate experimentalist who laid the foundations for the electromechanical theories that we use today. Faraday, through experimentation, demonstrated the relationships between the electric current and the magnetic field. In 1821, Faraday built a device that produced a continuous circular motion that was the basis for the first alternating current (ac) motor. Faraday published his electromagnetic work in three volumes entitled *Experimental Researches in Electricity*, from 1839 to 1855. These classical books have been reprinted several times and are still available in specialized bookstores. Faraday's work inspired the Scottish physicist James Clerk Maxwell (1831–1879) to develop the phasor relationships of the electric current, magnetic field, and mechanical force, which are known as *Maxwell's equations*. The unit of capacitance, farad, is named after Michael Faraday.

The next inventor is probably the least known, but had enormous impact on power systems. Hippolyte Pixii (1808–1835) was a French instrument maker who built the first generator (or dynamo). His machine consisted of a magnet rotated by a hand crank, and surrounded by a coil. Pixii found that the rotation of the magnet induced voltage across the coil. Pixii's machine, named by him as magnetoelectric, was later developed into the electrical generator we use today. Figure 1.2 shows one of the early models of Pixii's generators.

FIGURE 1.2 (See color insert following page 300.)   Pixii's generator. (Image courtesy of the Smithsonian Museum.)

FIGURE 1.3   Gaulard and Gibbs transformer. (Image courtesy of the Smithsonian Museum.)

The Italian inventor Antonio Pacinotti (1841–1912) invented a device that had two sets of windings wrapped around a common core. The windings were electrically isolated from the core and from one another. When an alternating voltage was applied across one winding, an induced voltage was observed across the second winding. The induced voltage had similar shape to the applied voltage but with different magnitude. This was the early invention of the transformer we use today. Westinghouse further developed the transformer and had several early models, among them are the Gaulard and Gibbs transformer developed in 1883 (Figure 1.3), and the Stanley transformer developed in 1886.

John Ambrose Fleming (1849–1945) was an English electrical engineer and physicist. In 1904, he invented the first electronic device that consisted of two electrodes inside a vacuum tube. One electrode was a heated filament (called cathode) that emitted electrons and the other (called anode) collected them. This way, the vacuum tube allowed the electrons to pass only in one direction, thus creating the first diode. His invention inspired others to create different types of vacuum tubes, Figure 1.4. One of them is the American scientist Lee DeForest who wrapped a thin grid of wires around the cathode. By applying a small negative voltage to the grid he managed to control the amount of electrons reaching the anode, thus controlling the electrical current. The vacuum tubes were then called valves as they act as a water valve where the flow of current can be reduced to zero by applying

FIGURE 1.4 Vacuum tube.

high enough negative voltage to the grid. This is the basic principle behind amplifiers. So it is not surprising that the vacuum tubes led to the inventions of radio, television, and phone communications. The current ratings of these vacuum tubes were made high enough to allow their use in power applications. For example, the first generations of direct current (dc) lines were based on the vacuum tube technology. The vacuum tubes were used in all electronic equipment for decades until solid-state technology materialized in the 1950s.

 Julius Edgar Lilienfeld (1881–1963) was born in Lemberg in Ukraine. He did his graduate studies in Germany and obtained his habilitation in 1910. He became a professor of physics at the University of Leipzig in Germany. Lilienfeld immigrated to the United States in 1927. In 1926, he discovered the field effect principles of the transistor, which was first disclosed in a U.S. patent in 1930. Later, he filed several other patents that describe the construction and operation of layered semiconductor materials. Although there is no evidence that Lilienfeld built a working transistor based on his discoveries, his patents describe many key theories for the bipolar transistors. The first practical bipolar junction transistor was later developed in 1947 by Bell Laboratories' physicists Walter Houser Brattain (1902–1987), John Bardeen (1908–1991), and William Bradford Shockley (1910–1989). These scientists shared a Nobel Prize in physics for their transistor in 1956.

The invention of the transistor opened the door wide to a range of semiconductor inventions such as the field effect transistor in 1952 and the metal oxide–semiconductor field effect transistor in 1963. In 1957, General Electric developed the silicon-controlled rectifier, which is one of the most important power electronic devices.

The semiconductor inventions led to what is known as the power electronics revolution. Because of these devices, we can now build electric equipment with new capabilities and better performance while drastically reducing their sizes, costs, and losses. The power capabilities of these devices range from the milliwatt to the gigawatt making them used in low power applications such as household chargers to high power applications such as transmission line converters. Today, it is almost impossible to find electric equipment without a power electronic device.

Most of the above marvelous innovations inspired two great scientists who are considered by many to be the fathers of our electric power systems: Thomas Alva Edison and Nikola Tesla. Before we discuss their work, let us first summarize their biographies.

1.1 THOMAS A. EDISON (1847–1931)

Thomas Alva Edison was born on February 11, 1847, in Milan, Ohio, and died in 1931 at the age of 84. Early in his life, Edison showed brilliance in mathematics and science, and was a very resourceful inventor with a wealth of ideas and knowledge. In addition to his engineering brilliance, Edison was also a businessman, although not highly successful by today's standard. During his career, he had a large number of patents issued under his name, a total of 1093 patents. His first patent was granted at the age of 21 in 1868, and his last one was at the age of 83. This makes him one of the most productive inventors ever with an average of about 1.5 patents per month. Of course, most of his patents were due to the teamwork with his research assistants.

His first patent was on an electric voting machine that tallied voting results fairly quickly. He tried to convince the Massachusetts legislatures to adapt his machine instead of the tedious manual process used at that time, but did not succeed. The politicians did not like his invention because the slow process at that time gave them a chance to influence, and perhaps change, the voting results.

From 1862 to 1868, Edison worked as a telegrapher in several places including New England, which was the Silicon Valley of that time; all the new and interesting innovations were made in New England. During this period, he invented a telegraphic repeater, duplex telegraph, and message printer. However, Edison could not financially gain from these inventions, so he decided to move to New York City in 1869.

While in New York City, the Wall Street stock trade was disrupted by a failure in the stock ticker machine. Coincidentally, Edison was in the area and managed to fix the machine. This landed him a high-paying job that allowed him to continue with his innovations during his off-working hours.

In 1874 at the age of 27, Edison opened his first research and development laboratory in Newark, New Jersey. In 1876, he moved the facility to a bigger laboratory at Menlo Park, New Jersey. These facilities were among the finest research and development laboratories in the world. Great scientists worked in his laboratories, such as Tesla, Lewis Howard Latimer (an African-American who worked on carbon filaments), Jonas Aylsworth (who pioneered plastics), Reginald Fessenden (a radio pioneer), William K.L. Dickson (movie developments), Walter Miller (sound recording), John Kruesi (phonograph), and Francis R. Upton (dynamo). At that time, the innovations and productivities of Edison's laboratories were unmatched by any other research facility in the world. By today's jargons, these laboratories were the core of the Silicon Valley of the nineteenth century. Because of his success with various innovations, Edison expanded his research and development laboratory in 1887, and moved it to West Orange, New Jersey. There, he created the Edison General Electric Company, which in 1892 became the famous General Electric Corporation.

The large number of Edison's inventions includes the silent movie, alkaline storage battery, electrical generator, ore separator, printer for stock ticker, printer for telegraph, electric locomotive, motion picture camera, loudspeakers, telephone, microphone, mimeograph, and the telegraph repeater.

Edison was a highly competitive scientist who was willing and able to compete with great scientists in several areas at the same time. Among his innovations is the carbon transmitter, which ultimately was developed into the telephone. However, Alexander Graham Bell was also working on the same invention, and he managed to beat Edison to the telephone patent in 1876. This specific event upset Edison tremendously, but did not slow him down. He worked on the phonograph, which was one of Bell's projects, but this time Edison patented the phonograph in 1878 before Bell could apply for his own patent.

One of Edison's most important inventions was the electric lightbulb. This simple device is probably one of his everlasting inventions. The first incandescent lightbulb was tested in Edison's laboratory in 1878 (see Figure 1.5). Although the filament of the bulb burned up quickly, Edison was very esthetic about the commercial prospects of his invention. Several developments were later made to produce filaments with longer lifetimes.

Before the invention of the electric lightbulb, New York streets were illuminated by oil lamps. So when Edison invented his bulb, the technology was resisted by Rockefeller, who was an oil tycoon. But soon after realizing that oil would be used to generate electricity, Rockefeller accepted this new technology and was one among its promoters.

Edison received the U.S. Congressional Gold Medal for career achievements in 1928. When Edison died in 1931, people worldwide dimmed their lights in honor of his achievements. Initially it was suggested that the proper action would be to turn off the electricity completely for a few minutes, but the suggestion was quickly ruled out because of its impracticality; power systems worldwide would have to shut down and restart, which was considered to be an inconceivable task.

Today, a number of museums around the world have sections dedicated to Edison and his inventions. Among the best in the United States are the Edison National Historic Sites in West Orange, New Jersey; and Menlo Park in Edison, New Jersey. Also, the National Museum of American History has a section dedicated to Edison.

FIGURE 1.5   Edison's lightbulbs. (Image courtesy of the Smithsonian Museum.)

1.2 NIKOLA TESLA (1856–1943)

Nikola Tesla was born in Smiljan, Croatia, on July 9, 1856. Tesla was a gifted person with an extraordinary memory and highly analytical mind. He had his formal education at the Polytechnic School of Gratz, Austria, and the University of Prague in the areas of mathematics, physics, and mechanics. In addition to his strength in science and mathematics, Tesla spoke six languages.

During his career, Tesla had more than 800 patents. Although lesser in number than Edison's

record, his patents were the direct result of his own work. Furthermore, unlike Edison, Tesla was broke most of his life and could not afford the cost of patent applications. Actually, his most resourceful era was his last 30 years, but because of his poor financial condition, he managed to apply for a very few patents.

Tesla moved from Europe to the United States in 1884 and worked for Edison in his laboratory as a research assistant. This was during the time Edison patented the lightbulb and was looking for a way to bring electricity to homes, offices, and factories. He needed a reliable system to distribute electricity and Tesla was hired to help in this area.

Among Tesla's inventions are the ac motor shown in Figure 1.6, the hydroelectric generator, radio, x-rays, vacuum tubes, fluorescent lights, microwaves, radar, the Tesla coil, the automobile ignition system, speedometer, and electronic microscope.

In the opinion of many scientists, historians, and engineers, Tesla invented the modern world. This is because he developed the concept of the power system we use today. By today's business standards, Tesla should have been among the richest in the world. But because of his unfortunate business failures, Tesla was broke when he died at the age of 86 on July 7, 1943.

Among the honors given to Tesla was the IEEE Edison medal in 1917, the most coveted prize in electrical engineering in the United States. Tesla was inducted into the Inventor's Hall of Fame in 1975. He received honorary degrees from institutions of higher learning including Columbia University, Yale University, University of Belgrade, and the University of Zagreb. He also received the Elliott Cresson medal of the Franklin Institute. In 1956, "tesla" was adopted as the unit of magnetic flux density in the meter-kilo-second-amp (MKSA) system in his honor. In 1975, the IEEE Power Engineering Society established the Nikola Tesla award for outstanding contributions in the field of electric power in his honor.

FIGURE 1.6 Tesla's ac motor. (Image courtesy of the Smithsonian Museum.)

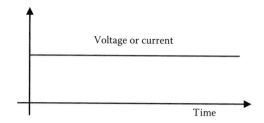

FIGURE 1.7   dc Waveform.

Among the best museums for Tesla are the Nikola Tesla Museum in Belgrade and the Nikola Tesla Museum of Science and Industry in Colorado Springs, Colorado.

1.3   BATTLE OF AC VERSUS DC

Edison continued to develop the lightbulb and made it reliable enough for commercial use by the end of the nineteenth century. Edison powered his bulbs by dc sources that were commonly used at that time. The voltage and current of a dc source are unchanging with time; an example is shown in Figure 1.7. The battery is an excellent example of a dc source.

The people in the northeast of the United States were very eager to use this marvelous electrical bulb in their homes because, unlike oil lamps, electrical bulbs did not emit horrible fumes. Edison, who was also a businessman, decided to build a dc power system and sell electricity to these eager customers. He erected poles and extended wires (conductors) above the city streets. In September 1882, his Pearl Street plant in lower Manhattan started operation as the world's first commercial electric lighting power station. One of the most significant events in technological history took place in 1883 when Edison electrified the city of Brockton, Massachusetts. The center of the city was the first place on earth to be fully electrified by a central power station.

Edison's dc generators were low voltage (100 V) machines. As a result of the increasing demand for electricity, the wires carried larger currents causing larger voltage drops across the wires, thus reducing the voltage at the customers' sites. To understand these relationships, consider the simple representation of Edison's system shown in Figure 1.8. The voltage across the load, V_{load}, can be expressed by the following equation:

$$V_{\text{load}} = V_{\text{s}} - IR_{\text{wire}}$$
$$V_{\text{load}} = IR = \frac{V_{\text{s}}}{R + R_{\text{wire}}} R = \frac{V_{\text{s}}}{1 + \left(\frac{R_{\text{wire}}}{R}\right)} \tag{1.1}$$

where
V_{s} is the voltage of the dc source (100 V)
R is the load resistance (large number of bulbs for large number of homes)
R_{wire} is the resistance of the wire (conductor) connecting the source to the load

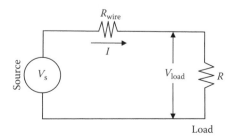

FIGURE 1.8   Simple representation of Edison's dc system.

Because more customers mean more load resistances are added in parallel, the total load resistance R decreases when more customers are added to the system. As you can see from Equation 1.1, higher loads (i.e., more customers and smaller R) lead to lower voltage at the customers' end of the line V_{load}.

EXAMPLE 1.1

Assume $V_s = 100$ V, $R = 1 \, \Omega$, and $R_{\text{wire}} = 0.5 \, \Omega$. Compute the following:

(a) Voltage at the load side
(b) Percentage of the load voltage with respect to the source voltage
(c) Energy consumed by the load during a 10 h period
(d) Maximum load (minimum resistance) if the load voltage cannot be reduced by more than 10% of the source voltage
(e) Energy consumed by the new load during a 10 h period

Solution

(a) By direct substitution in Equation 1.1, we can compute the voltage at the load side

$$V_{\text{load}} = \frac{V_s}{1 + \left(\frac{R_{\text{wire}}}{R}\right)} = \frac{100}{1 + \left(\frac{0.5}{1}\right)} = 66.67 \text{ V}$$

This low voltage at the load side may not be high enough to light the lightbulbs.

(b) $V_{\text{load}}/V_s = 66.67\%$
(c) Energy is the power multiplied by time. Hence, the energy consumed by the load is

$$E = Pt = \frac{V_{\text{load}}^2}{R} t = \frac{66.67^2}{1} 10 = 44.444 \text{ kW h}$$

(d) If the minimum load voltage is 90% of the source voltage, the load resistance can be computed from Equation 1.1 as

$$\frac{V_{\text{load}}}{V_s} = \frac{1}{1 + \left(\frac{R_{\text{wire}}}{R}\right)}$$

$$0.9 = \frac{1}{1 + \left(\frac{0.5}{R}\right)}$$

$$R = 4.5 \, \Omega$$

(e) Notice that the new load resistance is 4.5 times the original load resistance. This means that fewer customers are connected and less energy is consumed by the load.

$$E = Pt = \frac{V_{\text{load}}^2}{R} t = \frac{(0.9 \times 100)^2}{4.5} 10 = 18.0 \text{ kW h}$$

This is less energy than the one computed in step c.

As seen in Example 1.1, the load voltage, V_{load}, is reduced when more customers are added. The main reason is because the conductor resistance, R_{wire}, and the load current, I, cause a voltage drop across the conductor, thus reducing the voltage across the load resistance as given in Equation 1.1. The wire resistance increases when the length of the wire increases, or the cross section of the wire decreases as given in the following equation:

$$R_{\text{wire}} = \rho \frac{l}{A} \tag{1.2}$$

where

ρ is the resistivity of the conductor in ohm meter that is a function of the conductor's material

A is the cross section of the wire

l is the length of the wire

EXAMPLE 1.2

Assume $V_s = 100$ V and the load resistance $R = 1\ \Omega$. Compute the length of the wire that would not result in load voltage reduction by more than 10%. Assume the wire is made of copper and the diameter of its cross section is 3 cm.

Solution

From Equation 1.1, we can compute the wire resistance

$$\frac{V_{\text{load}}}{V_s} = \frac{1}{1 + \left(\frac{R_{\text{wire}}}{R}\right)}$$

$$0.9 = \frac{1}{1 + \left(\frac{R_{\text{wire}}}{1}\right)}$$

$$R_{\text{wire}} = 0.111\ \Omega$$

From the table of units in Appendix C, the resistivity of copper is $1.673 \times 10^{-8}\ \Omega$ m. Hence, the length of the wire can be computed by using Equation 1.2.

$$l = \frac{AR_{\text{wire}}}{\rho} = \frac{(\pi r^2)R_{\text{wire}}}{\rho} = \frac{\left(\pi \times (0.015)^2\right)0.111}{1.673 \times 10^{-8}} = 4.69\ \text{km}$$

A wire with length longer than 4.69 km will result in lower voltage across the load than needed.

As seen in Equation 1.2 and Example 1.2, long wires cause low voltage across the load. Because of this reason, the limit of Edison's system was about 3 mi, beyond which the wire resistance would be high enough to render the lightbulbs useless. To address this problem, Edison considered the following three solutions:

1. Reduce the wire resistance by increasing its cross section. This solution was expensive because
   (a) Conductors with bigger cross sections are more expensive.
   (b) Bigger conductors are heavier and would require higher poles to be placed at shorter spans.
2. Have several wires feeding areas with high demands. This was also an expensive solution as it would require adding more wires for long miles.
3. Place electrical generators at every neighborhood. This was also an impractical and expensive solution.

Edison was faced with the first real technical challenge in power system planning and operation. He looked at all possible solutions from anyone who could help. That was probably why he hired Tesla in his laboratory and promised him lucrative bonuses if he could solve this problem.

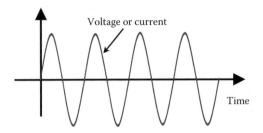

FIGURE 1.9   ac Waveform.

Tesla knew the problem was related to the low voltage (100 V) Edison was using in his dc system. The viable alternative was to increase the supply voltage so the current could be reduced, and consequently the voltage drop across the wire could be reduced as well. Since the power consumed by the load is the multiplication of voltage by current (Equation 1.3), an increase in the source voltage results in a decrease in current. If the current decreases, the voltage drop across the wire, IR_{wire}, is reduced, and the load voltage, $V_{load} = V_s - IR_{wire}$, becomes a higher percentage of the source voltage. Although logical, adjusting the voltage of the dc systems was unknown technology at that time.

$$P = VI \tag{1.3}$$

Instead of the dc system, Tesla promoted a totally different concept. Earlier in his life, Tesla conceived the idea of the ac whose waveform is shown in Figure 1.9. The ac is a sinusoidal waveform changing from positive to negative and back to positive in one cycle. Tesla wanted to use the ac to produce a rotating magnetic field that can spin motors (see Chapter 12). While working at the Continental Edison Company in Paris, Tesla actually built the first motor running on ac. The machine was later named the induction motor (see Chapter 12).

Tesla was also aware of the device invented by Pacinotti in 1860 that could adjust the ac voltages. This device was further developed by Tesla and became the transformer we use today (see Chapter 11). The transformer requires alternating magnetic field to couple its two separate windings. Therefore, it is designed for ac waveforms and is ineffective in dc systems.

Tesla molded these various technologies into a new design for the power system based on the ac (Figure 1.10). The voltage source of the proposed system was ac at generally low voltage. Transformer 1 was used to step up (increase) the voltage at the wire side of the transformer to a high level. The high voltage of the wire resulted in lower current inside the wire. Thus, the voltage drop across the wire, IR_{wire}, was reduced. At the customers' sites, the high voltage of the wire was stepped down (reduced) by Transformer 2 to a safe level for home use.

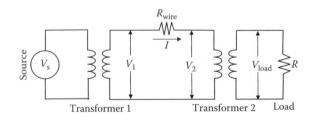

FIGURE 1.10   Tesla's proposed system.

Tesla stated three main advantages for his ac system:

1. Can be transmitted over long distances with little voltage drop across its wires.
2. No need to install generating plants at residential areas as suggested by Edison for his dc system.
3. Can create rotating magnetic fields for electric motors.

Edison was not impressed by Tesla's ac system because of its unsafe high voltage wires that would pass through residential areas. However, most historians believe that Edison's rejection to the ac system was because he had too much money invested in the dc infrastructure. This disagreement created an irreparable rift between these two great scientists. Tesla and Edison continued to disagree on various other issues, and Tesla left Edison's laboratory and eventually worked for Westinghouse.

The dc versus ac battle started between Edison and General Electric on one side, and Tesla and Westinghouse on the other side. Edison used unconventional methods to convince the public that Tesla's high voltage ac system was too dangerous. He made live demonstrations where he used high voltage to deliberately electrocute animals such as puppies, cats, horses, and even elephants. Edison went so far as to convince the state of New York to use an electric chair powered by high voltage ac system to execute condemned inmates on the death row. Most historians believe that his real motive was to further tarnish the safety of the ac system. Sure enough, New York City executed its condemned inmates by the proposed electric chair, and Edison captured the occasion by referring to the execution process as "Westinghoused," a clear reference to the ac system promoted by Tesla and Westinghouse. When Tesla signed a contract with Westinghouse to erect ac wires, posters were put on the power poles to warn residents from the danger of being Westinghoused. In addition to the negative publicity, Westinghouse ran into financial troubles and Tesla's contract with Westinghouse was terminated before the ac system was fully implemented. Instead of becoming the world's richest man, Tesla died broke.

Despite all the negative publicity created by Edison, the advantages of the ac systems were overwhelming. In fact, Edison later admitted that ac is superior to dc for power distribution, and was quick to exploit the economical advantages of the ac system through several inventions. Once when Edison met George Stanley, the son of the William Stanley who developed the transformer for Westinghouse, he told him "tell your father I was wrong."

Tesla was considered by many historians to be the inventor of the new world. This is probably true since social scientists use energy consumption as a measure of the advancements of modern societies.

1.4 TODAY'S POWER SYSTEMS

The great scientists mentioned in this chapter, and many more, have provided us with the power system we know today. The power engineers made use of the various innovations to deliver electricity safely and reliably to each home, commercial building, and factory. Indeed, it is hard to imagine our life without electricity.

In 2006, the world generated over 19×10^{12} kW h of electric energy; the data for a 12-year period are shown in Figure 1.11. As you see in the figure, in 10 year, from 1996 to 2006, the world generated 40% more electric energy. This awesome amount of energy is the main reason for the highly developed societies we enjoy today. To deliver this amount of power to the consumers, power engineers constructed several millions of transmission lines, power plants, and transformers. It is overwhelming to fathom how such a massive system is controlled and operated on continuous basis.

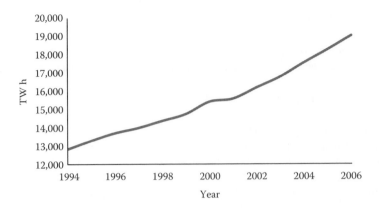

National Aeronautics and Space Administration (NASA) often publishes a fascinating panoramic photo of the earth at night, one of them is shown in Figure 1.12. The photo shows the earth's human-made lights, which comprises of hundreds of pictures made by the orbiting Defense Meteorological Satellites Program. As you see in Figure 1.12, man-made lights make it quite possible to identify the borders of cities and countries. Note that the developed or populated areas of the earth, such as Europe, eastern United States, and Japan, are quite bright reflecting their high consumption of electric energy. Also, for countries with high populations around rivers, such as Egypt, you can trace the path of the rivers by following the lights trail. The dark areas, such as in Africa, Australia, Asia, and South America, reflect low consumption of electricity or unpopulated regions. In North Africa, you can identify the great Sahara desert by the dark and expanded region. Similarly, you can identify the unpopulated outback region in Australia.

FIGURE 1.12   Earth at night. (Image courtesy of NASA.)

EXERCISES

1. In your opinion, identify 10 of the most important innovations in electrical engineering, and name the inventors of these innovations.
2. Thomas Alva Edison has several innovative inventions to his credit. Select one of them and write an essay on the history of its development and the impact of the invention on the society.
3. Nikola Tesla has made several innovative inventions. Select one of them and write an essay on the history of its development and the impact of the invention on the society.
4. Transformer is one of the major inventions in power systems. Why cannot we use it in dc systems?
5. State the advantages and disadvantages of using low voltage transmission lines.
6. State the advantages and disadvantages of using high voltage transmission lines.
7. In your opinion, what are the major developments to foresee in future power systems?
8. Simple power system consists of a dc generator connected to a load center via a transmission line. The load resistance is 10 Ω. The transmission line is 50 km copper wire of 3 cm in diameter. If the voltage at the generator terminals is 400 V, compute the following:
   (a) Voltage across the load
   (b) Voltage drop across the line
   (c) Line losses
   (d) System efficiency
9. Simple power system consists of a dc generator connected to a load center via a transmission line. The load power is 100 kW. The transmission line is 100 km copper wire of 3 cm in diameter. If the voltage at the load side is 400 V, compute the following:
   (a) Voltage drop across the line (V_{line})
   (b) Voltage at the source side (V_{source})
   (c) Percentage of the voltage drop (V_{line}/V_{source})
   (d) Line losses
   (e) Power delivered by the source
   (f) System efficiency
10. Repeat the previous problem assuming that the transmission line voltage at the load side is 10 kV.
11. Identify 10 household appliances that use power electronic devices and circuits. Discuss the advantages of using power electronics in two of these appliances.

2 Basic Components of Power Systems

Modern power networks are made up of three distinct systems: generation, transmission, and distribution. Figure 2.1 shows a sketch of a typical power system. The generation system includes the main parts of the power plants such as turbines and generators. The energy resources used to generate electricity in most power plants are combustible, nuclear, or hydropower. The burning of fossil fuels or a nuclear reaction generates heat that is converted into mechanical motion by the thermal turbines. In hydroelectric systems, the flow of water through the turbine converts the kinetic energy of the water into rotating mechanical energy. These turbines rotate the electromechanical generators that convert the mechanical energy into electric energy.

The generated electricity is transmitted to all customers by a complex network of transmission systems composed mainly of transmission lines, transformers, and protective equipments. The transmission lines are the links between power plants and load centers. Transformers are used to increase (step up) or decrease (step down) the voltage. At the power plant, a transmission substation with step-up transformers increases the voltage of the transmission lines to very high values (220–1200 kV). This is done to reduce the current through the transmission lines, thus reducing the cross section of the transmission wires and consequently reducing the overall cost of the transmission system. At the load centers, the voltage of the transmission lines is reduced by step-down transformers to lower values (15–25 kV) for the distribution of power within city limits. At the consumer sites, the voltage is further reduced to values from 100 to 240 V for household use depending on the standards of the country.

The power system is extensively monitored and controlled. It has several levels of protections to minimize the effect of any damaged component on the system ability to provide safe and reliable electricity to all customers. A number of key devices and equipment used in power systems are covered in detail in this chapter, and a summary of their functions and operations is given below.

2.1 POWER PLANTS

At the power plant, energy resources such as coal, oil, gas, hydropower, or nuclear power are converted into electricity. The main parts of the power plant are the burner (in fossil plants), the reactor (in nuclear power plants), the dam (in a hydroelectric plant), the turbine, and the generator. Power plants can be huge in size and capacity. For example, the Itaipu hydroelectric power plant located at the Brazilian–Paraguayan border produces 75 TW h annually (Figure 2.2). China is building the largest nuclear power plant in the southern province of Guangdong with a capacity of 6 GW.

Although power plants are enormous in mass, they are delicately controlled. A slight imbalance between the input power of the turbine and the output electric power of the generator may cause a blackout unless rapidly corrected. To avoid blackouts, the massive amount of water or steam inside the plant must be tightly controlled at all times, which is an enormous challenge to mechanical and structural engineers.

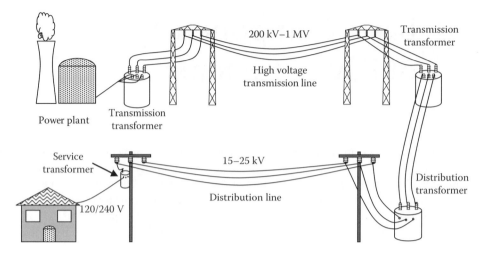

FIGURE 2.1   Main components of power systems.

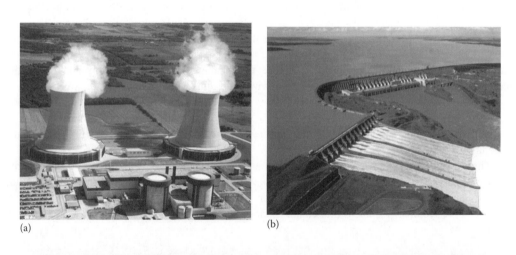

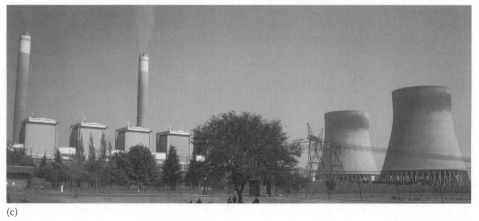

FIGURE 2.2 (See color insert following page 300.)   Power plants. (a) Byron nuclear power plant. (Image courtesy of the U.S. Department of Energy.) (b) Itaipu hydroelectric power plant. (c) Coal-fired thermal power plant.

FIGURE 2.3   Model of a hydroelectric turbine.

2.1.1 TURBINES

The function of the turbine is to rotate the electrical generator by converting the thermal energy of the steam or the kinetic energy of the water into rotating mechanical energy. There are two types of turbines: thermal and hydroelectric. Figure 2.3 shows a model of a hydroelectric turbine. It consists of blades mounted on a rotating shaft and curved to capture the maximum kinetic energy from the water. The angle of the blade can be adjusted in some types of turbines for better control on its output mechanical power.

In thermal power plants, fossil fuels or nuclear reactions are used to produce steam at high temperatures and pressures. The steam is passed through the blades of the thermal turbine, and causes the turbine to rotate. The steam flow is controlled by several valves at critical locations to ensure that the turbine is rotating at a precise speed.

A typical hydroelectric power plant consists of a dam that holds the water upstream at high elevations with respect to the turbine. The difference in height between the water surface behind the dam and the turbine blades is called head. The larger the head, the more potential energy is stored in the water behind the dam. When electricity is needed, the water is allowed to pass to the turbine blades through pipes called penstocks. The turbine then rotates and the valve of the penstock regulates the flow of water, thus controlling the speed of the turbine.

Since the generator is mounted on the shaft of the turbine, the generator rotates with the turbine and electricity is generated. To ensure that the voltage of the generator is at a constant frequency, the turbine must run at a precise and constant speed. This is not a simple task as we shall see later in Chapter 14.

2.1.2 GENERATORS

The generator used in all power plants is the synchronous machine (see Chapter 12). The invention of the synchronous generator goes back to Hippolyte Pixii (1808–1835) who was the first to build a dynamo. The synchronous machine has a magnetic field circuit mounted on its rotor and is firmly connected to the turbine. The stationary part of the generator, called a stator, has windings wrapped around the core of the stator. When the turbine rotates, the magnetic field moves inside the machine in a circular motion. As explained by Faraday, the relative speed between the stator windings and the magnetic field induces voltage across the stator windings. When an electrical load is connected across the stator windings, the load is energized as explained by Ohm's law.

The output voltage of the generator (5–22 kV) is not high enough for the efficient transmission of power. Higher voltage generators are not practical to build as they require more insulation, making the generator unrealistically large in size. Instead, the output voltage of the generators is increased by using step-up transformers.

2.2 TRANSFORMERS

The main function of the transformer is to increase (step up) or decrease (step down) the voltage (see Chapter 11). As explained in Chapter 1, the voltage of the transmission line must be high enough to reduce the current in the transmission line. When electric power is delivered to the load centers, the voltage is stepped down for safer distribution over city streets as seen in Figure 2.1. When the power reaches customers' homes, the voltage is further stepped down to the household level of 100–240 V depending on the various standards worldwide.

2.3 HIGH VOLTAGE TRANSMISSION LINES

Electrical lines (conductors) deliver electric power from the generating plant to customers as shown in Figure 2.1. The bulk power of the generating plant is transmitted to load centers over long distance lines called high voltage transmission lines. The lines that distribute the power within city limits are called medium voltage distribution lines. There are several other categories such as subtransmission and high voltage distribution lines. But these are fine distinctions that we should not worry about at this stage.

The transmission lines are high voltage wires (220–1200 kV) mounted on tall towers to prevent them from touching the ground, humans, animals, buildings, or equipment. High voltage towers are normally 25–45 m in height; one of these towers is shown in Figure 2.4. The higher the voltage of the wire, the taller is the tower.

High voltage towers are normally made of galvanized steel to achieve the strength and durability needed in harsh environments. Since the steel is electrically conductive, the high voltage wires cannot be attached directly to the steel tower. Instead, insulators made of nonconductive material mounted on the tower are used to hold the conductors away from the tower structure. The insulators

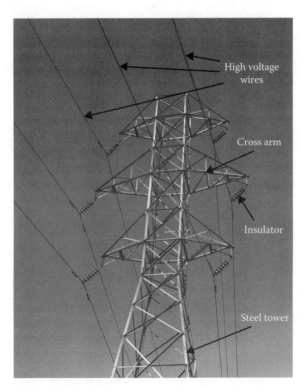

FIGURE 2.4 High voltage transmission tower.

withstand the static and dynamic forces exerted on the conductor during windstorms, freezing rain, or earth movements. Insulators come in various shapes and designs; one of them is shown in Figure 2.5a. It consists of mounting ends and insulated central rod with several disk-shaped insulating materials. In some types of insulators, the disk has a slightly conical shape where the top diameter is slightly smaller than the bottom diameter as shown in Figure 2.5b. The top end of the central rod is attached to the tower and the lower end is attached to the conductor. The disks have two main functions:

1. Increase the flashover distance between the tower and the conductor (called creepage distance).
2. When rain falls, the insulators are automatically cleaned since the disks are conical shaped or are slightly bent toward the earth.

An electrical discharge between any two metals having different potentials depends on the potential difference and the separation between the two metals. If an insulator is placed between two metals, a flashover may occur along the surface of the insulator if the potential difference between the two metals is high enough, or the length of the insulator is small enough. To protect against the flashover in high voltage transmission lines, the insulators must be unrealistically long. However, if a disk-shaped insulator is used, the distance between the tower and the conductor over the surface of the insulator is more than the actual length of the insulator. This is explained in Figure 2.5c where the

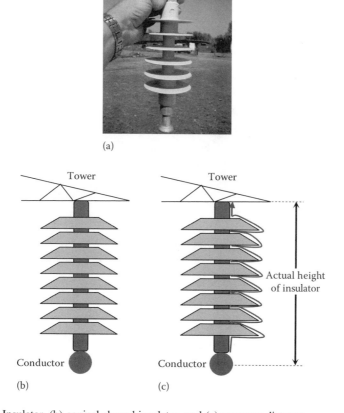

FIGURE 2.5 (a) Insulator, (b) conical-shaped insulator, and (c) creepage distance.

jagged arrow shows the flashover path which is longer than the actual height of the insulator. The flashover path is called the creepage distance and is defined as the shortest distance along the surface of the insulation material. But why are the disks slightly bent toward the ground or have their upper radii smaller than their lower radii? The simple answer is to make it easy to clean the insulators. When rain falls, the deposited salt and dust on the insulator are washed away. If it is mounted upside down, there will be pockets where dust is trapped, thus reducing the insulation capability of the insulators.

2.4 MEDIUM VOLTAGE DISTRIBUTION LINES

The conductors of the medium voltage distribution lines are either buried underground or mounted on poles and towers. In most cities, the distribution network is mainly underground for esthetic and safety reasons. However, it is also common to have overhead distribution lines as shown in Figure 2.6.

Since the voltage of the distribution lines is much lower than that for the transmission lines, the distribution towers are shorter and their insulators are smaller. The distribution towers are often made of steel, wood, concrete, or composite materials. Most of the commercial and industrial plants have direct access to the distribution network, and they use their own transformers to step down the voltage to the levels needed by their various equipment. In residential areas, utilities install transformers in vaults or on towers to reduce the distribution line voltage to any value between 100 and 240 V depending on the standard of the country.

2.5 CONTROL CENTERS

The stability and security of the power system must be maintained at all times to avoid any interruption of service to the customers or the collapse of the power system. The system must also be protected from being damaged due to the failure of any of its components or equipment.

FIGURE 2.6   Medium voltage distribution tower.

In addition, the system must operate efficiently and economically to ensure the best rates to the customers. These are enormous tasks whose responsibility lies mainly with the electrical engineers.

To effectively operate and control the power system, it must be extensively monitored at all times. All mechanical and electrical functions inside the power plants are monitored, evaluated, and controlled mostly in real time. All major equipments, such as substation transformers and transmission lines, are also extensively monitored and controlled. The various circuit breakers in the network that switch the various lines and components as well as protecting the equipment during fault conditions are continuously checked for their status and their operations are often controlled remotely. Besides the extensive monitoring, the engineers continuously evaluate the operation of the system, predict the future demand, and establish favorable energy trade conditions between utilities. These functions require the use of sophisticated algorithms that optimize the operation of the power system.

All the above tasks are managed in the control center, which is the brain of the electric power system. One of these control centers is shown in Figure 2.7. The control center has a large mimic board representing the main components of the power system such as transmission lines and circuit breakers. The mimic board displays the current condition of the system such as the power flow in all lines and the status of each circuit breaker. Several operators (called system operators) are present at all times to monitor the power system and implement corrective actions should a problem arise. They also shift flows on transmission lines and circuits to ensure that all power system components are operating within their specified voltage and current ranges. When the system is operating under normal conditions, the system operators run several software to optimize the operation of the system, predict future demand and power flows, assess system stability, schedule generation, etc.

FIGURE 2.7 (See color insert following page 300.) Control center. (Image courtesy of Tennessee Valley Authority.)

2.6 WORLDWIDE STANDARDS FOR HOUSEHOLD VOLTAGE AND FREQUENCY

The magnitude of the household voltage or frequency is not the same worldwide. This is mostly because of three reasons:

- Due to the competitions between the United States and Europe in the late 1800s and early 1900s, the manufacturing companies in both continents have independently developed their power system equipment without coordinating their efforts.
- Different safety concerns in Europe and the United States have led to various voltage standards.
- Wiring and equipment costs have led nations to select the most economical voltage standards.

Table 2.1 shows the voltage and frequency standards in select countries. The voltage standard is for single-pole outlets used in most household appliances. In the United States, 120 V is used everywhere inside the house, except for heaters, driers, and ovens where the receptacles are double-pole at 240 V.

The various voltages and frequencies standards worldwide have created confusions among consumers who wish to use equipment manufactured based on one standard in a country with a different standard. For example, the compressors of the refrigerators designed for 60 Hz standard slow down by 17% when used in a 50 Hz system. Also, electrical equipment designed for 100 V standard will be damaged if it is used in 240 V systems.

For small appliances, modern power electronics circuits have addressed this problem very effectively. Almost all power supplies for travel equipment (electric razors, portable computers, digital cameras, audio equipment, etc.) are designed to operate at all voltage and frequency standards.

TABLE 2.1
Standards for Voltage and Frequency in Several Nations

Country	Voltage (V)	Frequency (Hz)
Australia	230	50
Brazil	110/220	60
Canada	120	60
China	220	50
Cyprus	240	50
Egypt	220	50
Guyana	240	60
South Korea	220	60
Mexico	127	60
Japan	100	50 and 60
Oman	240	50
Russian Federation	220	50
Spain	230	50
Taiwan	110	60
United Kingdom	230	50
United States	120	60

2.6.1 Voltage Standard

It appears that the 120 V was chosen somewhat arbitrarily in the United States. Actually, Edison came up with a high-resistance lamp filament that operated well at 120 V. Since then, the 120 V was selected in the United States. The standard for voltage worldwide varies widely from 100 V in Japan to 240 V in Cyprus. Generally, a wire is less expensive when the voltage is high; the cross section of the copper wires is smaller for higher voltages. However, from the safety point of view, lower voltage circuits are safer than the higher voltage ones; 100 V is perceived to be less harmful than 240 V.

2.6.2 Frequency Standard

Only two frequencies are used worldwide: 60 and 50 Hz. The standard frequency in North America, Central America, most of South America, and some Asian countries is 60 Hz. Almost everywhere else, the frequency is 50 Hz.

In Europe, major manufacturing firms such as Siemens and AEG have established 50 Hz as a standard frequency for their power grids. Most of Asia, parts of South America, all of Africa, and the Middle East have adopted the same 50 Hz standard.

In the United States, Westinghouse adopted the 60 Hz standard. Nikola Tesla actually wanted to adopt a higher frequency to reduce the size of the rotating machines, but 60 Hz was eventually selected for the following reasons:

- It is a high enough frequency to eliminate light flickers in certain types of incandescent lamps.
- It is conveniently synchronized with time.
- Machines designed for 60 Hz can have less iron and smaller magnetic circuits than the ones designed for 50 Hz.

In Japan, both frequencies are used; in Eastern Japan (Tokyo, Kawasaki, Sapporo, Yokohama, and Sendai), the grid frequency is 50 Hz; and in Western Japan (Osaka, Kyoto, Nagoya, and Hiroshima), the frequency is 60 Hz. The first generator in Japan was imported from Germany for the Kanto area during the Meiji era, and the frequency was 50 Hz. Subsequent acquisitions of power equipment were mainly from Europe and were installed in the eastern side of Japan making the frequency of the eastern grid 50 Hz. After World War II, Japan imported their power plants from the United States and installed them in the western side of the island (the first power plant was in the Kansai area). The western grid is thus operating at 60 Hz. The two frequency standards created two separate grids with the dividing line between them going from the Fuji River in Shizuoka upward to the Itoi River in Niigata. These two grids are not directly connected by an alternating current (ac) line, but are connected by direct current (dc) lines. On each end of the dc line, a power electronic system converts the ac frequency into dc.

It is interesting to know that for stand-alone systems, such as aircraft power systems, the frequency is 400 Hz. This is selected to reduce the size and weight of the rotating machines and transformers aboard the aircraft.

2.6.2.1 Frequency of Generating Plants

The frequency of the power system's voltage is directly proportional to the speed of the generators in the power plants (which is the same as the speed of the turbines). The relationship between the frequency and the speed, which is explained in details in Chapter 12, is given in the following equation:

$$f = \frac{P}{120} n \tag{2.1}$$

where

   n is the speed of the generator (rpm)
   P is the number of magnetic poles of the field circuit of the generator
   f is the frequency of the generator's voltage

One of the earliest ac generators was a 10-pole machine running at 200 rpm. The frequency of this generator, as explained by Equation 2.1, was $16\frac{2}{3}$ Hz. This anomalous frequency was low enough to allow the series-wound dc motor to operate from ac supply without any modification. This was a justifiable reason because the series-wound motor was used extensively in locomotive tractions. However, this low frequency created noticeable flickers in incandescent lamps, and was therefore rejected as a standard.

When the Niagara Falls power plant was built, the engineers used 12-pole generators running at 250 rpm, and the frequency was 25 Hz. The power plant was built to produce compressed air, and the low frequency was not a major concern. In the following developments, the frequency was raised to 40 Hz, then 60 Hz to reduce the flickers.

2.6.2.2   Frequency of Power Grids

In early days, the generators operated independently without any connection between them. The typical system was composed of a single generator, feeders (transmission lines), and several loads. In such a system, the frequency can drift without any major impact on the stability of the system. In the 1940s, it became economically important to interconnect the generators by a system of transmission lines, and the power grid was born. The interconnections demanded that the frequency of the grid be fixed to a single value. If the frequencies of all generators are not exactly equal, the power system would collapse as explained in Chapter 14.

EXERCISES

   1. What is the function of a power plant turbine?
   2. What is the function of a power plant generator?
   3. What is the function of the hydroelectric dam?
   4. Why are transformers used with transmission lines?
   5. A two-pole generator is to be connected to a 60 Hz power grid. Compute the speed of its turbine.
   6. A two-pole generator is to be connected to a 50 Hz grid. Compute the speed of its turbine.
   7. Why are insulators used on power line towers?
   8. Why are tower insulators built as disk shapes?
   9. Why is the frequency of the airplane power system 400 Hz?
  10. Why are there different voltage and frequency standards worldwide?
  11. Why are the transmission line towers higher than the distribution line towers?
  12. What are the various forces that an insulator must withstand?
  13. What are the main functions of the control center?
  14. What is a mimic board?

3 Energy Resources

As shown in Figure 3.1, our energy resources are often divided into three loosely defined categories:

1. Fossil fuel
2. Nuclear fuel
3. Renewable resources

Fossil fuels include oil, coal, and natural gas. Renewable energy resources include hydropower, wind, solar, hydrogen, biomass, tidal, and geothermal. All these resources can also be classified as primary and secondary resources. The primary resources are

1. Fossil fuel
2. Nuclear fuel
3. Hydropower

The secondary resources include all renewable energy minus hydropower. Over 99% of all electric energy worldwide is generated from primary resources. The secondary resources, although increasing rapidly, have not yet achieved a level comparable to the primary resources.

In the subsequent statistics, three terminologies are used: generated electricity, consumed electricity, and generation capacity. Generated electricity is the amount of electrical energy produced at the power plant. This is equal to the consumed energy plus the losses in all power system equipment such as transformers and transmission lines. Generation capacity is the amount of energy that could be generated if all generators operate at their full capacity at all times. The generation capacity is always more than the generated energy.

The distribution of electricity generated worldwide by primary resources is shown in Figure 3.2. As you see, most electrical energy is generated by oil, coal, and natural gas. Hydroelectric energy is limited to about 6% of the world's electrical energy because of the limited water resources suitable for generating electricity. Nuclear energy is only 6% of the total electrical energy because of the public resistance to building new nuclear facilities in the past 20 years.

In the United States, most electrical energy is generated by fossil fuel as depicted in Figure 3.3. Because of its abundance, coal is by far the prevailing source of electric energy in the United States. Actually, the United States has the largest coal reserve among all countries. After coal, nuclear energy and natural gas each account for about 19% of generated electricity. Unfortunately, the share of secondary resources is only about 2% of total generated electricity. Some attribute this low percentage to the low cost of electricity in the United States.

The worldwide generation capacity in 2006 is shown in Figure 3.4. The world's total generation capacity was about 19,028 TW h (19.028×10^{12} kW h) in 2006. About 27% of this capacity was in North America; this is about 4,254 TW h. In the Asia–Pacific region, the capacity in 2006 was about 34% of the total world capacity. Four years earlier, in 2002, the Asian capacity was 29%. This tremendous increase in the Asian capacity is due to the accelerated installation of power plants to support the societal needs and the rapid industrial developments in the region, particularly in China and India.

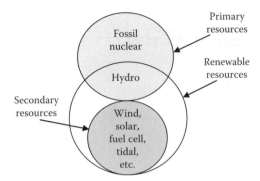

FIGURE 3.1   Energy resources.

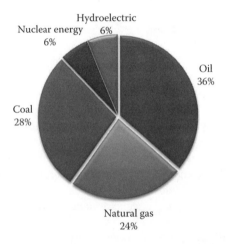

FIGURE 3.2   Primary resources used to generate electricity worldwide in 2006. (From *British Petroleum Statistical Review of World Energy*, 2007.)

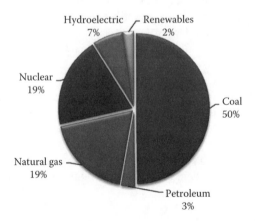

FIGURE 3.3   Primary resources used to generate electricity in the United States in 2006. (From U.S. Department of Energy, *Annual Energy Review*, 2007.)

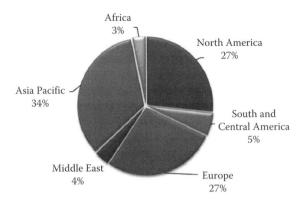

FIGURE 3.4 Generation capacity worldwide as of 2006. (From *British Petroleum Statistical Review of World Energy*, 2007.)

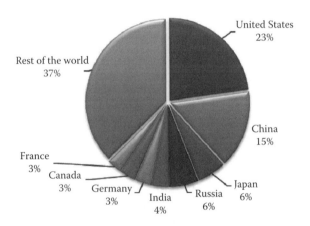

FIGURE 3.5 Consumption of electric energy from primary resources worldwide in 2006. (From USA—the *CIA World Factbook*, 2007.)

In 2006, the estimated worldwide consumption of electrical energy was about 16.28×10^{12} kW h; the distribution is shown in Figure 3.5. The U.S. share of this energy was about 3.72×10^{12} kW h, which is about 23% of the world's electric energy consumption. In second place was China with 15% of the world's electric energy consumption. In 2002, China's share was just 9%. This tremendous increase in energy consumption in China is due to its recent industrial developments and the enhancement in its standard of living.

EXAMPLE 3.1

Compute the annual electrical energy consumption per capita worldwide in 2006.

Solution

According to the U.S. Census Bureau, the world population by the end of 2006 was about 6.5×10^9.

Annual electric energy consumed per capita worldwide = Total world consumption/world population = $16.28 \times 10^{12}/6.5 \times 10^9 = 2.5$ **MW h**.

EXAMPLE 3.2

Repeat Example 3.1 excluding the consumption of the United States.

Solution

The world energy consumption – the U.S. consumption $= (16.28 - 3.72) \times 10^{12} = 12.56 \times 10^{12}$ kW h.

According to the U.S. Census Bureau, the U.S. population by the end of 2006 was about 3×10^{8} people.

World population – the U.S. population $= (6.5 - 0.3) \times 10^{9} = 6.2 \times 10^{9}$ people.

Annual electric energy consumed per capita worldwide outside the United States = total world consumption excluding the United States/world population excluding the United States $= 12.56 \times 10^{12}/6.2 \times 10^{9} = 2.03$ MW h.

EXAMPLE 3.3

Compute the annual electric energy consumed per capita in the United States in 2006. Compare the result with the average of the rest of the world.

Solution

Annual electric energy consumed per capita in the United States = total consumption in the United States/the U.S. population $= 3.72 \times 10^{12}/3 \times 10^{8} = 12.4$ MW h.

Annual electric energy consumed per capita in the United States/annual electric energy consumed per capita worldwide excluding the United States $= 12.4/2.03 = 6.1$.

In 2006, the average person in the United States consumed more than six times the electrical energy consumed by the average person in the rest of the world. This high consumption is not just attributed to the luxury living in the United States, but is also due to the United States' highly industrial base. Repeat the example for France and draw a conclusion.

3.1 FOSSIL FUEL

Fossil fuels are formed from fossils (dead plants and animals) buried in the earth's crust for millions of years under pressure and heat. They are composed of high carbon and hydrogen elements such as oil, natural gas, and coal. Further, because the formation of fossil fuels takes millions of years, they are considered nonrenewable.

Since the start of the Industrial Revolution in Europe in the nineteenth century, the world became extremely dependent on fossil fuels for its energy needs. The bulk of fossil fuels are used in transportation, industrial processes, generating electricity as well as residential and commercial heating.

The use of fossil fuel to generate electricity is always a subject of hot debate. Burning fossil fuels causes a wide range of pollution that include the release of carbon dioxide, sulfur oxides, and the formation of nitrogen oxides. These are harmful gases that cause some health and environmental problems as discussed in Chapter 5. Also, the availability and prices of fossil fuels are vulnerable to world politics. Oil production and distribution are often interrupted during wartimes and due to political tensions between nations.

3.1.1 OIL

Oil is the most widely used fossil fuel worldwide. It is also called petroleum, which is composed of two words *petro* and *oleum*. The first is a Greek word for rock and the second is a Latin word for oil.

Petroleum is extracted from fields with layers of porous rocks filled with oil. The Chinese discovered oil as early as AD 300. They used bamboo tubes to extract oil that might have been used as a medical substance. In the twelfth century, the famous traveler, Marco Polo, while in Persian near

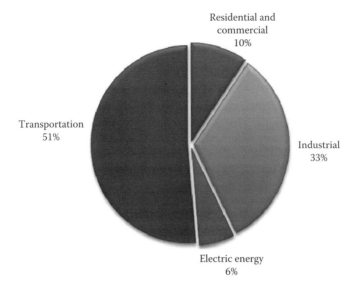

FIGURE 3.6 Consumption of oil worldwide by sectors in 2006. (From U.S. Department of Energy, Energy Information Administration, 2007.)

the Caspian Sea, observed a geyser gushing black substance used by the inhabitants to provide heat and light. However, it was not until the fifteenth century that oil was first commercialized in Poland and used to light streetlamps.

In the mid-eighteenth century, the first oil wells in North America were drilled in Ontario, Canada, and Pennsylvania. Around 1870, John D. Rockefeller formed Standard Oil of Ohio that eventually controlled about 90% of the U.S. refineries. In the early nineteenth century, oil was discovered in several Middle Eastern countries as well as Central and South America.

Today, oil is used mainly in the transportation and industrial sectors as seen in Figure 3.6. The generation of electrical energy uses only 6% of the total oil consumed annually. This is mainly because coal is a much cheaper option for generating electricity.

The estimated oil reserve worldwide, as of the end of 2006, was about 1.372×10^{12} barrel, mainly in the Middle East as seen in Figure 3.7. The world consumption of oil in 2006 was about

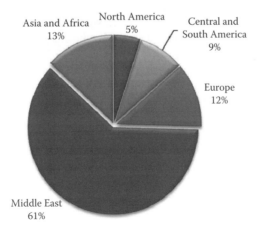

FIGURE 3.7 Distribution of known oil reserve as of 2006. (From *British Petroleum Statistical Review of World Energy*, 2007.)

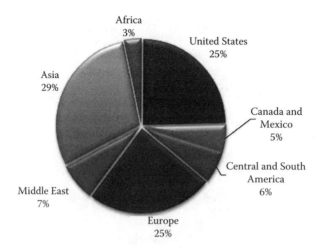

FIGURE 3.8 Consumption of oil worldwide in 2006. (From *British Petroleum Statistical Review of World Energy*, 2007.)

3.06×10^{10} barrel distributed as seen in Figure 3.8. This high rate of consumption is troubling, and new fields must always be discovered before the available supply is dried out.

EXAMPLE 3.4

Assume no new oil field is discovered, then for how long can we maintain the consumption of oil at the 2006 rate?

Solution

The world's known oil reserve is 1.372×10^{12} barrel, and the annual world consumption is 3.06×10^{10} barrel. Hence, the world reserve will last for $1.372 \times 10^{12}/3.06 \times 10^{10} = 44.84$ years. This is a disturbingly short period.

3.1.2 NATURAL GAS

Although natural gas was discovered in Pennsylvania, as early as 1859 by Edwin Drake, it was not utilized commercially because the process of liquefying natural gas at low temperatures was unknown before the nineteenth century. Natural gas was then considered a by-product of crude oil extraction, and was often burned out in the fields. After the discovery of large reservoirs of natural gas in Wyoming in 1915, the United States developed cryogenic liquefaction methods and reliable pipeline systems for the storage and transportation of natural gas. In the 1920s, Frank Phillips (the founder of Phillips Petroleum) commercialized natural gas in the forms of propane and butane.

The distribution of world's reserve of natural gas is shown in Figure 3.9. The world reserve as of 2006 is about 1.81×10^{14} m^3. The majority of this reserve is in Russia and the Middle East. The 2006 estimated reserve in the United States is 5.93×10^{12} m^3, which is about 3.27% of total world reserve.

In 2006, the world consumed about 2.85×10^{12} m^3 of natural gas; the distribution is shown in Figure 3.10. The majority of consumed natural gas was in North America and Europe; the United States alone consumed about 6.2×10^{11} m^3. About 31% of the natural gas production worldwide is used to generate electricity and 44% is used in the industrial sector.

EXAMPLE 3.5

Assume no new natural gas field is discovered in the United States, then for how long can the United States be self-sustained at the 2006 consumption rate?

Solution

The known reserve of natural gas in the United States as of 2006 is 5.93×10^{12} m^3

The U.S. consumption in 2006 is 6.2×10^{11} m^3

Hence, the reserve in the United States can sustain the 2006 consumption rate for $5.93 \times 10^{12}/6.2 \times 10^{11} = 9.56$ years.

EXAMPLE 3.6

Assume no new natural gas field is discovered worldwide, then for how long will the natural gas last?

Solution

The world known reserve as of 2006 is 1.81×10^{14} m^3.

The world consumption in 2006 is 2.85×10^{12} m^3.

Hence, the world reserve can sustain the 2006 consumption rate for $1.81 \times 10^{14}/2.85 \times 10^{12} = 63.5$ years.

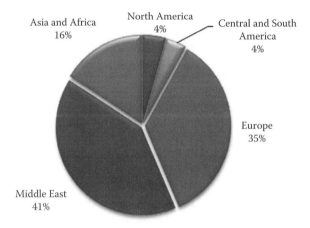

FIGURE 3.9   Known natural gas reserve as of 2006. (From *British Petroleum Statistical Review of World Energy*, 2007.)

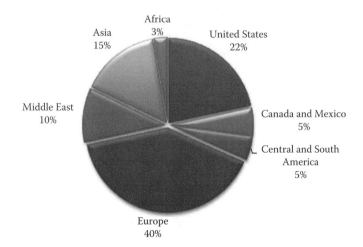

FIGURE 3.10   Consumption of natural gas worldwide in 2006. (From *British Petroleum Statistical Review of World Energy*, 2007.)

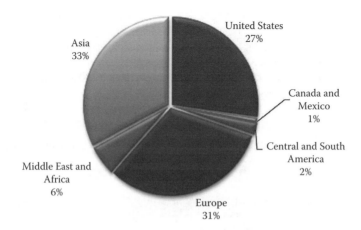

FIGURE 3.11 World's coal reserve as of 2006. (From *British Petroleum Statistical Review of World Energy*, 2007.)

3.1.3 COAL

Charcoal, the black porous carbon substance produced by burning wood, is one of the earliest sources of heat and light known to man. Five thousand years ago, the ancient Egyptians used charcoal for cooking, heating, baking, pottery, and for liquefying metals such as gold.

Coal is another form of the carbonized vegetations formed in the carboniferous period millions of years ago. Geological activity, heat, pressure, and millions of years are needed to form it. Coal was widely used in England during the twelfth century. However, as people burned coal in closed quarters without adequate ventilations, a large number of people were poisoned from the carbon monoxide released from burning coal. Consequently, King Edward I imposed the death penalty upon anyone caught burning coal. The ban lasted for two centuries.

In the seventeenth century, English scientists discovered that coal actually burned cleaner and produced more heat (two to four times) than wood charcoal. This discovery started intense exploration of coal worldwide to provide Europe with the energy needed to power its industrial revolution. Because of coal, the Scottish James Watt invented the steam engine that propelled ships, drove trains, and powered industrial machines. Coal was later used in the 1880s to generate electricity.

In the United States, the French explorers Louis Joliet and Jacques Marquette discovered coal in Illinois in 1673. This discovery was followed by many more in Kentucky, Wyoming, Pennsylvania, West Virginia, and Texas. Today, the largest sources of coal in the United States are in Wyoming and West Virginia.

The world reserve of coal as of 2006 is estimated about 9.1×10^{11} tons distributed as shown in Figure 3.11. The U.S. reserve is 27% of the world reserve, which is about 2.47×10^{11} tons.

The world consumption of coal in 2006 is about 3.1×10^{9} toe (ton oil equivalent). The distribution of the consumption is shown in Figure 3.12. As you see, although Asia has 33% of the world's coal reserve, it accounts for 58% of the world's consumption. In North America and Europe, the consumption of coal is very high, but the dependency on coal for the production of electricity is reduced because of the wider use of natural gas.

3.2 NUCLEAR FUEL

Nuclear fuel is heavy nuclear material that releases energy when its atoms are forced to split and in the process, some of its mass is lost. The nuclear fuel used to generate electricity is mostly uranium (U), but plutonium (Pu) is also used. Uranium is found in nature and contains several isotopes. Natural uranium is almost entirely a mixture of three isotopes: ^{234}U, ^{235}U, and ^{238}U, where the left superscripts indicate the atomic mass of the isotopes. The concentration of these isotopes in natural

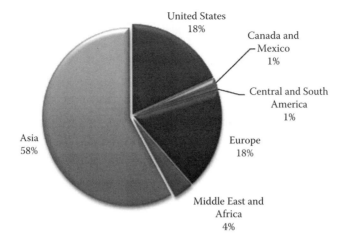

FIGURE 3.12 Consumption of coal worldwide in 2006. (From *British Petroleum Statistical Review of World Energy*, 2007.)

uranium is 99.2% for ^{238}U and 0.7% for ^{235}U. However, only ^{235}U can fission in nuclear reactors. Since the concentration of ^{235}U in uranium ores is very low (0.7%), an enrichment process is used to increase its concentration in nuclear fuel. For nuclear power plants, ^{235}U concentration is about 3%–5%, and for nuclear weapons, it is over 90%.

The data for uranium production are not freely available for political and security reasons. However, the world's production was estimated at 39,430 tons in 2006. The largest producers of uranium are Canada (about 9,860 tons), Australia (about 7,590 tons), Niger, Namibia, Russia and the former Soviet Union states, and the United States. The U.S. production of uranium is estimated at 2,010 tons annually in 2006.

Plutonium is the other nuclear fuel used to generate electricity, but it is less common than uranium. Minute amounts of Pu can also be found naturally. So it is mainly a man-made element, which was discovered by a group of scientists from the University of Berkeley in 1941. They found that plutonium is created when ^{238}U absorbs a neutron to become ^{239}U and ultimately decays to ^{239}Pu. Plutonium is produced in breeder reactors, and it has three common isotopes: ^{238}Pu, ^{239}Pu, and ^{240}Pu. The other Pu isotopes are created by different combinations of uranium and neutron.

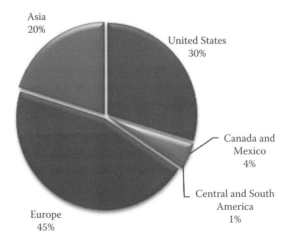

FIGURE 3.13 Consumption of nuclear fuel worldwide in 2006. (From *British Petroleum Statistical Review of World Energy*, 2007.)

The isotope ^{239}Pu is used in nuclear weapons and ^{238}Pu is used in nuclear power plants. In some cases, ^{238}Pu is mixed with uranium to form a mixed-oxide fuel that increases the power plant output.

Another use of Pu is in standalone power systems such as the ones installed on satellites. In addition, a minute amount of plutonium can provide long lasting power for medical equipment such as heart pacemakers.

The consumption of nuclear fuel worldwide is distributed as shown in Figure 3.13. The world consumption of nuclear fuel in 2006 was about 6.36×10^8 toe. Europe and the United States led the world in the amount of nuclear fuel used to generate electricity. The United States consumed about 1.88×10^8 toe of nuclear fuel in 2006.

EXERCISES

1. Exclude the United States and compute the generation capacity of the rest of the world. Find the world's per capita capacity.
2. Compute the generation capacity per capita in the United States. Compare the result with the world average.
3. Exclude the United States and compute the annual world consumption of electrical energy per capita. Compare the result with the U.S. consumption per capita.
4. Find the ratio of the electrical energy capacity to the electrical energy demand (consumption) in the United States. Identify the amount of surplus or deficit.
5. Exclude the United States and find the ratio of the electrical energy capacity to the electrical energy demand worldwide. Identify the amount of surplus or deficit.
6. Assume the demand in the United States is increasing at a rate of 5% annually. For how long can the United States be self-sustained with respect to electricity without constructing new generating plants?

4 Power Plants

The vast majority of electricity generated worldwide (about 99%) is generated from power plants using primary energy resources such as hydropower, fossil fuel, and nuclear fuel. The descriptions of the power plants that use primary resources are covered in this chapter, and the methods to generate electricity from secondary resources (solar, wind, geothermal, etc.) are discussed in Chapter 6.

The geological and hydrological characteristics of the area where the power plant is to be erected determine, to a large extent, the type of the power plant. For example, fossil fuel power plants in the United States are concentrated mainly in the east and midwest regions where coal is abundant. Similarly, hydroelectric power plants are concentrated in the northwest region where water and water storage facilities are available. Nuclear power plants, however, are distributed in all regions since their demand for natural resources is limited to the availability of cooling water.

The percentage of energy produced worldwide by primary resources is given in Chapter 3. The data in that chapter show that fossil fuels are the main source of electric energy (over 80%), and about 63% of the electric energy is produced by coal and oil-fired power plants. In the United States, coal counts for about 50% of the fuel used to generate electricity.

4.1 HYDROELECTRIC POWER PLANTS

"Hydro" is a Greek word meaning water, "hydropower" means the power in the moving water and hydroelectric is the process by which hydropower is converted into electricity. The hydroelectric power plant harnesses the energy of the hydrologic cycle. Water from oceans and lakes absorbs solar energy and evaporates into air forming clouds. When the clouds become heavy, rains and snows occur. The rain and melted snow travel through streams and eventually end up in oceans. The motion of water toward oceans is due to its kinetic energy, which can be harnessed by the hydroelectric power plant that converts it into electrical energy. If water is stored at high elevations, it possesses potential energy proportional to that elevation. When this water is allowed to flow from a higher elevation to a lower one, the potential energy is transformed into kinetic energy, which is converted into electrical energy by hydroelectric power plants.

The world's first hydroelectric power plant was constructed across the Fox River in Appleton, Wisconsin, and began its operation on September 30, 1882. The plant generated only 12.5 kW, which was enough to power two paper mills and the private home of the mill's owner. The latest and largest hydroelectric power plant, so far, is the one being built in China's Three Gorges, which has a capacity of 22.5 GW. Some of the world's largest hydroelectric power plants are given in Table 4.1.

4.1.1 TYPES OF HYDROELECTRIC POWER PLANTS

The common types of hydroelectric power plants are impoundment hydroelectric, diversion hydroelectric, and pumped storage hydroelectric power plants.

1. Impoundment hydroelectric: It is the most common type of hydroelectric power plant and is suitable for water bodies with high heads. The dam in these power plants creates a reservoir at a high elevation behind the dam. A good example is the Grand Coulee Dam shown in Figure 4.1.

TABLE 4.1
World's Largest Hydroelectric Power Plants

Name of Dam	Location	Capacity (GW)	Year of Completion
Three Gorges	China	22.5	2010
Itaipu	Brazil/Paraguay	14	1983
Guri	Venezuela	10	1986
Tucurui	Brazil	8.37	1984
Grand Coulee	Washington	6.5	1942
Sayano-Shushensk	Russia	6.4	1989
Krasnoyarsk	Russia	6	1968
Churchill Falls	Canada	5.43	1971
La Grande 2	Canada	5.33	1979
Bratsk	Russia	4.5	1961

2. Diversion hydroelectric: An example of a diversion hydroelectric power plant is shown in Figure 4.2. It is suitable for low heads and is based on diverting some water of a river with strong current through the turbines. This hydroelectric plant does not require a water reservoir at high elevation, so its generating capacity is less than that for the impoundment hydroelectric power plant.

3. Pumped storage hydroelectric: A pumped storage hydroelectric power plant operates as a dual-action water flow system. When the power demand is low, electricity is used to pump water from the lower water level in front of the dam to the high level behind the dam. This increases the potential energy behind the dam for later use.

4.1.2 IMPOUNDMENT HYDROELECTRIC POWER PLANTS

A typical impoundment hydroelectric power system has six key components: dam, reservoir, penstock, turbine, generator, and governor. A schematic of a hydroelectric power plant is shown in Figure 4.3.

FIGURE 4.1 (See color insert following page 300.) The Grand Coulee Dam and Franklin D. Roosevelt Lake. (Image courtesy of U.S. Army Corps of Engineers.)

FIGURE 4.2 (See color insert following page 300.) Fox River diversion hydroelectric power plant, Wisconsin. (Image courtesy of U.S. Army Corps of Engineers.)

1. Dam: It is a barrier that prevents water from flowing downstream, thus creating a lake behind the dam. The potential energy of the water behind the dam is directly proportional to the volume and height of the lake. The dam can be enormous in size; the Grand Coulee Dam in Washington State is 170 m in height, 1.6 km in length, and its crest is 9 m wide. Its base is 150 m wide, which makes the base about four times as large as the base of the Great Pyramid of Egypt. The volume of the concrete used to build the dam is almost 9.16×10^6 m^3. Although massive, the Grand Coulee is not the biggest dam in the world. The Three Gorges Dam in China is the biggest dam ever built so far, followed by the Itaipu Dam in Brazil. Table 4.2 shows some structural data of the Grand Coulee and the Three Gorges dams.

2. Reservoir: The dam creates a lake behind its structure called a reservoir and often covers a wide area of land. The Grand Coulee Dam created the Franklin D. Roosevelt artificial lake shown in Figure 4.1, which is about 250 km long, and has over 800 km of shore line. Its surface area is about 320 km^2, the depth of the lake ranging from 5 to 120 m.

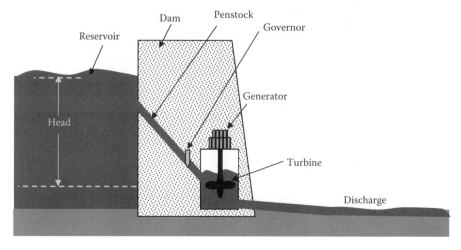

FIGURE 4.3 Simple schematic of a hydroelectric power plant.

TABLE 4.2

Comparative Data on Two Large Hydroelectric Dams

	Grand Coulee Dam	Three Gorges Dam
Length of dam	1.6 km	2.34 km
Height of dam	170 m	185 m
Width at river base	150 m	115 m
Width at crest	9 m	40 m
Volume of concrete	9.16×10^6 m^3	28×10^6 m^3
Surface area of reservoir behind dam	320 km^2	72,128 km^2

3. Penstock: It is a large pipeline that channels water from the reservoir to the turbine. Figure 4.4 shows the penstock of the Grand Coulee Dam during its construction. The water flow in the penstock is controlled by a valve called governor.
4. Turbine: A turbine is an advanced water wheel. The high-pressure water coming from the penstocks pushes against the blades of the turbine causing the turbine shaft to rotate. The electrical generator is mounted directly on the same shaft of the turbine, thus the generator rotates at the speed of the turbine. Turbine–generator units are shown in Figure 4.5.
5. Generator: It is an electromechanical converter that converts the mechanical energy of the turbine into electrical energy. The generators used in all power plants are the synchronous machine type. The generator is equipped with various control mechanisms such as the

FIGURE 4.4   Penstock of the Grand Coulee Dam. (Image courtesy of U.S. Bureau of Reclamation.)

FIGURE 4.5 (See color insert following page 300.) Hydroelectric turbine–generator units at the Lower Granite power plant, Walla Walla, Washington. (Image courtesy of U.S. Army Corps of Engineers.)

excitation control and various stabilizers to maintain the voltage constant and to ensure that the generator's operation is stable.

6. Governor: It is the valve that regulates the flow of water in the penstock. When it is fully open the kinetic energy of the water is at its maximum. If the system is to shut down, the valve is closed fully.

4.1.3 ANALYSIS OF IMPOUNDMENT HYDROELECTRIC POWER PLANT

The amount of the generated electric energy of the impoundment hydroelectric power plant depends on several parameters. The most important ones are

1. Water head behind the dam
2. Reservoir capacity
3. Flow rate of the water inside the penstock
4. Efficiencies of the penstock, turbine, and generator

4.1.3.1 Reservoir

The water behind the dam forms a reservoir (lake). The potential energy of the water in the reservoir PE_r is a linear function of the water mass and head.

$$PE_r = MgH \tag{4.1}$$

where
  M is the water mass (kg)
  g is the acceleration of gravity (m/s^2)
  H is the water head (average elevation) behind the dam (m)

The unit of PE_r is joule (W s). The mass of the water is a function of the water volume and water density

$$M = \text{vol} \times \rho \qquad (4.2)$$

where
vol is the volume of water (m^3)
ρ is the water density (kg/m^3)

At temperatures up to 20°C, ρ is 1000 kg/m^3

EXAMPLE 4.1

A hydroelectric dam forms a reservoir of 20 km^3. The reservoir average head is 100 m. Compute the potential energy of the water in the reservoir.

Solution

$$PE_r = \text{vol} \times \rho gH = 20 \times 10^9 \times 1000 \times 9.81 \times 100 = 1.962 \times 10^7 \text{ GJ}$$

This immense energy is the potential energy of the entire reservoir. Keep in mind that seasonal variations in rain falls or snow spills do change the volume of the reservoir; thus the potential energy of the reservoir is accordingly varied.

4.1.3.2 Penstock

The potential energy of the water entering the penstock, PE, is

$$PE = mgH \qquad (4.3)$$

where m is the mass of water entering the penstock. This potential energy is converted into kinetic energy as the water moves inside the penstock. The kinetic energy, KE, of the water leaving the penstock is

$$KE = \frac{1}{2}mv^2 \qquad (4.4)$$

where v is the velocity of water exiting the penstock (m/s). The PE and KE of the penstock are not equal unless the penstock is vertical and the water friction is ignored. In reality, the penstock is inclined; therefore, its length is longer than the water head, which results in some energy losses. Hence, the penstock efficiency, η_p, is defined as the ratio of its output energy KE to its input energy PE.

$$\eta_p = \frac{KE}{PE} = \frac{v^2}{2gH} \qquad (4.5)$$

Since power is energy divided by time, the mechanical power of the water exiting the penstock P_w (also known as hydropower) is

$$P_w = \frac{KE}{t} = \frac{1}{2}\frac{m}{t}v^2 = \frac{1}{2}fv^2 \qquad (4.6)$$

where f is the flow of water inside the penstock (kg/s) and is defined as

$$f \equiv \frac{m}{t} = \frac{\text{vol} \times \rho}{t} \tag{4.7}$$

EXAMPLE 4.2

The penstock of a hydroelectric dam allows $800 \text{ m}^3/\text{s}$ of water to flow at a speed of 30 m/s. Compute the mechanical power (hydropower) of the water exiting the penstock.

Solution

$$P_{\text{w}} = \frac{1}{2} \frac{\text{vol}}{t} \rho v^2 = \frac{1}{2} 800 \times 1000 \times 30^2 = 360 \text{ MW}$$

This hydropower is not completely converted into electrical power due to the losses inside the turbine and the generator.

The volume of water passing through the penstock during an interval time, t, is

$$\text{vol} = Avt \tag{4.8}$$

where
   A is the cross-sectional area of the penstock
   t is the time interval

Hence, the mechanical power of the water exiting the penstock can be rewritten as

$$P_{\text{w}} = \frac{1}{2} \frac{\text{vol}}{t} \rho v^2 = \frac{1}{2} A \rho v^3 \tag{4.9}$$

EXAMPLE 4.3

The diameter of the penstock of a hydroelectric dam is 4 m. The water velocity inside the penstock is 40 m/s. Compute the power of the water exiting the penstock.

Solution

$$P_{\text{w}} = \frac{1}{2} A \rho v^3 = \frac{1}{2} \left(\pi 2^2 \right) \times 1000 \times 40^3 = 402 \text{ MW}$$

4.1.3.3 Turbine

Hydroelectric turbines are specially designed water wheels that come in three main types; two of them are shown in Figure 4.6. Kaplan turbines, named after Viktor Kaplan, are used mainly in diversion power plants with small heads. Pelton turbines, invented by Lester Pelton, are used in high head impoundment power plants. Francis turbines, invented by James Francis, are used in either type of power plant.

   A schematic of a simple turbine is shown in Figure 4.7. Its main parts are the shaft and the blades. The shaft of the turbine is directly connected to the shaft of the generator, and its blades are designed to rotate the shaft similar to water wheels. The kinetic energy captured by the turbine is a function of the sweep area, A_{s}, of the blades.

$$A_{\text{s}} = \pi r^2 \tag{4.10}$$

(a)

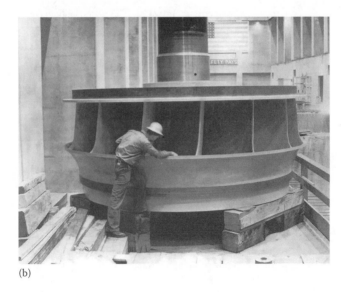

(b)

FIGURE 4.6 Two different types of hydroelectric turbines. (a) Kaplan turbine and (b) Francis turbine. (Images courtesy of the U.S. Army Corps of Engineers.)

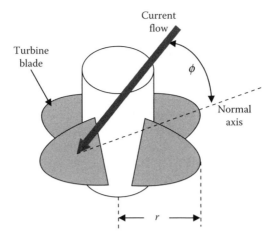

FIGURE 4.7 Simple hydroelectric turbine.

where r is the radius of the sweep area. If the water enters the turbine at an incident angle ϕ from the normal axis of the blade, the sweep area can be modified as follows

$$A_s = \pi r^2 \cos \phi \qquad (4.11)$$

The mechanical power of the water hitting the turbine P_t is

$$P_t = \frac{1}{2} A_s \rho v_t^3 \qquad (4.12)$$

where v_t is the velocity of water when it hits the blades of the turbine. It is slightly lower than the speed of water v exiting the penstock in Equation 4.9.

Because of the various mechanical losses of the turbine, the power P_t cannot entirely be converted into the mechanical power P_m entering the generator. The ratio of P_m to P_t is known as the coefficient of performance, C_p, of the turbine, which is essentially the turbine efficiency.

$$C_p = \frac{P_m}{P_t} \qquad (4.13)$$

Hence,

$$P_m = C_p \left(\frac{1}{2} A_s \rho v^3 \right) \qquad (4.14)$$

EXAMPLE 4.4

The sweep diameter of a turbine's blades is 3 m and the incident angle of the water is 10°. The water velocity at the surface of the blades is 30 m/s. The turbine has a coefficient of performance of 0.5. Compute the mechanical power at the turbine shaft.

Solution

$$P_m = C_p \left(\frac{1}{2} A_s \rho v^3 \right) = 0.5 \left(\frac{1}{2} \left(\pi \times 1.5^2 \cos 10 \right) \times 1000 \times 30^3 \right) = 46.99 \text{ MW}$$

EXAMPLE 4.5

The Grand Coulee Dam has six 40 ft diameter penstocks for its third powerhouse. Each penstock passes 250,000 gal of water per second when the average head of the water behind the dam is 380 ft.

(a) Compute the volume of the discharged water per second for all penstocks in metric units.
(b) Assume the penstock efficiency is 90%. Compute the hydropower exiting each penstock.
(c) Compute the speed of the water inside the penstock.
(d) Assume that the coefficient of performance of the turbine is 0.5 and the generator efficiency is 95%. Estimate the generated electrical power for each penstock.
(e) Compute the overall system efficiency.

Solution

The first step is to convert all data into metric values. Use the conversion table in Appendix A.

The diameter of the penstock is $40 \times 0.3048 = 12.2$ m.
The volume of water passing through the penstock per second is $250{,}000 \times 3.7854 = 9.46 \times 10^5$ L/s.
Since 1 L of water in volume is equal to 1 kg in weight, the penstock flow rate is 9.46×10^5 kg/s.
The water head is $380 \times 0.3048 = 115.82$ m.

(a) Volume of the discharged water per second is 9.46×10^5 L/s or 946 m^3/s
    The total volume of water discharged by the six penstocks per second is $946 \times 6 = 5676$ m^3/s
(b) The input hydropower to each penstock is

$$P_{\text{in}} = \frac{\text{PE}}{t} = \frac{m}{t}gH = 946 \times 9.81 \times 115.82 = 1.075 \text{ MW}$$

The output hydropower of each penstock

$$P_{\text{w}} = \eta_{\text{p}}P_{\text{in}} = 0.9 \times 1.075 = 967.5 \text{ kW}$$

(c) The speed of water inside the penstock can be computed by using Equation 4.6.

$$P_{\text{w}} = 0.5\,fv^2 = 0.5\frac{m}{t}v^2 = 0.5 \times 946\ v^2$$

$$v = \sqrt{\frac{967.5 \times 10^3}{473}} = 45.23 \text{ m/s}$$

(d) The generator's output power for each penstock P_{g} is

$$P_{\text{g}} = P_{\text{w}} \times C_{\text{p}} \times \eta_{\text{g}}$$

where η_{g} is the efficiency of the generator

$$P_{\text{g}} = 967.5 \times 0.5 \times 0.95 = 459.56 \text{ kW}$$

(e) Overall efficiency $\eta_{\text{total}} = \eta_{\text{p}}\, C_{\text{p}}\, \eta_{\text{g}} = 0.9 \times 0.5 \times 0.95 = 0.4275 = 42.75\%$

4.2 FOSSIL FUEL POWER PLANTS

Fossil power plants (coal, oil, or natural gas) utilize the thermal cycle described by the laws of thermodynamics to convert heat energy into mechanical energy. This conversion, however, is highly inefficient as described by the second law of thermodynamics where a large amount of the

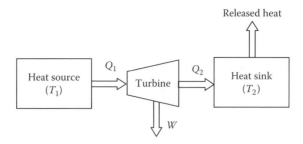

FIGURE 4.8 Second law of thermodynamics.

heat energy must be wasted to convert the rest into mechanical energy. The conversion process is depicted in Figure 4.8. Assume that the energy source in the figure produces heat energy Q_1 at a temperature T_1. Since heat flows only from high temperature to low temperature, a heat sink of temperature $T_2 < T_1$ is needed to facilitate the flow of heat. The second law of thermodynamics shows that the ideal efficiency, η_{ideal}, of a heat engine (turbine, internal combustion engine, etc.) is

$$\eta_{ideal} = \frac{T_1 - T_2}{T_1} \tag{4.15}$$

Equation 4.15 shows that the engine efficiency is increased when T_2 is decreased. In other words, the lower the heat sink temperature, the higher is the efficiency of the heat engine. Keep in mind that this efficiency does not include friction and other mechanical losses or heat leakages. Therefore, the real efficiency is less than η_{ideal}.

As seen in Figure 4.8, the turbine of the power plant is installed between a heat source and a heat sink (known as a cooling tower). The turbine is a thermomechanical device (heat engine) that converts heat energy into mechanical energy. It extracts some of the thermal energy in Q_1 and converts it into mechanical energy W. The rest is dissipated in the heat sink (cooling tower), without which no heat travels through the turbine.

The mechanical energy, W, is the difference between the source energy Q_1 and the energy dissipated in the heat sink Q_2.

$$W = Q_1 - Q_2 \tag{4.16}$$

The ideal efficiency of the turbine, η_{ideal}, can be written in terms of heat energy as

$$\eta_{ideal} = \frac{W}{Q_1} = \frac{Q_1 - Q_2}{Q_1} \tag{4.17}$$

Note that if $T_2 = T_1$, the heat sink does not dissipate any heat energy and $Q_2 = Q_1$. In this case, no mechanical energy is produced by the turbine, and the turbine efficiency is zero.

The cooling towers can be divided into two types: dry and wet. The dry type (Figure 4.9) uses a closed-loop water system where the steam generated during the extraction of Q_2 is condensed and reused. This type is suitable for areas with limited water resources. The wet type allows the steam to be vented in the air, which is suitable for areas near oceans or large lakes where water is abundant.

FIGURE 4.9 Cooling tower.

4.2.1 THERMAL ENERGY CONSTANT

Thermal energy constant (TEC) is defined as the amount of thermal energy produced per 1 kg of burned fuel. The unit of TEC is also called the British thermal unit (BTU); one BTU is equivalent to 252 cal or 1.0544 kJ. Table 4.3 shows typical TEC values for various fossil fuels. As seen in the table, oil and natural gas produce the highest BTU among all fossil fuels.

EXAMPLE 4.6

The cooling tower of a coal-fired power plant extracts 18,000 BTU/kg of burned coal. Compute the mechanical energy of the turbine and the overall system efficiency.

Solution

According to Table 4.3, coal has a TEC of 27,000 BTU/kg or 28.469×10^3 kJ/kg
 The condenser extracts 18,000 BTU/kg or 18.979×10^3 kJ/kg
 The mechanical energy of the turbine is

$$W = Q_1 - Q_2 = 28.469 - 18.979 = 9.49 \times 10^3 \text{ kJ/kg}$$

The ideal efficiency of the thermal turbine η_{ideal} is

$$\eta_{\text{ideal}} = \frac{\text{Output mechanical energy}}{\text{Input thermal energy}} = \frac{W}{Q_1} = \frac{9.49 \times 10^3}{28.469 \times 10^3} = 33.3\%$$

It is normal that the efficiency of the thermal cycle is below 50% because the heat sink dissipates (wastes) a large amount of the thermal energy to complete the thermal cycle.

TABLE 4.3
Thermal Energy Constant for Various Fossil Fuels

Fuel Type	Thermal Energy Constant (BTU/kg)
Petroleum	45,000
Natural gas	48,000
Coal	27,000
Wood (oven dry)	19,000

4.2.2 DESCRIPTION OF THERMAL POWER PLANT

Generally, most fossil fuel power plants have similar designs. The main differences among them are the burners, fuel feeders, and stack filters. Nevertheless, these differences are not essential for the description of the operation of any thermal power plant, and therefore, we shall discuss the coal-fired type only.

As explained in Chapter 3, coal is a combustible black rock that is abundant in the midwest and the east of the United States. Coal comes from organic matter that built up in swamps millions of years ago during the Carboniferous period (about 30 million years ago) and became fossilized.

A typical view of the power house of a thermal power plant is shown in Figure 4.10. It consists mainly of a turbine and a generator. The turbine consists of blades mounted on a shaft. The angles and contours of the blades are designed to capture the maximum thermal energy from the steam. Figure 4.11 shows a large thermal turbine under construction.

The schematic of the main components of a coal-fired power plant is shown in Figure 4.12, and a photograph of one plant is shown in Figure 4.13. The process starts when the coal is delivered to the plant by trucks and railroad trains. The coal is first crushed and delivered to the burner via conveyor belts. It is then burned to generate heat that is absorbed by water pipes inside the boiler.

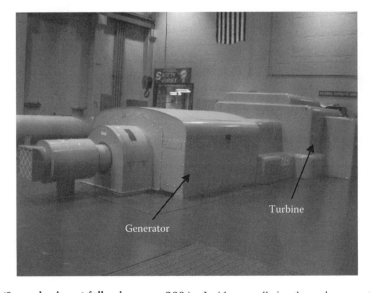

Generator

Turbine

FIGURE 4.10 (See color insert following page 300.)   Inside a small-size thermal power plant.

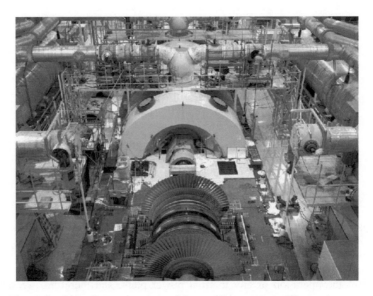

FIGURE 4.11   Thermal turbine. (Image courtesy of Oregon Department of Energy.)

The water turns into high-pressure steam at high temperature. The steam leaves the boiler at a temperature higher than 500°C and enters the turbine at a velocity greater than 1600 km/h. The high speed steam hits the blades of the turbine and causes the turbine to rotate. The turbine's shaft is connected to the shaft of the generator, thus causing the generator to rotate at the same velocity and electricity is generated. Due to the presence of the cooling tower, the thermal cycle is completed as

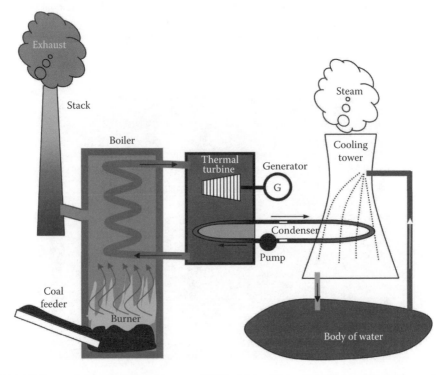

FIGURE 4.12 (See color insert following page 300.)   Main components of a coal-fired power plant.

FIGURE 4.13 (See color insert following page 300.)   Weston coal-fired power plant. (Image courtesy of the U.S. Department of Energy.)

described by the second law of thermodynamics. In the cooling tower, the steam is turned into liquid and goes back to the boiler to complete the thermal cycle. Inside the cooling tower, the condenser uses water from nearby lakes or oceans to cool down the steam inside the pipes.

Although coal-fired power plants are simple in design and easy to maintain, they are major producers of pollution as shown in Chapter 5. Carbon dioxide (CO_2), carbon monoxide (CO), sulfur dioxide (SO_2), nitrous oxides (NO_x), soot, and ashes are some of the by-products of coal combustion. In fact, coal burning by power plants and industries is responsible for 30%–40% of the total CO_2 in the air. In older and unregulated plants, most of these pollutants are vented through the stack. However, with newer technologies, large amounts of the pollutants are trapped by filters or removed from the coal before it is burned. Examples of the pollution reduction measures that are taken in most coal-fired plants include the following:

- Coal is chemically treated to remove most of its sulfur before it is burned.
- Filters are used to remove the particulate (primarily fly ash) and some of the exhaust gases from the boilers. There are various types of filters; among them are the wet scrubber system and the fabric filter system. With the wet scrubber, the exhaust gas passes through liquid, which traps flying particulate and SO_2 before the gas is vented through the stacks. The fabric filter system works like a vacuum cleaner where the particulates are trapped in bags.
- SO_2 is removed by its own scrubber system.
- NO_x are reduced by upgrading the boilers to low NO_x burners.
- CO_2 removal is too expensive to implement and not all power plants use a CO_2 scrubber system.

4.3   NUCLEAR POWER PLANTS

From the availability point of view, nuclear fuel is the most abundant source of energy. As given in Chapter 3, the common fuel for nuclear power plants is uranium; an atom of uranium produces about 10^7 times the energy produced by an atom of coal. In the United States, there are over 100 commercial nuclear power plants in operation generating about 20% of the total electric energy. Worldwide, there are about 400 power plants generating as much as 70% of the energy demand in nations such as France. However, because of public concern, few nuclear power plants have been constructed in the United States, and several are expected to be mothballed (retired) in the near future.

Nuclear power plants can generate electricity by one of two methods:

- "Fission" is the splitting of heavy nuclei element such as uranium, plutonium, or thorium into many lighter elements. By this process, mass is converted into energy. Fission power plants have two main designs: boiling water reactor (BWR) and pressurized water reactor (PWR). About two third of the nuclear reactors are pressurized water reactors, and almost all of the commercial nuclear power plants worldwide are fission reactors.
- "Fusion" is a process by which two lighter elements are combined into a heavier element. The fusion technique is not yet fully developed for commercial power plants.

4.3.1 Nuclear Fuel

An atom comprises subatomic particles known as protons, neutrons, and electrons. Atoms with a different atomic number (number of protons in the nucleus) are called elements. Atoms with the same atomic number, but with a different number of neutrons are called isotopes. Different isotopes of the same element can have different masses due to the difference in the number of neutrons they possess.

Natural uranium is almost entirely a mixture of three isotopes, ^{234}U, ^{235}U, and ^{238}U; the superscripts indicate the atomic masses of the isotopes. The concentration of these isotopes in natural uranium is 99.2% for ^{238}U and 0.7% for ^{235}U. Unfortunately, only ^{235}U can fission in nuclear reactors. Since the concentration of ^{235}U in uranium ores is very low, an enrichment process is used to increase its concentration in nuclear fuel. In nuclear power plants, ^{235}U concentration is about 3%–5%, and for nuclear weapons it is over 90%.

Another way to develop a fissionable nuclear fuel is through breeder reactors. A breeder reactor uses the widely available, nonfissionable uranium isotope ^{238}U, together with small amounts of fissionable ^{235}U, to produce a fissionable isotope of plutonium, ^{239}Pu. Plutonium is a man-made element and cannot be found in nature.

Commercial uranium fuel comes in the form of ceramic pellets. The pellets are made of enriched Uranium containing 3%–5% ^{235}U isotopes by mass. The pellet is cylindrical and about 2–3 cm in height, containing the energy equivalent of 1000 kg of coal, 700 L of oil, or 1500 m^3 of natural gas. The pellets are loaded into zirconium alloy metal tubes. Zirconium alloy is chosen because of its ability to resist radiation and thermal stresses. About 800 tube assemblies are placed inside the reactor.

4.3.2 Fission Process

The fission process is depicted in Figure 4.14. When a nuclear power plant starts up, neutrons are let loose to strike uranium atoms at certain speed causing it to split (fission). The fission process produces the following:

- Fission fragments: They are the leftover materials after the uranium atoms have split. They are cesium-140 and rubidium-93, which are radioactive materials.
- Released neutrons: After each fission action, three free neutrons are released. If they hit three other uranium atoms, they cause more fission to occur, which produces nine free neutrons and so on. This is known as a chain reaction. The neutrons produced by nuclear fission are fast-moving and must be slowed down to initiate further fission. The material used for this purpose is called a moderator; the common moderators are regular water (H_2O) and heavy water (D_2O), as well as graphite. Heavy water is chemically similar to

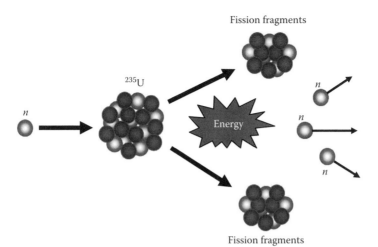

FIGURE 4.14 Fission reaction.

regular water, but the hydrogen atoms are replaced by an isotope of hydrogen called deuterium. Deuterium has one neutron more than hydrogen, which makes D_2O heavier than H_2O by about 10%.

- Energy: The mass of the original uranium atom is more than the combined masses of the fission fragments plus the released neutrons. The lost mass is converted into energy as described by Albert Einstein's formula

$$E = mC^2 \qquad (4.18)$$

where
   E is the released energy
   m is the lost mass
   C is the speed of light

Each fission event releases approximately 3.2×10^{-11} J of energy. To put the number into perspective, 1 J of energy requires approximately 31×10^9 fission events. One kilogram of ^{235}U can have approximately 25.4×10^{23} fission events.

EXAMPLE 4.7

A nuclear reactor produces an average of 1 GW of thermal power. Compute the number of its fission events per year.

Solution

   1 J requires 31×10^9 fission events.
   1 W requires 31×10^9 fission events per second.
   1 GW requires $10^9 (31 \times 10^9) = 31 \times 10^{18}$ fission events per second.
   1 GW for 1 h requires $3600 (31 \times 10^{18}) = 11.16 \times 10^{22}$ fission events.
   1 GW for 1 year requires $8760 (11.16 \times 10^{22}) = 9.7762 \times 10^{26}$ fission events.

The number of fission events in 1 year is staggering. However, in terms of nuclear fuel mass, the mass of ^{235}U required to produce 1 GW of thermal energy for the entire year is very small, which is explained in Example 4.8.

EXAMPLE 4.8

Compute the mass of ^{235}U to produce an average of 1 GW of thermal power. Compare the mass of the nuclear fuel with the equivalent mass of coal.

Solution

In Example 4.7, the number of annual fission events is 9.7762×10^{26}.

Since 1 kg of ^{235}U can have 25.4×10^{23} fission events, the mass of fuel needed for the reactor annually is

$$\text{Mass of } ^{235}\text{U annually} = \frac{9.7762 \times 10^{26}}{25.4 \times 10^{23}} = 385 \text{ kg}$$

About 1 kg of coal produces $27{,}000 \times 1.0544 = 28{,}469$ kW s of thermal energy as given in Table 4.3.

To produce 1 GW of thermal power for 1 s, we need to burn $10^6/28{,}469 = 35.13$ kg of coal.

To produce 1 GW of thermal power annually, we need to burn $(60 \times 60 \times 8760) \times 35.13 = 1.107 \times 10^9$ kg of coal.

Mass of coal/mass of uranium $= 2.88 \times 10^6$.

Keep in mind that the high ratio of coal to uranium is because we assumed that the entire mass of uranium is pure ^{235}U. In a nuclear power plant the uranium is just 3%–5% enriched. Therefore, 1 kg of this enriched nuclear fuel will have much less than 25.4×10^{23} fission events. Nevertheless, the example shows that a small amount of nuclear fuel can unleash a tremendous amount of thermal energy.

4.3.3   FISSION CONTROL

The fission process can be sustained indefinitely if the fuel is available and the chain reaction is maintained. To control the amount of the energy released, the chain reaction must be regulated by controlling the number of neutrons available for fission. This is done by using control rods made of material absorbent to neutrons and inserted inside the reactor between the fuel tubes as shown in Figure 4.15. The common material used for control rods are hafnium, cadmium, or boron. The control rods can be inserted or removed from the reactor by a motion control mechanism. When they are inserted, the rods absorb neutrons, so fewer are available for fission. The reactor can be completely shut down by fully inserting the control rods into the core. In emergency situations, the rods are released and the gravity pulls them down inside the core to their fully inserted positions. This is called safety control rod axe man (SCRAM) process; the process was developed for the early generation of test reactors when the rods were suspended by a rope. In an emergency, the rope was cut by an axe, and the person in charge of this process was called SCRAM.

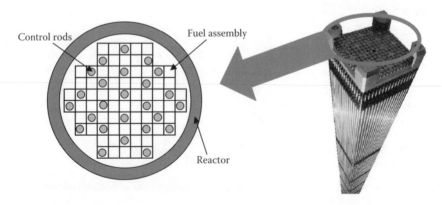

FIGURE 4.15   Fuel assembly and control rods.

4.3.4 BOILING WATER REACTOR

The BWR used in commercial power plants was originally designed by the General Electric Company. The first installation was at Humboldt Bay, California, in 1963. The BWR typically boils the water inside the reactor vessel itself. The operating temperature of the reactor is approximately 300°C, and its steam pressure is about 7×10^5 kg/m^2. Current BWR reactors can generate as much as 1.4 GW with an overall efficiency of 33%.

The schematic of a BWR nuclear power plant is shown in Figure 4.16. The plant has the following key components:

- Containment structure: The dome structure houses the reactor and has multiple barriers of thick steel and concrete to contain the radiations inside the structure.
- Reactor vessel: The reactor vessel houses the reactor, nuclear fuel, and fuel rods. The vessel is filled with water, which acts as a moderator for the chain reaction and also extracts the heat energy generated by the nuclear reaction. The walls of the vessel are made of thick barriers of steel and concrete to guard against any radiation leakage or accidental meltdown.
- Reactor: The reactor contains the nuclear fuel assembly and the control rods. The control rods can be inserted or removed from the reactor by an actuation system to control the amount of heat generated. In case of an emergency shutdown, the rods are released and dropped to the fully inserted position to halt the chain reaction.
- Turbine: The energy generated by the nuclear reaction heats the water inside the reactor vessel, and steam is produced at high temperature and pressure. The steam passes through the turbine, which is a thermomechanical converter that converts the thermal energy into mechanical energy, much like the turbine of the thermal power plant.
- Generator: The turbine is connected to the generator via a common drive shaft. The generator is an electromechanical converter, where the mechanical energy of the turbine is converted into electrical energy.

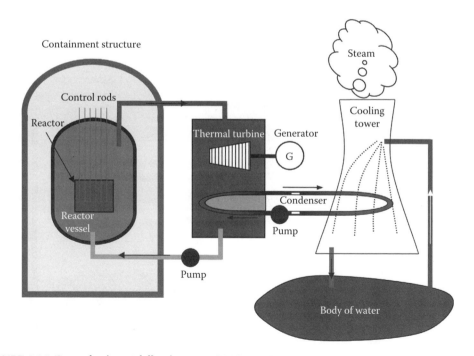

FIGURE 4.16 (See color insert following page 300.) Main components of a boiling water reactor.

- Condenser: It is the heat exchanger that extracts heat from the steam exiting the turbine. It acts as a thermal link between the turbine steam and the cooling tower.
- Cooling tower: The function of the cooling tower is to act as the heat sink of the thermal cycle. It uses cold water from a nearby reservoir, lake, river, or ocean. The cold water is poured over the condenser pipes to cool them down. The cooling water extracts the heat from the condenser and turns it into steam, which is vented from the top of the cooling tower.

Among the advantages of the BWR are its simple design and high efficiency. However, because the fuel rods are in direct contact with the steam, the radioactive material from the nuclear reaction is carried over by the steam and eventually reaches the turbine. This radioactive steam pollutes the turbine and poses a challenge to the maintenance crew working on the turbine–generator system.

4.3.5 Pressurized Water Reactor

A pressurized water reactor was originally designed by Westinghouse as the power source for navy ships and submarines. The first commercial PWR plant in the United States was the Shippingport Atomic Power Station near Pittsburgh, which operated from 1958 to 1982. The PWR is the most widely used type of nuclear reactor worldwide.

The PWR design is distinctly different from that of the BWR because, in PWR, a heat exchanger is placed between the water of the reactor and the steam entering the turbine. This is done to prevent the radioactive water from contaminating the turbine. This heat exchanger is called a steam generator. A schematic of a PWR power plant is shown in Figure 4.17 and several photographs of PWR power plants are shown in Figure 4.18. The water of the PWR reactor is under enough pressure to remain in liquid form even when it reaches 300°C or more; thus the system is called a pressurized water reactor. Raising the temperature of the water under pressure makes it

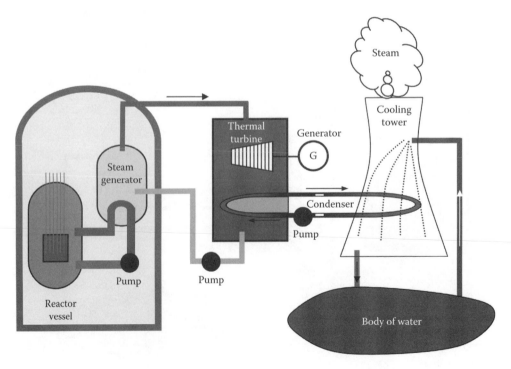

FIGURE 4.17 (See color insert following page 300.) Pressurized water reactor.

(a)   (b)

FIGURE 4.18 (See color insert following page 300.) Pressurized water reactor nuclear power plants.

absorb more energy. The PWR has three separate heat exchange loops, or water loops, but the water in these loops never mix. In the first loop, called primary loop, the pressurized water is pumped through the reactor to extract the thermal energy generated by the nuclear reaction and then pass through extremely strong pipes that lead to a steam generator (heat exchanger). The water in the first loop is radioactive. The secondary loop includes the heat exchanger, the turbine, and the generator. The water in the second loop is free from any significant radioactive material. The third loop is the cooling loop that includes the turbine–generator, the condenser, and the cooling tower.

4.3.6 Safety Features in Nuclear Power Plants

Safety has been an important consideration from the very beginning of the nuclear era. However in the 1940s, the developments were focused on demonstrating the potential of nuclear power as a reliable and cheap source of energy. Unfortunately, several nations sacrificed safety measures by adopting cheap and substandard designs.

The world enjoyed relatively safe nuclear power without any major accident until 1979. On March 28, 1979, a reactor at the Three Mile Island nuclear power facility near Harrisburg, Pennsylvania, suddenly overheated, releasing radioactive gases. Equipment failure and human error were the main reasons for the worst nuclear accident in the history of the United States. Although the plant engineers managed to shut down the reactor without damaging the core, public concerns regarding the safety of nuclear power plants was greatly intensified.

On April 1986, unit 4 of the Chernobyl nuclear power plant in the Ukraine was destroyed because of a combination of weak design and a series of human errors. The accident killed 31 people almost immediately and about 7 tons of uranium dioxide products escaped and began to spread across the surrounding areas.

These two accidents highlighted the need for effective safety measures to prevent future nuclear accidents. Until today, the nuclear industry is struggling to reverse the ever-growing skeptical public opinion regarding the integrity and safety of nuclear power plants. Among the reasons for public concern are the safety of the fuel assembly, the effectiveness of the control rods, and the loss of water inside the reactor. Each of these components is discussed in the following paragraphs:

Fuel rods: The fuel pellets are enriched uranium in the size of jellybeans. Most of the pellets are enriched with 3%–5% ^{235}U. The pellets are placed in tubes made of an alloy of zirconium and niobium. This alloy resists corrosion and has a low neutron absorption property. It is capable of maintaining its property even at high-temperature, high-pressure, and high-irradiation environments. The fuel assembly is often replaced every 3–5 years.

Control rods: The control rods are made of material such as boron, which absorbs neutrons. The boron, therefore, impedes the chain reaction and controls the rate of the fission reaction. The control rods are inserted between the fuel tubes or mixed with the fuel rods. During emergency conditions, the control rods would drop between the fuel rods and stop the nuclear reactions completely.

Reactor water: Reactor water removes excess heat from the reactors to prevent any meltdown. Water also slows down the neutrons and increases the probability of fission. Without water, neutrons would become too energized to initiate fission. Therefore, any loss of water would slow down the fission process dramatically.

4.3.7 DISPOSAL OF NUCLEAR WASTE

The fission fragments are radioactive with a half-life of thousands of years. The spent fuel rods from a nuclear reactor are the most radioactive of all nuclear wastes. The spent fuel rods are stored in special storage facilities that prevent radiation leakages. The storage facilities are of two types: temporary and permanent. The United States lacks permanent storage facilities, and its temporary facilities store the nuclear waste for a very long time. The temporary facilities are of two types: wet storage and dry storage.

4.3.7.1 Wet Storage

When the spent fuel rods are removed from the core of the reactor, they are extremely hot and must be cooled down. The rods are, therefore, placed in a pool filled with boric acid to cool them down and to allow the boric acid to absorb some of the radiation of the fission fragments. A diluted solution of boric acid is quite safe and is commonly used as antibacterial and eye drops. The spent fuel rods are immersed in the boric acid fluid for at least 6 months before they are transported to permanent storage facilities. As an additional safety measure, control rods are placed amongst the spent fuel rods to inhibit any fission action of leftover ^{235}U.

4.3.7.2 Dry Storage

After the spent fuel rods are cooled down in the wet storage facilities, they can be placed in temporary dry storage made of reinforced casks or buried in concrete bunkers. The casks are steel cylinders that are welded or bolted closed. Each cylinder is surrounded by an additional steel and concrete as a further measure against radiation leaks. The casks can be used for both storage and transportation.

4.3.7.3 Permanent Storage

Nuclear waste is composed of low- and high-level radiation waste. The low-level waste loses its radioactivity in a few hundred years and is often buried in shallow sites. The high-level waste, such as the spent fuel rod, is harder to dispose because it contains fission fragments that are radioactive for thousands of years. Therefore, they are buried in deep geological permanent storage facilities. This deep site must have no or little groundwater to prevent the erosion of the containments (steel cylinders). The site must also be stable geologically, so that earthquakes do not damage the containments. In the United States, a permanent storage site has been selected at Yucca mountain, Nevada. The site, which is expected to be ready by 2010, is about 460 m underground and is far from any population center.

EXERCISES

1. Name three types of hydroelectric power plants.
2. Why are cooling towers used in thermal power plants?
3. What are the two nuclear reactions?

4. What is the enrichment process of nuclear fuel?
5. How is the chain reaction controlled in nuclear power plants?
6. What is the main difference between BWR and PWR?
7. What is heavy water? Why is it used in nuclear power plants?
8. Write an essay on the latest technology used to dispose spent fuel rods.
9. A man owns a land that includes a low head water fall and wants to build a small hydroelectric plant. He used a tube of 3 m length and 1 m diameter as a penstock. He computed the speed of the water inside the penstock by dropping a small ball at the entrance of the tube and measuring the time it took to reach the other end of the tube. The ball traveled the penstock in 2 s.
   (a) Compute the mechanical power of the water.
   (b) If the coefficient of performance of the turbine is 0.5 and the efficiency of the generator is 90%, compute the expected generation of the site.
   (c) If the cost of building the small hydroelectric system is $20 K, compute the payback period if the cost of electricity from the neighboring utilities is $0.2/kW h. Assume that the water flow is always constant.
   (d) Is building this small hydroelectric system a good investment?
10. A hydroelectric dam creates a reservoir of 10 km^3. The average head of the reservoir is 100 m. Compute the potential energy of the reservoir.
11. A penstock is used to bring water from behind a dam into a turbine. The penstock is 6 m in diameter and moves water at a rate of 500 m^3/s. Compute the mechanical power entering the turbine.
12. The penstock of a hydroelectric power plant is 4 m in diameter. The penstock efficiency is 95% and the water head is 60 m. Compute the mechanical power of the water at the exit of the penstock. Also, compute the water flow inside the penstock.
13. A hydroelectric dam has a penstock that discharges 10^5 kg/s of water. The head of the dam is 80 m.
   (a) Compute the volume of the discharged water per second.
   (b) Assume that the penstock efficiency is 85%; compute the power of the water entering the turbine.
   (c) Compute the speed of the water inside the penstock.
   (d) Assume that the coefficient of performance of the turbine is 0.5 and the generator efficiency is 92%; estimate the generated electrical power.
   (e) Compute the overall system efficiency.
14. A natural gas power plant has a condenser that extracts 18,000 BTU/kg. Compute the mechanical energy of the turbine and the overall system efficiency.
15. An oil-fired power plant has a condenser that extracts 18,000 BTU/kg. Compute the mechanical energy of the turbine and the overall system efficiency.
16. Compare natural gas to oil in terms of thermal power and efficiency. Use the results of the previous two problems to verify your assessment.
17. Why are heat sinks used in thermal power plants?
18. Write a report on the general design of oil-fired power plants.
19. Write a report on the storage of nuclear waste.
20. Write a report on the disposal of the contaminated structure of a nuclear power plant.
21. Write a report on the nuclear accident of the Three Mile Island power plant. Show the sequence of events and comment on the safety measures taken to prevent any subsequent catastrophic failure.
22. Write a report on the Chernobyl nuclear accident. Show why such an accident is unlikely to happen in the United States.
23. Estimate the amount of nuclear energy produced by 10 kg of ^{235}U.

5 Environmental Impact of Power Plants

Beginning from the Stone Age when fire was used for heating, humans have continually been involved in activities that have harmful effects on their health and on the environment. Unfortunately, as we become more industrially advanced, the earth's atmosphere has become more polluted, water resources further contaminated, and the earth's crust more acidic.

A good starting point in the history of pollution is the eleventh century when wood was extensively used in Europe as a heating source. The pollution from burning wood was relatively limited, but wood became scarce and was replaced by coal at the beginning of the twelfth century. The dense smoke created by burning coal was initially viewed as just a minor discomfort. However, when coal was burned inside closed quarters with inadequate ventilation, the carbon monoxide released from the combustion suffocated a large number of people. This led King Edward I in the twelfth century to impose the death penalty upon anyone who was caught burning coal. The ban lasted for two centuries.

In the late seventeenth century, the Industrial Revolution started in Europe and spread all over the world. During this period, coal propelled the heavy industries as it was the only reliable source of energy. In the early twentieth century, oil started to replace coal and the automobile industry flourished along with several other heavy industries such as steel and rubber. This was the period of unchecked assault on the environment with automobiles responsible for 60% of all atmospheric pollution.

Until recently, pollution was primarily an urban phenomenon in industrial countries, but now it has spread all over the world with more than 20% of the world population living in communities that do not meet the World Health Organization (WHO) air quality standards. Only recently humans have begun to comprehend the severity of the problems created by pollution. In the early 1960s, several public and scientific organizations intensified their efforts to enforce various regulations to maintain the delicate ecological balance of nature. In the United States, President Richard M. Nixon signed into law the National Environmental Policy Act on January 1, 1970. The law led to the formation of the Environmental Protection Agency (EPA), which is chartered with the protection of human health and the environment.

Because air pollution respects no national boundaries, international cooperation and treaties have been established to reduce the flow of pollution across borders. The United Nations Commission on Sustainable Development is an example of an international umbrella organization that promotes control measures leading to a cleaner and healthier environment.

In the electric energy sector, over 99% of the electric energy generated worldwide is produced by the primary resources: fossil fuels, hydro, and nuclear. In 2006, the estimated worldwide consumption of electric energy from primary resources was about 16×10^{12} kW h. The extensive use of these primary resources is because they are readily available, produce enormous amount of energy, and are cheaper than alternative resources (solar, wind, etc.). From the viewpoint of the environment, each of these primary resources is associated with air, water, and land pollution. Although there is no pollution-free method for generating electricity (from primary or secondary resources), the primary resources are often accused of being responsible for most of the negative environmental impact. In this chapter, the various pollution problems associated with the generation of electricity are presented and discussed.

5.1 ENVIRONMENTAL CONCERNS RELATED TO FOSSIL FUEL POWER PLANTS

The earth's atmosphere is a mixture of gases and particles that surrounds the planet. Clean air is generally composed of the following mixture of gases. Parts per million (ppm) of the substance in air by volume is the number of molecules of the substance in a million molecules of air.

- Nitrogen (N_2), 78.1%
- Oxygen (O_2), 21%
- Argon (Ar), 0.9%
- Carbon dioxide (CO_2), 330 ppm
- Neon (Ne), 18 ppm
- Water vapor (H_2O)
- Small amounts of krypton (Kr), helium (He), methane (CH_4), hydrogen (H), nitrous oxide (N_2O), xenon (Xe), and ozone (O_3)

This delicate balance must be maintained for the air to be healthy for humans, animals, and vegetation. Unfortunately, the combustion of large quantities of fossil fuels could alter this balance, and could introduce other polluting gases and particles in local areas. Indeed, it is alleged that fossil fuels are probably the most air-polluting energy resources the worst among them being raw coal. A single raw coal power plant without adequate filters can pollute an entire city of the size of New York. Although pollution from burning fossil fuels can have a direct effect on the environment, it can also combine with other gases and particles to create more potent effects. This is known as a synergistic effect among pollutants.

The question often asked is what level of pollution is considered harmful? The U.S. EPA has established national ambient air quality standards (NAAQS) (Table 5.1) for six pollutants: carbon monoxide, lead, nitrogen dioxide, particulate matter, ozone, and sulfur dioxide. The table shows two types of national air quality standards: primary and secondary. The primary standards set limits to protect the health of sensitive population such as asthmatics, children, and the elderly. The secondary standards set limits to protect general public health as well as to guard against decreased visibility and damage to animals, crops, vegetation, and buildings.

When the level of a pollutant in an area exceeds a particular standard, the area is classified as nonattainment for that pollutant. In this case, the EPA imposes federal regulations on the polluters and gives deadlines, by which time the area must satisfy the standard.

5.1.1 SULFUR OXIDES

Fossil fuels such as coal, oil, and diesel are not pure substances, but are often mixed with other minerals such as sulfur (S) and nitrogen (N). Mined coal, in particular, contains more than 6% sulfur. When such a fossil fuel is burned, the released sulfur is combined with oxygen to form sulfur oxides (sulfur dioxide, SO_2 and sulfur trioxide, SO_3).

$$S + O_2 \Rightarrow SO_2 \tag{5.1}$$

Sulfur dioxide (SO_2) is a corrosive, acidic, and colorless gas with a suffocating odor that can cause severe health problems. When the concentration of the gas reaches 2 ppm, the suffocating odor of the gas can be easily detected. Inhaling large amounts of SO_2 can damage the upper respiratory tract and lung tissues. This can be more severe for the very young and the very old. In addition, asthmatic patients are more sensitive to SO_2 and their health status can deteriorate quickly. What makes SO_2 particularly dangerous is its rapid effect on people, usually within the first few minutes of exposure.

TABLE 5.1
EPA's NAAQS

Pollutant	Primary Standards	Averaging Times	Secondary Standards	Additional Restrictions
Carbon monoxide	9 ppm (10 mg/m³)	8 h	None	Not to be at or above this level more than once per year
	35 ppm (40 mg/m³)	1 h	None	Not to be at or above this level more than once per year
Lead	1.5 μg/m³	Quarterly	Same as primary	
Nitrogen dioxide	0.053 ppm (100 μg/m³)	Annual	Same as primary	
Particulate matter 2.5–10 μm	50 μg/m³	Annual	Same as primary	
	150 μg/m³	24 h		Not to be at or above this level for more than 3 days over a 3-year period
Particulate matter <2.5 μm	15 μg/m³	Annual	Same as primary	
	65 μg/m³	24 h		
Ozone	0.08 ppm	8 h	Same as primary	The average of the annual fourth highest daily 8 h maximum over a 3-year period is not to be at or above this level
	0.12 ppm	1 h	Same as primary	Not to be at or above this level for more than 3 days in a 3-year period
Sulfur oxides	0.03 ppm	Annual	—	
	0.14 ppm	24 h	—	Not to be at or above this level more than once per year
	—	3 h	0.5 ppm (1300 μg/m³)	Not to be at or above this level more than once per year

Table 5.2 summarizes various epidemiological studies linking SO_2 to respiratory and cardiovascular problems. Note that even a small value of parts per million can increase the health risk.

It is hard to estimate the exact amount of sulfur dioxide released after burning fossil fuels. However, a coal-fired power plant can produce as much as 7 kg/MW h of SO_2, while a natural gas power plant emits about 5 g/MW h. It is estimated that 20 million tons of SO_2 are released by power plants worldwide.

TABLE 5.2
Health Effects of SO_2

Concentration of SO_2 (ppm)	Exposure Time	Effect
3	3 min	Increased airway resistance
0.2	4 days	Increased cardiorespiratory diseases
0.04	1 year	Increased cardiovascular diseases

History has several examples of human tragedies due to excessive release of SO_2 from industrial plants other than power plants. In 1930, in Meuse Valley, Belgium, industrial pollution in the form of sulfur dioxide killed 63 people. In 1948, in Donora, Pennsylvania, sulfur dioxide emissions from industrial plants caused various respiratory illnesses in about 6000 people, and eventually caused the death of 20 people in a few days. In 1952, London experienced the worst air pollution disaster ever reported from burning coal during a dense foggy day; about 4000 people died mainly because of sulfur dioxide. During the first Gulf War, high concentrations of sulfur dioxide were released when oil fields were set on fire. Soldiers and civilians suffered from severe cardio-respiratory ailments.

To reduce sulfur emissions, most governments have imposed various regulations (such as the NAAQS in the United States), and are closely monitoring the level of emissions released by industrial plants. Penalties are often imposed on those who exceed the government set "quota" of sulfur emissions. Since fossil fuel power plants produce almost half of all sulfur emissions worldwide, the penalties are severe for utilities with inadequate sulfur-filtering systems. These regulations have led to the decline in sulfur emissions in the United States, Canada, and many cities in western Europe. The regulations also encourage owners of coal-fired plants to remove sulfur from coal before it is burned, or even convert their facilities to natural gas plants. However, unfortunately, sulfur emissions are still very high in a number of cities in eastern Europe, Asia, Africa, and South America.

5.1.2 NITROGEN OXIDES

Other harmful gases produced by burning fossil fuels are the nitrogen oxides NO_x (O_x represents various stages of oxidation). Coal or natural gas power plants produce about 2 kg/MW h of NO_x.

Nitrogen oxides are toxic gases with NO_2 being a highly reactive oxidant and corrosive element. NO_2 irritates the eye, nose, throat, and respiratory tract. Very high concentrations of NO_2 over long periods can cause respiratory infections and chronic bronchitis. In addition, NO_2 plays major roles in the formation of smog and acid rain. It absorbs sunlight resulting in the brownish color of smog.

5.1.3 OZONE

Ozone (O_3) can be found at two different altitudes: the troposphere (up to 10 km altitude) and the stratosphere (10–50 km altitude). The stratosphere has a high concentration of ozone of about 30,000 ppm.

The stratosphere ozone protects the earth by absorbing the dangerous ultraviolet radiation of the sun. However, because of gases such as the chlorofluorocarbons (CFCs), the ozone in the stratosphere area can be depleted. When CFC is decomposed by ultraviolet radiation, it releases chlorine atoms that react with ozone to create chlorine oxide and oxygen. By this process, one chlorine atom can destroy as many as 100,000 ozone molecules. Keep in mind that CFC is not released by power plants, but it is used as a refrigerant and can also be found in aerosol products. In 1978, CFC products were banned in the United States and more recently in most countries.

The troposphere ozone is formed when nitrogen dioxide (NO_2) is released by industrial plants. This pollutant, when excited by solar radiation, is converted into nitric oxide (NO) and in the process releases free oxygen atoms (O) that can combine with oxygen molecules (O_2) to form ozone at low elevations (up to 10 km).

$$NO_2 + \text{solar energy} \Rightarrow NO + O \tag{5.2}$$

$$O + O_2 \Rightarrow O_3 \tag{5.3}$$

Nitric oxide, NO, can also be formed during lightning storms, which facilitate the reaction of nitrogen and oxygen.

Although the stratosphere ozone is beneficial, the troposphere ozone is a pollutant with harmful effects on the respiratory system and can make asthmatic patients more sensitive to SO_2. It is also one of the main ingredients of smog that irritates lungs and causes damage to vegetation.

Fortunately, in nature, the troposphere ozone can be recycled back into nitrogen dioxide (NO_2) and oxygen molecules (O_2) when nitric oxide (NO) is available and sun energy is not high.

$$NO + O_3 \Rightarrow NO_2 + O_2 \tag{5.4}$$

Equations 5.2 through 5.4 form a delicately balanced process where ozone is formed, then destroyed. The reaction in Equation 5.4 is rapid, causing the concentration of O_3 to remain low as long as NO is available. However, when hydrocarbons are present (e.g., from automobile emissions), they react with NO to form organic radicals. These hydrocarbons then compete for the NO in air, so less of them are available to destroy the troposphere ozone. The result is an increase in the concentration of ozone at the troposphere level.

5.1.4 ACID RAIN

Sulfur and nitrogen dioxides produced by burning fossil fuels are the main ingredients of acid rain. Sulfur dioxide can further react with oxygen to form sulfur trioxide.

$$2SO_2 + O_2 \Rightarrow 2SO_3 \tag{5.5}$$

When sulfur trioxide reaches the clouds, it reacts with water to form sulfuric acid (H_2SO_4).

$$SO_3 + H_2O \Rightarrow H_2SO_4 \tag{5.6}$$

Similarly, when nitrogen dioxide (NO_2) reaches the clouds, it reacts with water and nitric acid (HNO_3) is formed.

$$3NO_2 + H_2O \Rightarrow 2HNO_3 + NO \tag{5.7}$$

Acid rain is the precipitation from clouds impregnated with these acids. Acid rain can be very damaging to crops, agricultural lands, and structures. When it reaches lakes, acid rain increases the acidity of water, which can have severe effects on fish populations. Acid rain can also damage limestone, historical buildings, and statues. Figure 5.1 shows two examples of acid rain damage. Figure 5.1a shows a statue in Santiago, Spain, washed away by acid rain and Figure 5.1b shows a tree that died from exposure to acid rain.

The acidity in the acid rain depends on the concentration of its hydrogen ions. Scientists have developed a scale called the potential of hydrogen (pH) to quantify the degree of acidity in a solution. It is a negative logarithmic measure of hydrogen ion H^+ concentration in moles per liter of solution. The mole is a chemical unit used to measure the amount of substance that contains as many elementary entities (atoms, molecules, etc.) as there are atoms in 12 g of the isotope carbon-12 (6.023×10^{23}).

$$pH = -\log H^+ \tag{5.8}$$

Since pH is a negative logarithmic scale, a change in just one unit from pH 3 to 2 would indicate a 10-fold increase in the acidity. The hydrogen ion concentration in pure water is about 1.0×10^{-7} mol; hence its pH is 7. Increasing the concentration of hydrogen ions increases the acidity of the liquid, thus pH < 7 is considered acidic, while pH > 7 is considered alkaline or basic. The pH for milk is about 7, while battery acid has a pH from 0 to 1. Clean rain usually has a pH of 5.6. Rain measuring

(a)   (b)

FIGURE 5.1 (See color insert following page 300.)   Effects of acid rain on (a) stones and (b) trees.

less than 5 on the pH scale is considered acid rain. The pH of the most acidic rain reported in the United States was about 4.3.

In the United States, the areas that suffer the most from acid rain are the Northeast and Midwest regions, which are known for their coal-based industries and coal-fired power plants.

5.1.5 CARBON DIOXIDE

The temperature of the earth's surface is determined by the difference between the solar energy reaching the earth and the radiated energy from the earth back to space. Because some naturally occurring gases form thermal blankets at various altitudes, not all solar energy is radiated back to space. This phenomenon is loosely known as the greenhouse effect. The balance between the solar energy reaching the earth and the radiated energy is delicately maintained in nature to keep the temperature on earth at a level that can sustain life. Some scientists believe that the release of greenhouse gases from industrial plants reduce the amount of the radiated energy, and therefore increase the temperature on earth. This phenomenon is known as global warming. The greenhouse gases include CO_2, CFCs, CH_4, N_2O, and O_3.

Global warming could lead to partial melting of glaciers and ice sheets, which can result in the sea level rising and flooding dry lands. Also, an increase in sea temperature could result from changes in the pattern of rain and wind. Some scientists believe that greenhouse gases have been increasing in concentration since the beginning of the Industrial Revolution in the seventeenth century. These scientists estimated that the global temperature on earth has increased by as much as

1°C during the last century. A counterargument made by other scientists is that while greenhouse gases can increase the temperature, other gases such as the stratospheric ozone and sulfate cause the atmosphere to cool down. Therefore, global warming may not be as severe as some have thought. This debate is not likely to end anytime soon.

CO_2, which is one of the greenhouse gases, is a colorless, odorless, and slightly acidic gas. Nature recycles CO_2 through water, animals, and plants. Humans exhale CO_2 as they breathe, and plants absorb it during photosynthesis.

$$CO_2 + H_2O + \text{solar energy} \Rightarrow O_2 + \text{carbohydrates} \tag{5.9}$$

CO_2 is also absorbed in oceans. The process of generating and absorbing CO_2 is precisely sustained in nature resulting in the right amount of CO_2 in air necessary to keep the earth's temperature at the current level. However, industrial activities that require burning of material containing carbon (coal, wood, and oil) can increase CO_2 concentration in air thus contributing to the warming of the atmosphere.

Coal-fired power plants produce as much as 1000 kg/MW h of CO_2 and natural gas power plants produce about half this amount. But keep in mind that CO_2 is only one of the greenhouse gases with relatively limited effect since its presence in the atmosphere is for short periods. Other gases with more sinister effects include CFCs, such as freon, and N_2O. One molecule of CFC has the same effect as 10,000 molecules of CO_2. What makes these gases much worse than CO_2 is that they remain in the atmosphere for very long periods.

5.1.6 Ashes

Ashes are small particles (0.01–50 μm) that are suspended in air. The combustion process in fossil fuel power plants produces a large amount of ash that could stay suspended in air for days before it reaches the earth. About 7 million tons of ash are released each year by electric power plants and industrial smelters. Ash from power plants may include a variety of metals such as iron, titanium, zinc, lead, nickel, arsenic, and silicon.

Ash affects breathing, weakens the immune system, and worsens the conditions of patients with cardiovascular diseases. Smaller ash (less than 10 μm) can reach the lower respiratory tract causing severe respiratory problems.

Most power plants install several filtering devices to eliminate or substantially reduce discharged ash. There are various types of filters, among them are the wet scrubber system and the fabric filter system. With the wet scrubber, the exhaust gas is passed through a liquid that traps flying particulates and sulfur dioxide before the gas is vented through the stacks. The fabric filter system works like a vacuum cleaner where the particulates are trapped in bags.

5.1.7 Legionnaires' Disease and Cooling Towers

A harmful group of bacteria known as Legionella can live inside cooling towers as well as in hot water systems, warm freshwater ponds, and creeks. One of these bacteria causes Legionnaires' disease, a form of infection that could lead to pneumonia.

For the cooling towers that vent their steam in air, hot water droplets carrying Legionella can be mixed with the steam. The droplets drift with wind and can cause infections in people several kilometers away from the cooling towers. To protect the public from these harmful bacteria, power plants install mist eliminators to prevent water droplets from escaping the cooling towers. In addition, biocides are used inside the cooling towers to prevent the growth of Legionella. Although these measures can be very effective in preventing the disease, not all utilities worldwide implement them.

5.2 ENVIRONMENTAL CONCERNS RELATED TO HYDROELECTRIC POWER PLANTS

Hydroelectric power is a major form of renewable energy. Although it does not emit any gas, the hydroelectric plant impacts the environment in other ways as discussed below.

Flooding: This is the most obvious impact of an impoundment-type hydroelectric power plant. A dam is constructed to form a reservoir by flooding the upstream land. The decaying vegetations submerged by the flooding emits greenhouse gases. Flooding can also release dangerous substances embedded in rocks, such as mercury, which accumulates in fish resulting in health hazards when consumed. Socially, people can also be affected by the dam; when China constructed the Three Gorges Dam, millions of people were evicted from areas behind the dam.

Water flow: Hydroelectric dams alter the flow of water downstream, which may change the quality of water in the river as well as its flow rate.

Silt: Dams and reservoirs can trap silt that would normally be carried downstream to fertilize lands and prevent shore erosions. In addition, trapped silt behind the dam can reduce the amount of water stored in the reservoir, which can reduce the amount of generated electricity.

Oxygen depletion: In deep reservoirs, cooler water sinks to the bottom because of its higher density. This could reduce the oxygen at the bottom of the reservoir, thus altering the biota and fish population that live at the bottom of the lake.

Nitrogen: When water spills over a dam, air is trapped in the water creating turbulence. This makes the water more dissolvent to nitrogen, which could be toxic to fish.

Fish: Dams alter the downstream flow of the river, which may affect the migration of certain species of fish. Hydroelectric turbines also kill some of the fish that pass through the penstocks.

5.2.1 CASE STUDY OF THE ASWAN DAM

After Egypt built the Aswan Dam in the 1960s, inexpensive electricity was generated and the flow of the Nile was controlled all year around for the first time in history. However, the dam created several ecological and social problems. One of the most tragic effects of the Aswan Dam was that the rising water submerged many valuable ancient monuments that are over 5000 years old. Some of these monuments, such as the temple of Abu Simbel built by Ramses II in the thirteenth century BC, were raised to higher dry lands. But the monument in its new location lost some of its miraculous engineering features such as sunlight penetration of the sanctuary on October 20 and February 20 of each year and the daily melodies played in various quarters.

One of the most important problems of the Aswan Dam is the entrapment of the annual flood sediment behind the dam creating a number of problems such as

- Slow filling of the lake behind the dam by sediment, resulting in spilling of the water over its shores. The increase in the surface area of the lake increases the evaporation, which leads to undesired increase in humidity in the otherwise dry climate. This increased humidity negatively impacts valuable ancient monuments in the area.
- Virtual elimination of silt that used to make fertile the agricultural land downstream, thus, negatively impacting farming in Egypt.
- Lack of sediment downstream exposes the banks of the Nile to steady erosion.
- Continuous recession of the shoreline of the Nile delta since the sediment that used to reach the Mediterranean shores counteracted the effects of coastal erosion.
- Mediterranean saltwater reaching upstream of the Nile, thus contaminating farmland and groundwater.
- Lack of sediment in the river allows more sunlight to penetrate the water as it is now clearer. This increases the population of several biota and plants such as phytoplankton,

which affects the fish population in the river, and makes the water more polluted. Thus more chlorine is added to the drinking water.

- Elimination of certain types of fish such as sardines and shrimps because the influx of nutrients associated with the annual silt rich flood can no longer reach the Mediterranean Sea to feed the local biota and fish.

- Frequent weak tremors in the area, thought by some scientists to occur as a result of the sediment trapped behind the dam.

5.3 ENVIRONMENTAL CONCERNS RELATED TO NUCLEAR POWER PLANTS

The public is highly concerned about nuclear power plants. Although some of this concern is justifiable, most is based on misinformation or the erroneous association of nuclear power plants with nuclear weapons. Nuclear power plants operated safely for almost 40 years until the accident at the Three Mile Island nuclear power plant that was followed by the Chernobyl nuclear disaster (see Chapter 4). These two accidents intensified public pressure to set stricter safety regulations on nuclear power plants or to abandon nuclear power altogether. The debate is still going strong and is not likely to end anytime soon. Among the public concerns are the following:

Radioactive release during normal operation: In boiling water reactors (BWRs), the fuel rods are in direct contact with steam. Hence, some of the radioactive elements from the nuclear reaction are carried by steam and eventually reach the turbine and condenser. Although the BWR has several layers of protections that hold the radioactive steam for several half-life cycles before it is released, accidental release is still a major public concern. In pressurized water reactors (PWRs), this problem does not exist since the fuel rods are separated from the steam cycle.

The release of a small amount of radioactive gas is not necessarily a health hazard. The limit on radiation set by the EPA is 170 milliREM/year above the background radiation from all natural resources. Radiation equivalent man (REM) is a measure of biological damage to tissues by specific amounts of radiation. High doses of REM can cause immediate radiation sickness such as hypodermal bleeding and hair loss. Under 25 REM, no short-term effects are observed, but long-term exposure to these small amounts of REM may lead to serious health issues such as cancer and may even cause genetic abnormality in offspring.

Loss of coolant: One of the major public concerns is the loss of coolant inside the reactor. If the flow of water is interrupted due to breaks in pipes or pumps, fuel rods could become excessively hot and could melt down. This scenario is sensationalized in movies and is given the name "China syndrome," where melted rods at the bottom of the reactor go through the reactor bottom into earth and all the way to China on the other side of the globe.

Needless to say this problem is not as severe as it is being depicted in some circles. The water inside the reactor is both a coolant and a moderator. The moderator slows down the neutrons to make fission possible. If the water inside the reactor is lost, the chain reaction is substantially curtailed. In addition, nuclear power plants have emergency core cooling and emergency core shutdown systems that prevent the meltdown of the fuel rods.

Reactor explosion: Nuclear material is wrongly associated with nuclear bombs, and nuclear explosion inside the reactor is often stated as a worry. This concern is unfounded since the uranium used in nuclear power plants is about 3%–5% enriched, while that used in nuclear bombs is over 90% enriched. Furthermore, the chain reaction in nuclear power plants is slowed down by the moderator; when the temperature of the reactor increases above the design limits, the moderator density increases, thus slowing down the chain reaction.

Disposal of radioactive waste: This is one of the crucial issues faced by the nuclear power industry. The spent fuel is a high-level radioactive waste with a half-life of thousands of years. Besides being a radioactive hazard, the spent fuel can stay hot for hundreds of years.

In the United States, about 2000 tons of spent fuel is produced annually from nuclear power plants, and much more is produced by several nuclear defense programs. The storage facilities of the nuclear waste must prevent for long periods the radioactive material from leaking. One type of storage vessel used in the industry is a container composed of multiple layers of steel and concrete barriers. These containers are eventually buried in areas that are geologically stable and are surrounded by rocks to prevent any leakage from reaching the water table. Some nations store the containers on the deep ocean floor. Some scientists have proposed some wild ideas such as shooting the radioactive waste into the sun, or storing the radioactive material in space. This is obviously a highly sensitive issue for the public.

A more realistic option is to reduce or eliminate the long-life radioactive waste by transmutation. Nuclear transmutation is a process by which the long-life fuel waste is converted into shorter-life waste that is easier to store. Transmutation occurs when radioactive waste is bombarded with charged material in accelerators.

EXERCISES

1. Name three types of major power plants.
2. How is acid rain formed?
3. What is the greenhouse effect?
4. State four drawbacks of a hydroelectric power plant.
5. What are the health effects of SO_2?
6. Fill in the blanks in the following sentences:
   (a) Sulfur dioxide emitted from power plants reacts with oxygen to form _____.
   (b) When sulfur trioxide reaches the clouds, it reacts with water to form _____.
   (c) When nitrogen dioxide reaches the clouds, it reacts with water and _____ is produced.
7. A coal-fired power plant produces an average of 100 MW annually. Estimate the sulfur dioxide released daily and annually if no filtering system is used.
8. A natural gas power plant produces an average of 100 MW annually. Estimate the sulfur dioxide released daily and annually if no filtering system is used.
9. If a 100 MW coal-fired power plant is converted into natural gas, compute the annual percentage reduction of SO_2.
10. In 2006, the world consumed 16.28×10^{12} kW h of electricity. Estimate the maximum amount of sulfur dioxide released from the coal-fired power plant worldwide.
11. Write a report on an area with severe acid rain. Discuss the pH values and their effect on the local environment.
12. How many moles of hydrogen ions are in a liquid with pH 5?
13. Choose a case study for a hydroelectric power plant. Identify the pros and cons of the plant.
14. Write a report on the impact of hydroelectric power plants on the fish populations.
15. Identify a few ideas to lessen the effect of hydroelectric power plants on fish migrations.
16. Identify a few ideas to increase silt in the downstream flow from hydroelectric power plants.
17. Write a report on the Three Mile Island accident. Identify the sequence of events that led to the accident. In your opinion, were the steps taken to correct the accident adequate?
18. Write a report on the Chernobyl nuclear disaster. Identify the sequence of events that led to the disaster. In your opinion, can this accident occur in the United States? Why?
19. Write a report on permanent storage of nuclear waste worldwide.

6 Renewable Energy

As seen in the previous chapters, the world relies heavily on fossil fuels (coal, oil, and natural gas) for its ever-growing appetite for energy. In the early 1950s, public concern regarding the negative environmental impact of burning fossil fuels encouraged engineers and scientists to develop reliable alternative energy resources. The efforts were accelerated in the 1970s when oil prices soared. Many countries began investing in renewable energy through various programs that encouraged the development and testing of reliable renewable energy systems. Tax credits, investments in research and development, subsidies, and developing favorable regulations are some of the various support measures taken by governments to accelerate the development of renewable energy technologies. However, unfortunately, the development efforts are still largely dependent on the price of crude oil as well as the intensity of societal pressures.

Renewable energy is a phrase that is loosely used to describe any form of electric energy generated from resources other than fossil and nuclear fuels. Renewable energy resources include hydropower, wind, solar, wave and tide, geothermal, and hydrogen. The sun is the source of all those renewable energies with the exception of geothermal and tidal energy. These resources produce much less pollution than burning fossil fuels and are constantly replenished, and thus called renewable.

Since the late nineteenth century, scientists and engineers have worked on developing technologies to generate electric energy using various forms of renewable resources. In 1842, Sir William George Armstrong invented a hydroelectric machine that produced frictional electricity. However, the world's first hydroelectric power plant was constructed across the Fox River in Appleton, Wisconsin, in 1882. The Italian Prince, Piero Ginori Conti, invented the first geothermal power plant in 1904. The plant is located at Larderello, Italy.

The developments in the renewable energy field during the past few years have led to more efficient and more reliable systems. Some of these systems, such as solar and fuel cells, are already used in aerospace and transportation. Also, wind energy is used to generate electricity all over the world. In this chapter, we shall discuss various renewable energy systems, but the reader should be aware that the area is growing fast, and new technologies and systems are invented every year.

6.1 SOLAR ENERGY

The sun is the primary source of energy in our solar system; the earth receives 90% of its total energy from the sun. Other natural forms of energy on earth include geothermal energy, volcanoes, and earthquakes.

Sunrays are packed with high energy. An area of $1 \, \text{m}^2$ in space can receive as much as $1.366 \, \text{kW}$ of power from sunrays. This translates into an extraterrestrial power density of about $1.366 \pm 4\%$ kW/m^2. This tremendous power is weakened at the earth's surface due to several reasons, such as the following:

- Various gases and water vapor in the earth's atmosphere absorb some of the solar energy.
- Distance from the sun, which can be measured by the projection angle of the sunrays (zenith angle); see Figure 6.1.
- Various reflections and scatterings of sunrays.

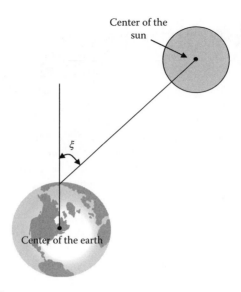

FIGURE 6.1 Zenith angle ξ.

The computation of solar power density on the earth, also called solar irradiance, is complex and requires the knowledge of several hard to find parameters. However, approximate models can be used, such as the one developed by Atwater and Ball and given in the following equation:

$$\rho = \rho_o \cos \xi (\alpha_{dt} - \beta_{wa})\alpha_p \tag{6.1}$$

where

ρ is the solar power density on the earth's surface (kW/m^2)

ρ_o is the extraterrestrial power density (the number often used is 1.353 kW/m^2)

ξ is the zenith angle (angle from the outward normal on the earth's surface to the center of the sun as shown in Figure 6.1)

α_{dt} is the direct transmittance of gases except for water vapor (the fraction of radiant energy that is not absorbed by gases)

α_p is the transmittance of aerosol

β_{wa} is the water vapor absorptions of radiation

The term "aerosol" refers to atmospheric particles suspended in the earth's atmosphere such as sulfate, nitrate, ammonium, chloride, and black carbon. The size of these particles normally ranges from 10^{-3} to 10^3 μm. As explained in Chapter 5, some aerosols such as CO_2, N_2O, and troposphere ozone cause the atmosphere's temperature to rise; and some such as stratospheric ozone, sulfate, and the by-product of biomass burning cause it to cool down.

Due to reflection, scattering, and absorption, the solar power at the earth's surface is a fraction of the extraterrestrial solar power. The ratio of the two solar power densities is known as the solar efficiency η_s.

$$\eta_s = \frac{\rho}{\rho_o} = \cos \xi (\alpha_{dt} - \beta_{wa})\alpha_p \tag{6.2}$$

The solar efficiency varies widely from one place to another and is also a function of the season and the time of the day. It ranges between 5% and 70%. The zenith angle has a major effect on the efficiency; for the same absorption conditions, the maximum efficiency occurs at noon in the equator when $\xi = 0°$.

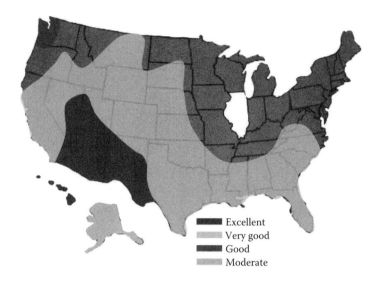

Excellent
Very good
Good
Moderate

FIGURE 6.2 (See color insert following page 300.) Solar power density in the United States.

EXAMPLE 6.1

A person wants to install a solar energy system in Nevada. At a certain time in the afternoon, the zenith angle is 30°, the transmittance of all gases is 70%, the water vapor absorption is 5%, and the transmittance of aerosols is 90%. Compute the power density and the solar efficiency at that time.

Solution

By directly substituting the parameters in Equation 6.1

$$\rho = \rho_o \cos \xi \, (\alpha_{dt} - \beta_{wa})\alpha_p = 1353 \cos 30 \, (0.7 - 0.05)0.9 = 685.5 \text{ W/m}^2$$

$$\eta_s = \frac{\rho}{\rho_o} = \frac{685.5}{1353} = 50.7\%$$

Figure 6.2 shows the map of solar power density in various regions in the United States. In the Mohave Desert and Nevada, the solar power density can be as much as 750 W/m². In the Northwest region, it can be as high as 300 W/m². Figure 6.3 shows a typical daily average of the world solar

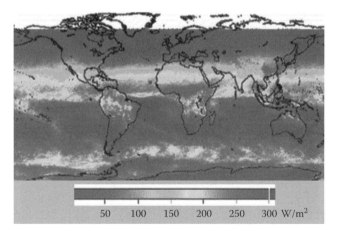

50   100   150   200   250   300 W/m²

FIGURE 6.3 (See color insert following page 300.) Sample of solar power density worldwide in December. (Image courtesy of NASA.)

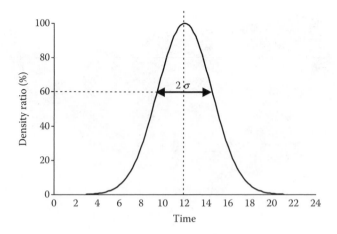

FIGURE 6.4 Typical solar distribution function (solar power density in a 24 h period).

power density in December. Central and South Africa, Argentina, Chile, Indonesia, and most of Australia can have as much as a daily average of 300 W/m². In these areas, the peak solar power density during the day can reach values higher than 700 W/m².

The solar power density during a typical day with stable weather follows the bell-shaped curve expressed by the normal distribution function in Equation 6.3 and Figure 6.4.

$$\rho = \rho_{max} \exp\left[\frac{-(t - t_o)^2}{2\sigma^2}\right] \tag{6.3}$$

where

t is the hour of the day using the 24 h clock
ρ_{max} is the maximum solar power density of the day at t_o (noontime in the equator)
σ is the standard deviation of the normal distribution function

The density ratio in Figure 6.4 is the percentage of the ratio ρ/ρ_{max}. Notice that when $t = 12 \pm \sigma$, $\rho/\rho_{max} = 0.607$.

A large σ means wider areas under the distribution curve, i.e., more solar energy is acquired during the day. In high latitudes, σ is smaller in the winter than in the summer. This is because daylight time is shorter during the winter as you move north.

EXAMPLE 6.2

An area located near the equator has the following parameters:

$$\alpha_{dt} = 80\%, \quad \alpha_p = 95\%, \quad \beta_{wa} = 2\%$$

Assume that the parameters are unchanged during the day, and the standard deviation of the solar distribution function is 3.5 h. Compute the solar power density at 3:00 PM.

Solution

The first step is to compute the maximum solar power density. Since the location is near the equator, the maximum solar power density occurs around noon when the zenith angle is almost zero.

$$\rho_{max} = \rho_o \cos \xi \, (\alpha_{dt} - \beta_{wa})\alpha_p = 1353 \cos 0 \, (0.8 - 0.02)0.95 = 1.0 \text{ kW/m}^2$$

The solar power density at 3:00 PM is then

$$\rho = \rho_{\max} \exp\left[\frac{-(t - t_0)^2}{2\sigma^2}\right] = 1.0 \exp\left[\frac{-(15 - 12)^2}{2(3.5)^2}\right] = 0.693 \text{ kW/m}^2$$

Solar energy is typically harnessed by two methods: passive and active. A passive solar energy system uses the sunrays directly to heat liquid or gas. The heated liquid or gas can then be used for air and water conditioning and industrial processes. An active system converts the sun's energy into electrical energy by using a photovoltaic (PV) semiconductor material called solar cell. The electricity generated can be used locally, exported to the power grid, or both.

6.1.1 Passive Solar Energy System

An example of the passive solar system is the thermosiphon hot water appliance shown in Figures 6.5 and 6.6. The system consists of a solar collector, water tank, and water tubes. The tank is located above the collector. The collector has an outer lens (or transparent glass) facing the sun and houses a long zigzagged water tube. The lens concentrates the sunrays, thus increasing the temperature of water inside the tubes. The warm water moves naturally upward to the tank (hot water rises above cold water). Since the water in the upper part of the tank is warmer than the water in the lower part, the cold water at the bottom of the tank goes back to the collector. When warm water is needed inside the house, it is extracted from the top of the tank. This water is replaced by the cold water from the main water feeder of the house.

The passive solar system is simple, inexpensive, and requires little maintenance. However, it demands enough solar power density to make it viable, and it is most effective during the daytime.

On a larger scale, passive solar systems can be used to concentrate solar radiation to increase the temperature of any fluids to the level needed in industrial applications. Figure 6.7a shows one of these systems, which is known as a solar farm or thermosolar system. The solar farm can be integrated with a thermal power plant to produce electricity on a continuous basis. The system is also known as Integrated Solar Combined Cycle System (ISCCS). It consists of a large number of parabolic trough mirrors called collectors, which concentrate the solar radiation energy on a pipe system called receiver located at the focal area of the collector mirrors (Figure 6.7b). The fluid of the receiver, which is often oil, is circulated through the receiver and is heated to about 400°C.

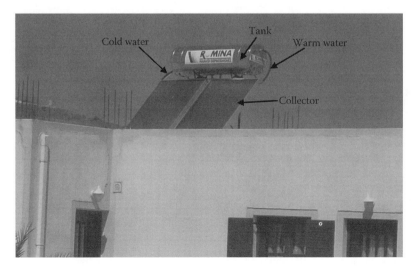

FIGURE 6.5 Passive thermosiphon hot water solar system.

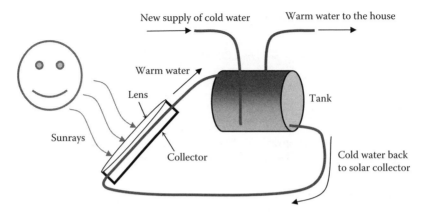

FIGURE 6.6 (See color insert following page 300.)   Thermosiphon hot water system.

The fluid is pumped to a heat exchanger to generate steam that can be used to generate electricity in a nearby thermal power plant. During the night, natural gas is used to produce the steam needed by the plant. Hence, the power generation is continuous.

6.1.2 ACTIVE SOLAR ENERGY SYSTEM (PV)

Light consists of particles called photons, which are the energy by-products of the nuclear reactions in the sun. Each photon is a packet of energy, but not all photons have the same amount of energy. Photons with shorter wavelengths (higher frequencies) such as gamma rays (about 10^{20} Hz) have more energy than photons with longer wavelengths such as visible light. The frequency range of visible sunlight is $4.5–7.5 \times 10^{14}$ Hz.

The photoelectric phenomenon was discovered in 1839 by the French physicist Edmond Becquerel (1820–1891) and was further refined by the German physicist Heinrich Hertz (1857–1894) in 1887. They found that when light photons hit certain materials, the electrons in the material absorb the energy of the photons. When this acquired energy is higher than the binding energy of the electrons, the electrons break away from their atoms. If the acquired energy of these electrons is harnessed, we have achieved photoelectric conversion of sunlight into electricity. The device to do this process was invented in the mid-twentieth century by the Bell Laboratories in

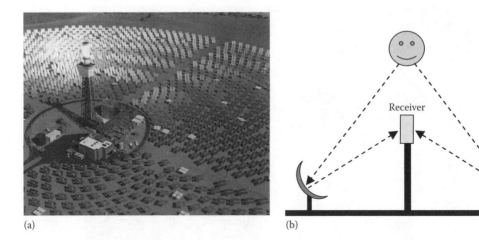

(a)   (b)

FIGURE 6.7 (See color insert following page 300.)   Solar farm. (Images courtesy of the U.S. Department of Energy.)

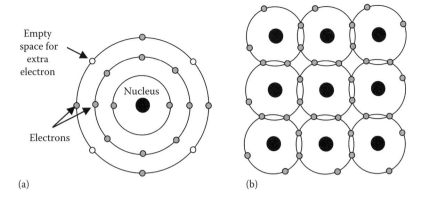

FIGURE 6.8   Silicon: (a) atom and (b) its crystal structures.

the United States and was called solar cell. It was designed to collect the breakaway electrons to form electric current. They used their solar cells to replace some of their dry cells (batteries) in their telephone network. The solar cell is often called PV, a compound word derived from two words: photo, which is a Greek word for light, and volt, a unit for measuring electric potential.

PV cells are made of semiconductor materials that include silicon (Si). A silicon atom has 14 electrons arranged in three energy levels (shells). The first two shells have 10 electrons, and the third (outer) shell has only 4 electrons, as shown in Figure 6.8. Although the atom is electrically neutral, the outer shell is only half full and there is enough space for four more electrons. Consequently, each silicon atom combines itself with four other atoms to form the silicon crystal structure shown in Figure 6.8b. Now, each atom has eight electrons in its outer shell.

Since all electrons are used in bonding, the crystal structure of silicon is a good insulator. To make silicon more conductive electrically, additives (impurities) are added. Common additives are phosphorus (P) and boron (B). Phosphorus has five electrons in its outer shell while boron has only three. The process of adding these two impurities is called doping.

When phosphorus is added to silicon, they bond as shown in Figure 6.9a. However since phosphorous (P) has five electrons in its outer shell, one electron in the compound will not bond to another atom. When this electron is given extra energy, it breaks free and roams inside the material

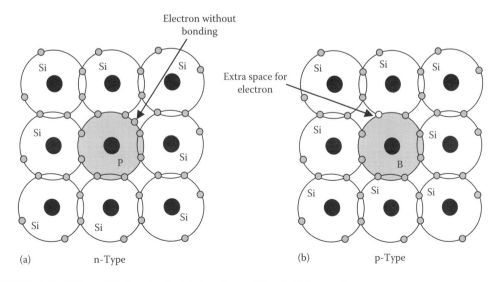

FIGURE 6.9   Silicon (Si) doped with (a) phosphorus (P) and (b) boron (B).

looking for a positive charge (hole) to attach to. This electron is known as a free carrier. Because of the presence of these negatively charged free carriers, this phosphorous–silicon compound is called n-type. Keep in mind that this n-type material is still electrically neutral since the free electrons are still inside the compound.

When silicon is doped with boron as shown in Figure 6.9b, the mix will have free holes because boron has only three electrons in its outer layer. This material is called p-type. Again, the p-type material is electrically neutral; the missing electrons (holes) are balanced out by the missing protons in the boron.

When n-type silicon is attached to p-type silicon, they turn into a device known as a diode or p–n junction. When the two types are attached, free electrons in the n-type silicon move toward the free holes in the p-type silicon and vise versa. They join in the junction between the two materials, which is known as the depletion zone. At equilibrium condition, the depletion zone creates a barrier that makes it harder for any more free electrons in the n-type material to move into the holes in the p-type and vice versa. This p–n junction device is the main component of the PV cell.

There are two major types of PV cells: concentrating and flat-plate. The structure of the concentrating PV cell is shown in Figure 6.10. The cell consists of a lens mounted on top of the n-type material. The p-type material is at the base of the cell. When the cell is illuminated, the electrons of the p–n junction acquire some energy from the light photons, which helps them to break free from their atoms. If we connect the two terminals of the p–n junction to a resistive load, the electrons move from the n-type side of the junction to the load and back to the p-type terminal. In this process, the energy acquired by the electrons is discharged in the load resistance. When the electron goes back to the p–n junction, it acquires new energy from the sunrays and the process is repeated. Notice that the direction of the current is opposite to the direction of the electrons; this is the convention adopted over 100 years ago.

The flat-plate PV cell, as the name implies, is rectangular and flat. This is the most common type of PV array used in commercial applications. Flat-plate cells are often mounted at fixed angles that maximize the exposure to the sun throughout the year; in the United States, it is the southern direction. In more flexible systems, the angle of the solar panel changes to track the optimal sun exposure during the day.

Because convex lenses concentrate lights, concentrating PV cells require less material for the same power output than the flat-plate cells; thus they are smaller in size. However, concentrating cells operate best when the sky is clear of clouds. On cloudy days, diffused light through the clouds can still produce electricity in flat-plate PV cells while concentrating PV cells generate less power with diffused light.

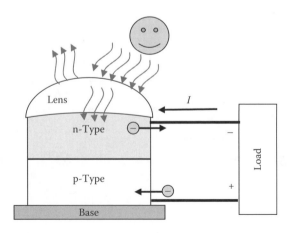

FIGURE 6.10 Concentrating PV cell.

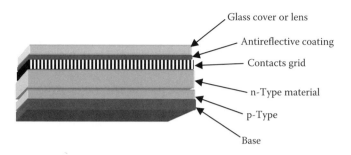

FIGURE 6.11   Main parts of a PV cell.

The construction of a solar cell requires several added components such as a cover glass, antireflective coating, and connecting grid. These components are shown in Figure 6.11. The cover glass is mounted on top of the cell to protect it from the harsh environment (dust, scratch, bird dropping, etc.). Antireflective coating is used to reduce the reflection losses of silicon because silicon is a very shiny material. A contact grid (mesh) is used to collect the electrons from the top of the n-type material. Of course, we cannot use solid plate instead of the mesh as it will prevent the sun's energy from reaching the p–n junction. Also, we cannot collect the electrons from the side of the junctions as this would require the electrons to travel long distances through the material, which will increase the internal losses of the PV. The base at the bottom is made of a solid plate if the cell is of a single-layer design.

Newer designs of PV cells consist of dual and triple junctions. These types of PV cells are made of two or three cells vertically stacked. Although more expensive, they produce more power, have higher efficiency and higher voltage than single-layer PV cells.

6.1.2.1   Ideal PV Model

The p–n junction diode is represented by the symbol shown in Figure 6.12. The p-type material (also known as anode) is represented by a triangle, and the n-type (cathode) is represented by a line at the top of the triangle (Figure 6.12b). To obtain the current–voltage (*I–V*) characteristics of the diode, the simple circuit shown in Figure 6.13a is used. The anode of the diode is connected to a direct current (dc) adjustable voltage source and the cathode is connected to a resistive load. When the anode-to-cathode voltage V_d is positive (forward biased), the current flows through the diode with little resistance. This is because the positive potential on the p-type material helps its holes to acquire enough energy to cross the barrier of the depletion zone and travel into the n-type side. When the voltage is reversed

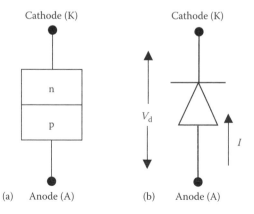

FIGURE 6.12   Representation of a p–n junction diode.

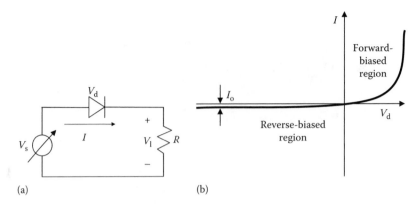

FIGURE 6.13 Characteristic of a p–n junction diode.

(reverse biased), almost no current flows though the diode. This is because the negative potential of the anode forces all holes to stay inside the p-type material. The characteristic is shown in Figure 6.13b. When the diode is forward biased, the voltage drop across the diode is very small (about 0.6 V), and when the diode is reverse biased, the current of the diode I_o (reverse saturation current) is very small.

The characteristic in Figure 6.13 can be obtained by using the diode model developed by the German physicist, Walter H. Schottky (1886–1976).

$$I = I_o \left(e^{\frac{V_d}{V_T}} - 1 \right)$$

$$V_T = \frac{kT}{q}$$
(6.4)

where
   I_o is the reverse saturation current of the diode
   V_d is the voltage across the diode
   V_T is known as thermal voltage
   q is the elementary charge constant (1.602×10^{-19} C)
   k is the Boltzmann's constant (1.380×10^{-23} J/K)
   T is the absolute temperature (K) (to convert from Celsius to kelvin, 273.15 is added to the
      Celsius value)

Since the solar cell is just a diode whose electrons acquire their energy from light photons, it seems reasonable that we use the equivalent circuit in Figure 6.14. The sun in the figure represents the source of energy, and the cell is represented by a diode whose current flows from the n-junction to the p-junction and then into the load, as explained earlier in Figure 6.10. This current is the reverse-biased

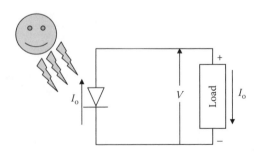

FIGURE 6.14 Modeling an ideal solar cell—first step.

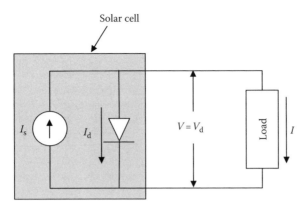

FIGURE 6.15   Modeling of an ideal cell with a current source.

current of the diode I_o, so we should expect it to be very small. However, the flow of this current creates an interesting situation; it makes the upper terminal of the load positive with respect to the lower terminal. Therefore, the diode now has a positive voltage on its anode with respect to the cathode. This is a forward-biased voltage, which causes a forward current to flow back into the diode. Now it seems that we have two currents in the circuit at the same time: current coming out of the diode due to the acquired energy by the PV from light, and current going into the diode due to the positive polarity across the load. How can we resolve this confusion? The answer is to separate the source of energy (light) from the diode model and represent the PV's acquired energy by an electric current source whose magnitude depends on the solar power density (irradiance), as shown in Figure 6.15. The diode current is then the resultant current through the p–n junction as given in Equation 6.4. The remaining current is the load current.

$$I = I_s - I_d \tag{6.5}$$

where
  I is the output current of the solar cell (load current)
  I_s is the solar current
  I_d is the current through the diode

The solar current I_s is a nonlinear variable that changes with light density (irradiance)
  The voltage across the load V is equal to the forward voltage across the diode V_d.

$$V = V_d \tag{6.6}$$

If we assume that I_s is independent of the load voltage, we can draw the current–voltage (I–V) characteristics as shown in Figure 6.16.

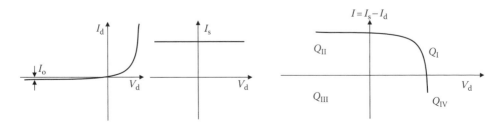

FIGURE 6.16   Current–voltage characteristics of the PV cell.

In Figure 6.16, the PV cell operates in the first quadrant Q_I only. Q_{II} and Q_{IV} are not realistic operational regions because in these quadrants, the load would be sending power to the solar cell.

The output power of the solar cell P is obtained by multiplying the output current I by the output voltage V.

$$P = VI \tag{6.7}$$

Since the solar cell is essentially a p–n junction, we can compute the current I_d using the Schottky model in Equation 6.4.

$$\begin{aligned} I_d &= I_o\left(e^{\frac{V_d}{V_T}} - 1\right) \\ V &= V_d \\ I &= I_s - I_d \end{aligned} \tag{6.8}$$

where

I_o is the reverse saturation current
V_d is the voltage across the diode
V_T is the thermal voltage, whose value is given in Equation 6.4

Substituting the current in Equation 6.8 into Equation 6.7 yields

$$P = VI = V_dI_s - V_dI_o\left(e^{\frac{V_d}{V_T}} - 1\right) \tag{6.9}$$

Equation 6.9 represents the power–voltage characteristic P–V and Equation 6.8 represents the I–V characteristic. Both characteristics are shown in Figure 6.17. When the load current is zero, the PV voltage is at its maximum value, which is known as open circuit voltage V_{oc}. When the load current increases, the voltage stays almost constant initially, then substantially decreases until it reaches zero. At zero voltage, the output current of the PV is called short circuit current I_{sc}. It is the current of the PV when its terminals are shorted.

To compute the open circuit voltage V_{oc}, we set the load current I in Equation 6.8 to zero, as shown in Figure 6.18.

$$\begin{aligned} I_d &= I_s = I_o\left(e^{\frac{V_{oc}}{V_T}} - 1\right) \\ V_{oc} &= V_T \ln\left(\frac{I_s}{I_o} + 1\right) \end{aligned} \tag{6.10}$$

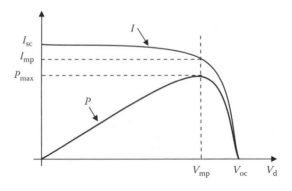

FIGURE 6.17 Current–voltage and power–voltage characteristics of a PV cell.

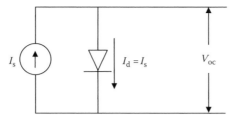

FIGURE 6.18   PV under open circuit condition.

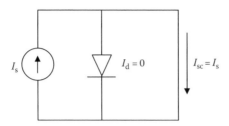

FIGURE 6.19   PV under short circuit condition.

If we short-circuit the terminals of the PV cell as shown in Figure 6.19, the short circuit current of the cell I_{sc} is equal to the solar current I_s. This is because the voltage across the diode is zero; hence the diode current is also zero.

$$I_{sc} = I_s \tag{6.11}$$

The P–V curve shows that the power is zero at the short circuit and open circuit conditions. This is because either the voltage or current at any of these two points is zero. The power reaches its maximum P_{max} at some point near the knee of the I–V curve. At P_{max}, the current is I_{mp} and the voltage is V_{mp} (Figure 6.17).

Example 6.3

An ideal PV cell with a reverse saturation current of 1 nA is operating at 30°C. The solar current at 30°C is 1 A. Compute the output voltage and output power of the PV cell when the load draws 0.5 A.

Solution

Let us first compute the thermal voltage V_T.

$$V_T = \frac{kT}{q} = \frac{1.38 \times 10^{-23}\,(30 + 273.15)}{1.602 \times 10^{-19}} = 26.11 \times 10^{-3}\ \text{V}$$

Use Equation 6.8 to compute the voltage.

$$I = I_s - I_o \left(e^{\frac{V}{V_T}} - 1 \right)$$

$$0.5 = 1 - 10^{-9} \left(e^{\frac{V}{0.02611}} - 1 \right)$$

Hence, the voltage of the PV cell is

$$V = \ln\left[(1 - 0.5) \times 10^9 + 1\right] V_T = 0.523\ \text{V}$$

The output power of the PV cell is

$$P = VI = 0.523 \times 0.5 = 0.2615 \text{ W}$$

EXAMPLE 6.4

An ideal solar cell with a reverse saturation current of 1 nA is operating at 20°C. The solar current at 20°C is 0.8 A. Compute the voltage and current of the solar cell at the maximum power point.

Solution

The output power of the PV cell is

$$P = VI$$

The voltage at maximum power can be obtained by setting the first derivative of the power equation to zero.

$$\frac{\partial P}{\partial V} = V \frac{\partial I}{\partial V} + I$$

The derivative of the current in Equation 6.8 is

$$\frac{\partial I}{\partial V} = -\frac{I_o}{V_T} e^{\frac{V}{V_T}}$$

Combining the above two equations yields

$$\frac{\partial P}{\partial V} = (I_s + I_o) - \left(1 + \frac{V}{V_T}\right) I_o e^{\frac{V}{V_T}}$$

where V_T is given by

$$V_T = \frac{kT}{q} = \frac{1.38 \times 10^{-23} \, (20 + 273.15)}{1.602 \times 10^{-19}} = 25.25 \times 10^{-3} \text{ V}$$

Setting the derivative of power equal to zero yields

$$\left(1 + \frac{V_{mp}}{V_T}\right) e^{\frac{V_{mp}}{V_T}} = \frac{I_s + I_o}{I_o}$$

$$\left(1 + \frac{V_{mp}}{25.25}\right) e^{\frac{V_{mp}}{25.25}} = 0.8 \times 10^9$$

The above equation is nonlinear, but we can still find the value of V_{mp} by iterating the solution. While doing so, keep in mind that the slope of the power curve is sharp between the maximum power point and the open circuit power. Therefore, you should use a small step size in your iteration. Solving the above equation iteratively yields the voltage at the maximum power point.

$$V_{mp} \approx 443.8479 \text{ mV}$$

Substituting the value of V_{mp} into Equation 6.8 yields I_{mp}

$$I_{mp} = I_s - I_o \left(e^{\frac{V_{mp}}{V_T}} - 1 \right)$$

$$I_{mp} = 0.8 - 10^{-9} \left(e^{\frac{443.8479}{25.25}} - 1 \right) = 0.7569 \text{ A}$$

Hence, the maximum output power of the cell is

$$P_{max} = V_{mp} I_{mp} = 443.8479 \times 0.7569 = 335.948 \text{ mW}$$

The operating point of the solar cell depends on the magnitude of the load resistance R, which is the output voltage V divided by the load current I. The intersection of the PV cell characteristic with the load line is the operating point of the PV cell. Consider the characteristics shown in Figure 6.20 where the straight lines are load lines representing the resistances of the load. The slope of the load line $[\tan(I/V)]$ is the inverse of the load resistance (i.e., load conductance). When the load resistance is adjusted to a value equal to R_1, the system operates at point 1 as shown in the figure. When the load resistance increases, the output voltage of the solar cell increases as well.

EXAMPLE 6.5

For the solar cell in Example 6.4, compute the load resistance at the maximum output power.

Solution

The voltage and current at the maximum output power computed in Example 6.4 are

$$V_{mp} \approx 443.8479 \text{ mV}$$
$$I_{mp} = 0.7569 \text{ A}$$

Hence, the load resistance at the maximum power point R_{mp} is

$$R_{mp} = \frac{V_{mp}}{R_{mp}} = \frac{443.8479}{756.9} = 0.5864 \ \Omega$$

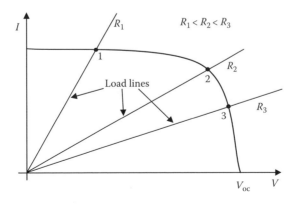

FIGURE 6.20 Operating points of the solar cell connected to a resistive load.

EXAMPLE 6.6

An ideal PV cell with a reverse saturation current of 1 nA is operating at 30°C. The solar current at 30°C is 1 A. The cell is connected to a 10 Ω resistive load. Compute the output power of the cell.

Solution

Equation 6.8 shows the relationship between the diode current I_d and voltage V.

$$I_d = I_o\left(e^{\frac{V}{V_T}} - 1\right)$$

where

$$V_T = \frac{kT}{q} = \frac{1.38 \times 10^{-23}\,(30 + 273.15)}{1.602 \times 10^{-19}} = 26.1 \times 10^{-3}\ \text{V}$$

The output current of the cell I is

$$I = I_s - I_d = I_s - I_o\left(e^{\frac{V}{V_T}} - 1\right)$$

In addition, the output current is the output voltage divided by the load resistance.

$$I = \frac{V}{R}$$

Combining the above two equations yields

$$I = \frac{V}{R} = I_s - I_o\left(e^{\frac{V}{V_T}} - 1\right)$$

or

$$V = I_s R - I_o R\left(e^{\frac{V}{V_T}} - 1\right)$$
$$V = 10 - 10^{-8}\left(e^{\frac{V}{0.0261}} - 1\right)$$

In the above equation, the voltage is the only unknown. However, the equation is nonlinear and can be solved iteratively to compute the voltage.

$$V \approx 0.4722\ \text{V}$$

Hence, the output power of the cell is

$$P = \frac{V^2}{R} = \frac{0.4722^2}{10} = 22.29\ \text{mW}$$

6.1.2.2 Effect of Irradiance and Temperature on Solar Cells

There are two key variables that affect the output power of a PV cell: (1) the light power density ρ (irradiance) and (2) the temperature T of the cell.

Increasing irradiance increases the magnitude of the solar current I_s and consequently increases the short circuit current and the open circuit voltage as shown in Figure 6.21. The effect of irradiance on the output power of the solar cell is shown in Figure 6.22. Notice that all the operating points are located at the intersections of the PV characteristics with the load line. Also, notice that the maximum power increases with an increase in irradiance.

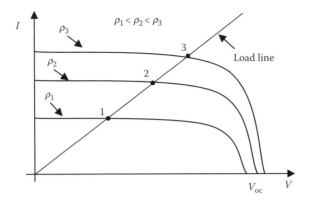

FIGURE 6.21 Effect of irradiance on the operating point.

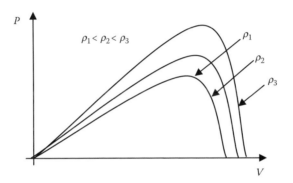

FIGURE 6.22 Effect of irradiance on PV output power.

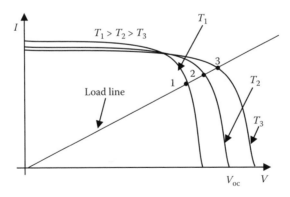

FIGURE 6.23 Effect of temperature on the operating point.

For the effect of temperature, let us examine Equation 6.10. The thermal voltage V_T is linearly dependent on the temperature as given in Equation 6.4. However, the saturation current I_o and the solar current I_s are also nonlinearly dependent on the temperature. The net result is that the open circuit voltage is reduced when the temperature increases. This relationship is shown in Figure 6.23. The effect of temperature on output power is shown in Figure 6.24.

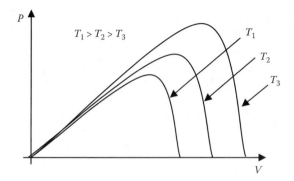

FIGURE 6.24 Effect of temperature on PV output power.

EXAMPLE 6.7

An ideal solar cell with a reverse saturation current of 0.1 μA is operating at 30°C. Find the short circuit current and the open circuit voltage of the cell assuming that the solar current at 30°C is 500 mA.

Solution

The short circuit current is given in Equation 6.11.

$$I_{sc} = I_s = 500 \text{ mA}$$

The open circuit voltage V_{oc} is given in Equation 6.10.

$$V_{oc} = V_T \ln\left(\frac{I_s}{I_o} + 1\right)$$

where

$$V_T = \frac{kT}{q} = \frac{1.38 \times 10^{-23} \, (30 + 273.15)}{1.602 \times 10^{-19}} = 26.1 \times 10^{-3} \text{ V}$$

Hence, the open circuit voltage is

$$V_{oc} = V_T \ln\left(\frac{I_s}{I_o} + 1\right) = 26.1 \times 10^{-3} \ln\left(\frac{500}{0.0001} + 1\right) = 403 \text{ mV}$$

6.1.2.3 Model of PV Module

A PV cell of 10 cm diameter produces about 1 W of power, which is enough to run a low-power calculator. Since the PV cell is essentially a diode, its operating voltage is about 0.6 V, which is the forward-biased voltage of a diode. To increase the power rating, the PV cells are connected together in parallel and series arrangements. The parallel connection increases the overall current, and the series connection increases the overall voltage. These interconnected PV cells are called a module or panel; a typical module is shown in Figure 6.25. For higher power needs, several modules are connected together to form a PV array as shown in Figure 6.25. Several of these arrays form a PV system. More sophisticated PV arrays are mounted on tracking devices that follow the sun throughout the day. The tracking devices tilt the PV arrays to maximize the exposure of the cells to the sunrays, thus increasing the output power of the system.

FIGURE 6.25 (See color insert following page 300.)   PV module (a), array (b), and system (c). (Images courtesy of the U.S. Department of Energy.)

EXAMPLE 6.8

An ideal PV cell produces 2.5 W at 0.5 V during certain environmental conditions. Compute the output power, current, and voltage if the cells are connected in the following arrangements:

(a) When the PV cells are connected as a panel of four parallel columns and each column has 10 series cells.
(b) When several panels are connected as an array of two parallel columns and each column has four series modules.
(c) When several arrays are connected as a solar system consisting of 10 parallel columns and each column has 20 series arrays.

Solution

(a) The total voltage of the panel is

$$V_{panel} = 0.5 \times 10 = 5 \text{ V}$$

When the power of each cell is 2.5 W, the total power of the panel is

$$P_{panel} = P_{cell} (10 \times 4) = 2.5 \times 40 = 100 \text{ W}$$

The total current of the panel is

$$I_{panel} = \frac{P_{panel}}{V_{panel}} = \frac{100}{5} = 20 \text{ A}$$

(b) The total voltage of the array is

$$V_{\text{array}} = V_{\text{panel}} \times 4 = 5 \times 4 = 20 \text{ V}$$

The total power of the array is

$$P_{\text{array}} = P_{\text{panel}} \ (4 \times 2) = 100 \times 8 = 800 \text{ W}$$

The total current of the array is

$$I_{\text{array}} = \frac{P_{\text{array}}}{V_{\text{array}}} = \frac{800}{20} = 40 \text{ A}$$

(c) The total voltage of the system is

$$V_{\text{system}} = V_{\text{array}} \times 20 = 20 \times 20 = 400 \text{ V}$$

The total power of the system is

$$P_{\text{system}} = P_{\text{array}} \times (20 \times 10) = 800 \times 200 = 160 \text{ kW}$$

The total current of the system is

$$I_{\text{system}} = \frac{P_{\text{system}}}{V_{\text{system}}} = \frac{160{,}000}{400} = 400 \text{ A}$$

EXAMPLE 6.9

A 2 m^2 panel of solar cells is installed in the Nevada area, discussed in Example 6.1. The overall efficiency of the PV panel is 10%.

(a) Compute the electrical power of the panel.
(b) Assume the panel is installed on a geosynchronous satellite. Compute its electrical power output.

Solution

(a) In Example 6.1, the computed power density is 685.5 W/m^2. The power of sunrays on the solar panel P_s is the power density ρ multiplied by the area of the panel A.

$$P_s = \rho A = 685.5 \times 2 = 1.371 \text{ kW}$$

The electrical power output of the panel P_{panel} is the solar power input P_s multiplied by the efficiency η of the panel.

$$P_{\text{panel}} = \eta P_s = 0.1 \times 1371 = 137.1 \text{ W}$$

As seen, because of the low efficiency of this solar panel, the electric power obtained from a 2 m^2 panel in one of the best areas in the United States produces only 137 W. This is enough to power two lightbulbs. Higher power can be obtained if the solar panel is larger or the efficiency of the panel is higher.

(b) If the same solar panel is mounted on a satellite, the solar power density on the panel is ρ_o. Hence,

$$P_s = \rho_o A = 1353 \times 2 = 2.706 \text{ kW}$$

The electrical power output P_{panel} of the panel is the solar power input P_s multiplied by the panel efficiency η.

$$P_{\text{panel}} = \eta P_s = 0.1 \times 2706 = 270.6 \text{ W}$$

As seen, the electric power of the same panel doubles when it is moved to outer space.

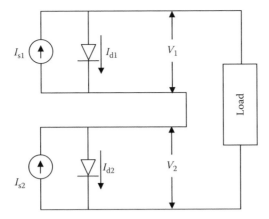

FIGURE 6.26   Solar cells in series.

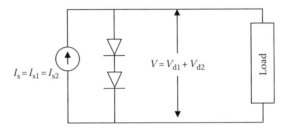

FIGURE 6.27   Equivalent circuit for solar cells in series.

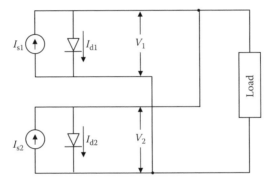

FIGURE 6.28   Solar cells in parallel.

The model in Figure 6.15 can also be used to compute the values of a PV module, array, and system if the cell parameters and the environmental conditions are known. However, before we connect the solar cells in series or parallel, we have to assume that the parameters and environmental conditions are uniform for every cell in the system. This is a reasonable assumption that simplifies the modeling of the solar systems.

For the series connection, let us examine the two solar cells in Figure 6.26. Because of the uniformity assumption, the solar currents of all cells are equal. The equivalent circuit for this case is shown in Figure 6.27. Because we now have two diodes in series, the output voltage of the module is double that of a single cell. The output current, however, is the same as that of a single cell.

In Figure 6.28, the two solar cells are connected in parallel. In this case, the voltage across the load is the same as the voltage of any one of the two cells. However, the output current of the

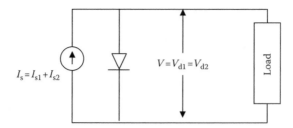

FIGURE 6.29 Equivalent circuit of solar cells in parallel.

module is the sum of the output currents of all cells. The equivalent circuit in the parallel connection is shown in Figure 6.29.

EXAMPLE 6.10

An ideal PV module is composed of 50 solar cells connected in series. At 20°C, the solar current of each cell is 1 A and the reverse saturation current is 10 nA. Draw the *I–V* and *I–P* characteristics of the module.

Solution

For each cell, the thermal voltage V_T is

$$V_T = \frac{kT}{q} = \frac{1.38 \times 10^{-23}\,(20 + 273.15)}{1.602 \times 10^{-19}} = 25.25 \text{ mV}$$

The diode current I_d of the cell is

$$I_d = I_o \left(e^{\frac{V}{V_T}} - 1 \right) = 10^{-8} \left(e^{\frac{V}{0.02525}} - 1 \right)$$

and the load current of the cell I_{cell} is

$$I_{cell} = I_s - I_d = 1 - 10^{-8} \left(e^{\frac{V_{cell}}{0.02525}} - 1 \right)$$

The power of the cell P_{cell} is

$$P_{cell} = V_{cell} I_{cell}$$

Using these two equations, we can find the characteristics of a single cell as shown in Figure 6.30.

The voltage of the module can be obtained by multiplying the single-cell voltage V_{cell} by the number of series cells *n*.

$$V_{module} = n V_{cell}$$

Similarly, the power of the module is the power of a single cell P_{cell} multiplied by the number of cells in the module.

$$P_{module} = n P_{cell}$$

Doing so, we can obtain the module characteristics in Figure 6.31.

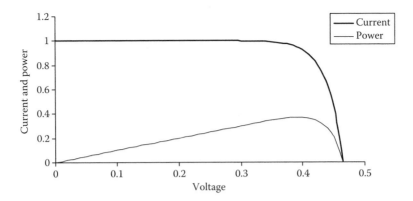

FIGURE 6.30   *I–V* and *P–V* characteristics of a single cell.

6.1.2.4   Real PV Model

The efficiency of most solar cells ranges between 2% and 20%, depending on the material and the structure of the cell. In a recent development, multilayer solar cells have achieved efficiencies as high as 40%. The efficiency of a PV cell is dependent on many factors; the main ones are

1. Reflection of the solar radiation at the top of the PV cell. The more the reflection, the less is the solar energy that reaches the p–n junction.
2. Light has photons of a wide range of energy levels; some do not have enough energy to excite the electrons and allow them to escape. Still other photons have too much energy that is hard to capture by the electrons. These two scenarios account for the loss of about 70% of the solar energy.
3. Resistances of the collector trace at the top of the cell. This resistance is due to the relatively thin traces forming a grid at the top of the cell. Remember that we cannot cover the top with a wide metal plate to reduce the resistance as this would prevent the light from penetrating the cell. In some modern PV cells, the contacts on the top of the cell are made of somewhat transparent material to reduce their blocking of sunlight. Remember that we cannot put the contacts on the side of the cell as this would increase the internal losses of the cell, as discussed earlier.
4. Resistance of the wires connecting the solar cell to the load.
5. Resistance of the semiconductor crystal.

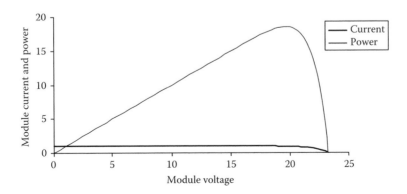

FIGURE 6.31   *I–V* and *P–V* characteristics of a module.

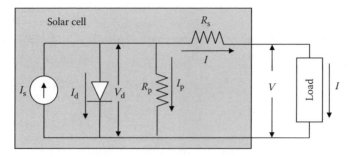

FIGURE 6.32   Model of a real PV cell.

The first two losses are called irradiance losses and the rest are electrical losses. To account for the electrical losses, we can modify the model in Figure 6.15 by including the resistances of the collector traces and the external wires as well as the resistance of the crystal itself. The wires and traces can be represented by a series resistance R_s, while the internal resistance of the crystal can be represented by a parallel resistance R_p as shown in Figure 6.32. The series resistance R_s is in the range of a few milliohm and the parallel resistance R_p is in the range of a few kiloohm.

The new model requires the modification of Equation 6.8 to account for the added components. The load current should now be represented by

$$I = I_s - I_d - I_p \tag{6.12}$$

And the load voltage is

$$V = V_d - IR_s \tag{6.13}$$

To compute the efficiency of the solar cell, we need to compute the irradiance efficiency $\eta_{\text{irradiance}}$ and the electrical efficiency η_e

$$\eta_{\text{irradiance}} = \frac{\text{sun power converted to electricity}}{\text{sun power}} = \frac{P_{se}}{P_s} = \frac{V_d I_s}{\rho A} \tag{6.14}$$

$$\eta_e = \frac{\text{output power of the cell}}{\text{sun power converted to electricity}} = \frac{P_{out}}{P_{se}} = \frac{VI}{V_d I_s} \tag{6.15}$$

where
P_s is the power of sunrays reaching the solar cell
P_{se} is the power of sunrays that is converted into electricity
P_{out} is the output electric power of the solar cell that is consumed by the load
ρ is the solar power density at the PV surface, as given in Equation 6.1
A is the area of the PV cell facing the sun

Then, the total efficiency η of the solar cell is

$$\eta = \eta_{\text{irradiance}} \eta_e = \frac{P_{se}}{P_s} \frac{P_{out}}{P_{se}} = \frac{P_{out}}{P_s} = \frac{VI}{\rho A} \tag{6.16}$$

EXAMPLE 6.11

A 100 cm^2 solar cell is operating at 30°C with an output current of 1 A; the load voltage is 0.4 V and the saturation current of the diode is 1 nA. The series resistance of the cell is 10 mΩ and the parallel resistance is 1 kΩ. At a given time, the solar power density is 200 W/m^2. Compute the irradiance efficiency.

Solution

The diode current can be computed if the thermal voltage, the voltage across the diode, and the saturation current are known. The thermal voltage at 30°C is

$$V_T = \frac{kT}{q} = \frac{1.38 \times 10^{-23} \, (30 + 273.15)}{1.602 \times 10^{-19}} = 26.1 \times 10^{-3} \text{ V}$$

The voltage across the diode is

$$V_d = V + IR_s = 0.4 + (1 \times 0.01) = 0.41 \text{ V}$$

The diode current can be computed using Equation 6.8.

$$I_d = I_o \left(e^{\frac{V}{V_T}} - 1 \right) = 10^{-9} \left(e^{\frac{0.41}{0.0261}} - 1 \right) = 6.64 \text{ mA}$$

The current in the shunt resistance is

$$I_p = \frac{V_d}{R_p} = \frac{0.41}{1000} = 0.41 \text{ mA}$$

The solar current is

$$I_s = I + I_d + I_p = 1 + 0.00664 + 0.00041 = 1.00705 \text{ A}$$

The irradiance efficiency can be calculated by Equation 6.14.

$$\eta_{\text{irradiance}} = \frac{V_d I_s}{\rho A} = \frac{0.41 \times 1.00705}{200 \times 0.01} = 0.205$$

EXAMPLE 6.12

For the solar cell in Example 6.11, compute the overall efficiency of the PV cell.

Solution

The electrical loss $P_{\text{e-loss}}$ of the PV cell is

$$P_{\text{e-loss}} = I^2 R_s + I_p^2 R_p = 1.0^2 \times 0.01 + \left(0.41 \times 10^{-3} \right)^2 1000 = 10.168 \text{ mW}$$

The electrical efficiency η_e as given in Equation 6.15 is

$$\eta_e = \frac{P_{out}}{P_{se}} = \frac{P_{out}}{P_{out} + P_{e\text{-loss}}} = \frac{VI}{VI + P_{e\text{-loss}}} = \frac{0.4 \times 1.0}{(0.4 \times 1.0) + 0.010168} = 0.975$$

As seen in the above calculations, electrical losses are relatively small compared with irradiance losses. The total efficiency η of the cell as given in Equation 6.16 is

$$\eta = \eta_{irradiance}\,\eta_e = 0.205 \times 0.975 = 0.20$$

6.1.2.5 Daily Power Profile of PV Array

The output power of a solar cell P_{out} can be computed using Equation 6.16.

$$P_{out} = \eta P_s = \eta \rho A \tag{6.17}$$

Hence, if we assume that the efficiency of the solar cell is approximately constant, the output power of the cell is linearly related to the solar power density ρ. Since the solar power density over a period of 1 day is almost a bell-shaped curve as shown in Figure 6.4, the output power of the cell is also a bell-shaped curve that can be expressed by

$$P_{out} = P_{max}e^{\frac{-(t-t_0)^2}{2\sigma^2}} \tag{6.18}$$

where
P_{out} is the electric power produced by the solar cell at any time t
P_{max} is the maximum power produced during the day at t_o (noon in the equator)

The energy E_{out} produced by the PV cell in 1 day is the integral of Equation 6.18.

$$E_{out} = \int_0^{24} P_{max}e^{\frac{-(t-t_0)^2}{2\sigma^2}}\,dt \approx P_{max}\sqrt{2\pi}\,\sigma \tag{6.19}$$

EXAMPLE 6.13

The solar power density at a given site is $\rho = \rho_{max} \exp[-(t-12)^2/12.5]$. A solar panel is installed at the site and its maximum electric power measured during the noontime is 100 W. Compute the daily energy produced by a PV panel.

Solution

From the equation of the solar power density, $2\sigma^2 = 12.5$

$$\text{Hence, } \sigma = \sqrt{\frac{12.5}{2}} = 2.5$$

Using Equation 6.19 we get

$$E_{panel} = P_{max}\sqrt{2\pi}\,\sigma = 100\sqrt{2\pi}\,2.5 = 0.627 \text{ kW h}$$

EXAMPLE 6.14

The solar power density at a given site is represented by the equation

$$\rho = 0.7 e^{\frac{-(t-12)^2}{18}} \text{ kW/m}^2$$

A 4 m^2 solar panel is installed at the site. The maximum electric power of the panel is 320 W.

(a) Compute the overall efficiency of the panel.
(b) Compute the output power of the panel at 4:00 PM.

Solution

(a) Maximum solar power density from the equation given above is 700 W/m^2.
Assuming the efficiency to be constant, we can compute the efficiency at the maximum operating point.

$$\eta = \frac{P_{max}}{\rho_{max}A} = \frac{320}{700 \times 4} = 11.43\%$$

(b) The output power of the panel at 4:00 PM can be directly computed using Equation 6.18.

$$P_{panel} = P_{max}e^{\frac{-(t-12)^2}{2\sigma^2}} = 320e^{\frac{-(16-12)^2}{18}} = 131.56 \text{ W}$$

6.1.2.6 PV System

Generating a dc at low voltage is not useful for most power equipment and appliances designed for alternating currents at 120/240 V. Therefore, a converter is needed to change the low voltage dc waveform of the PV array to an alternating current (ac) waveform at the frequency and voltage levels required by the load equipment. The solar array and the converter are the main components of the PV system. These PV systems are quite popular for applications such as household and commercial buildings; some are shown in Figure 6.33.

(a)

(b)

FIGURE 6.33 (See color insert following page 300.) Various PV systems. (Images courtesy of the U.S. Department of Energy.)

(*continued*)

(c)

(d)

FIGURE 6.33 (continued)

PV systems typically have two designs: storage and direct systems. The storage PV system shown in Figure 6.34 consists of four main components: solar array, charger/discharger, battery, and converter. The function of the battery is to store the excess energy of the PV system during the daytime so that it can be used during the night. These batteries are 12 V deep-cycle types that can be fully discharged without being damaged. They are more expensive than normal car batteries and often used in boats and recreation vehicles. The converter is used to transform the dc power of the PV array, or the battery, into ac for household or commercial use. Several of these

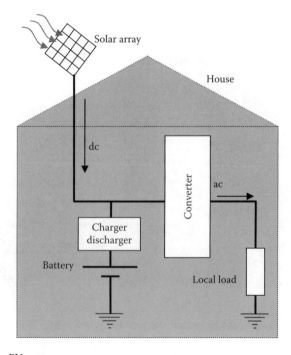

FIGURE 6.34   Storage PV system.

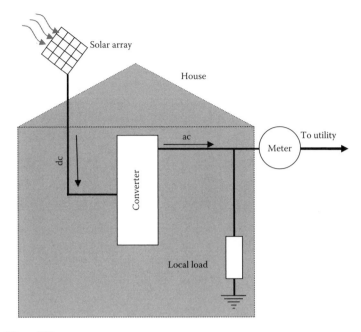

FIGURE 6.35   Direct PV system.

converters are discussed in Chapter 10; the commercial ones are in the range of 1000–6000 W, which are adequate for most household and small commercial applications. The output of the converter is used to power household loads. This system is also known as a stand-alone PV system.

The use of storage batteries is the major drawback of the storage PV system. This is because the batteries decrease the overall efficiency of the PV system; about 10% of the energy stored is normally lost. They add to the total expense of the PV system, their life span is about 5 years, and they occupy considerable floor space. In addition, the leak from the batteries is very acidic and causes corrosion and damage to surrounding areas.

A less expensive system that requires a direct connection to the utility grid is shown in Figure 6.35. In this system, the storage batteries and their charger are eliminated. The excess power in this solar system is pumped back to the power grid. The meter measures the net energy (PV energy minus the consumed energy of the house).

EXAMPLE 6.15

A house is equipped with the solar system shown in Figure 6.35. The following table shows the daily load of the house P_L and the average output of the PV system P_{out}. Assume that the daily pattern is repeated throughout 1 month. Compute the saving in energy cost due to the use of the PV system for 1 month. Assume that the cost of energy from the utility is $0.20/kW h.

Time	Average Output of PV Array P_{out} (W)	Average House Load P_L (W)
8:00 PM–6:00 AM	0	200
6:00–8:00 AM	200	800
8:00–11:00 AM	300	100
11:00 AM–5:00 PM	500	100
5:00–8:00 PM	200	1200

Solution

The energy consumed by the load and the energy traded with the utility are shown in the following table:

Time	House Energy (kW h)	Average Power Exported to Utility (W)	Energy Exported to Utility (kW h)	Average Power Imported from Utility (W)	Energy Imported from Utility (kW h)
T	$E_L = P_L T$	$P_{export} = P_{out} - P_L$	$E_{export} = P_{export} T$	$P_{import} = P_L - P_{out}$	$E_{import} = P_{import} T$
8:00 PM–6:00 AM	2.0	0	0	200	2.0
6:00–8:00 AM	1.6	0	0	600	1.2
8:00–11:00 AM	0.3	200	0.6	0	0
11:00 AM–5:00 PM	0.6	400	2.4	0	0
5:00–8:00 PM	3.6	0	0	1000	3.0
Total daily energy	8.1		3.0		6.2

Without the solar system, the daily cost of energy C_1 is

$$C_1 = 8.1 \times 0.2 = \$1.62$$

With the solar system, the daily cost of energy C_2 is

$$C_2 = (6.2 - 3.0) \times 0.2 = \$0.64$$

Saving in 1 month S is given by

$$S = (C_1 - C_2) \times 30 = \$29.4$$

Keep in mind that this saving does not reflect the cost of the PV system.

6.1.2.7 Assessment of PV Systems

The most common applications of PV cells are in consumer products such as calculators, watches, battery chargers, light controls, and flashlights. The larger PV systems are extensively used in space applications (such as satellites) where their usage has increased 1000-fold since 1970. In higher-power applications, three factors determine the applicability of PV systems: (1) the cost and the payback period of the system, (2) the accessibility to a power grid, and (3) the individual inclination to invest in environmentally friendly technologies.

The price of electricity generated by a PV system varies widely and is based on the size of the PV system, the various design options, the installation cost, the lifetime of the system, the maintenance cost, and the solar power density at the selected site. The estimated average prices of 1 kW h of PV energy using engineering economic models over a period of 40 years are shown in Figure 6.36. From this figure, we see that the price decreases rapidly, although it is still more expensive than most utility rates; the cost of generating 1 kW h of PV electricity is more than twice the price charged by the utilities in the Northwest region of the United States in 2007.

In remote areas without access to power grids, the PV system is often the first choice among the available alternatives. In developing countries and remote areas, thousands of PV systems have already been installed and have improved the quality of life tremendously. These solar power systems operate clean water pumps, provide power for lighting, keep medicine refrigerated, keep equipment sterilized, operate emergency radios, and are used in many other critical

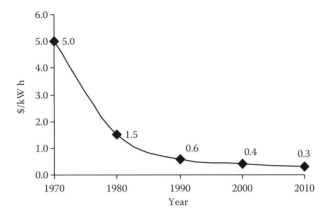

FIGURE 6.36 Price of kilowatt hour from PV systems.

applications. By the end of the twentieth century, PV systems worldwide had the capacity of more than 900 GW h annually; some of the large systems are shown in Figure 6.37. To put the number into perspective, this PV energy is enough for about 70,000 homes in the United States, or about 4 million homes in developing countries.

The dream of solar energy enthusiasts is to replace a large amount of the energy that is currently produced by primary resources with PV systems. This is because 1 kW of PV power can reduce the annual discharge of sulfur dioxide (SO_2) from coal-fired power plants by as much as 30 kg.

(a)   (b)

FIGURE 6.37 (See color insert following page 300.) High-power PV systems. (Images courtesy of the U.S. Department of Energy.)

Although solar energy is very promising, the PV technology is not free from drawbacks. The following list identifies the main issues associated with PV systems:

1. With the current technology, manufacturing silicon PV demands energy that is almost as much as the energy produced by an average solar cell in its lifetime.
2. Although the power generated by the PV system is pollution free, toxic chemicals are used to manufacture the solar cells such as arsenic and silicon compounds. Arsenic is generally an odorless and flavorless semimetallic chemical that is highly toxic and can kill humans quickly if inhaled in large amounts. Long-term exposure to small amounts of arsenic with through the skin or by inhalation can lead to slow death and a variety of illnesses. Silicon, by itself, is not toxic. However, when additives are added to make the PV semiconductor material from silicon, the compound can be extremely toxic. Since water is used in the manufacturing process, the runoff could cause the arsenic and silicon compounds to reach local streams. Furthermore, should a PV array catch fire, these chemicals can be released into the environment.
3. Solar power density can be intermittent due to weather conditions (e.g., heavy cloud cover). It is also limited exclusively to use in daytime.
4. For high-power PV systems, the arrays spread over a large area. This makes the system difficult to accommodate in or near cities.
5. PV systems are considered by some to be visually intrusive (e.g., rooftop arrays are clearly visible to neighbors).
6. Efficiency of the solar panel is still low, making the system more expensive and large in size.
7. Solar systems require continuous cleaning of their surfaces. Dust, shadows, falling leaves, and bird droppings can substantially reduce the efficiency of the system.

6.2 WIND ENERGY

Wind energy is one of the oldest forms of energy known to man; it dates back more than 5000 years in Egypt, when sailboats were used for transportation. Although no one knows exactly who built the first windmill, archaeologists discovered a Chinese vase with a painting that resembles a windmill, which dates back to the third millennium BC. By the second millennium BC, the Babylonians used windmills extensively for irrigation. They were constructed as a revolving doors system, similar to the vertical-axis wind machine used today. In the twelfth century, windmills were built in Europe to grind grains and pump water. They were also used in Holland to drain lands below the water level of the Rhine River. During this era, working in windmill plants was one of the hazardous jobs in Europe. The workers were frequently injured because the windmills were constructed of a huge rotating mass with little or no control on its rotation. The grinding or hammering sounds were so loud that many workers became deaf; the grinding dust of certain materials such as wood caused respiratory health problems, and the grinding stones often grind against each other causing sparks and fires.

In addition to producing mechanical power, windmills were also used to communicate with neighbors by locking the windmill sails in a certain arrangement. During World War II, the Netherlanders used to set windmill sails in certain positions to alert the public of a possible attack by their enemies.

Nowadays, windmills are mostly used to generate electricity by converting the kinetic energy (KE) in the wind into electrical energy. These types of windmills are often called wind turbines. The first wind turbine was built by the American inventor Charles F. Brush in 1888. It was just a 12 kW machine with an enormous structure, but lasted for several years. Because power grids did not reach farmlands until the second quarter of the twentieth century, farmers in the United States relied on wind turbines for their electric energy needs.

The design and operation of wind turbines have improved substantially in the last two decades making them a very viable technology. The size of wind turbines has increased from just a few kilowatt machines to up to 5 MW single unit systems. One of these machines is large enough to power about 2500 homes in Europe.

6.2.1 KINETIC ENERGY OF WIND

The role of wind turbines is to harness the KE of the wind and convert it into electrical energy. The KE of an object is the energy it possesses while in motion. Starting with Newton's second law, the KE of a moving object can be expressed as

$$KE = \tfrac{1}{2} mv^2 \tag{6.20}$$

where
   m is the mass of the moving object (kg)
   v is the velocity of the object (m/s)

The unit of KE is watt second (Ws). To calculate the KE of the wind, we need to calculate the mass of air passing through a given area A. The mass of air depends on the area A, the speed of the wind v, the air density δ, and the time t.

$$m = Av\delta t \tag{6.21}$$

where
   Unit of A is m^2
   Unit of v is m/s
   Unit of δ is kg/m^3
   Unit of t is s

Substituting the mass in Equation 6.21 into Equation 6.20 yields

$$KE = \tfrac{1}{2} A\delta t v^3 \tag{6.22}$$

Since energy is power multiplied by time, the wind power P_{wind} in watt is

$$P_{wind} = \frac{KE}{t} = \frac{1}{2} A\delta v^3 \tag{6.23}$$

Notice that the KE and the power of wind are proportional to the cube of the wind's speed. Hence, when the wind speed increases slightly by 10%, the KE of wind increases by 33.1%.

   Wind power density ρ in watt per square meter (W/m^2) is often used to judge the potential of a given site for wind energy installation. Figure 6.38 shows the average wind power density in the United States at an elevation of 10 m. There are areas in the western United States with as much wind power density as 1 kW/m^2.

   From Equation 6.23, the wind power density can be written as

$$\rho = \frac{P_{wind}}{A} = \frac{1}{2} \delta v^3 \tag{6.24}$$

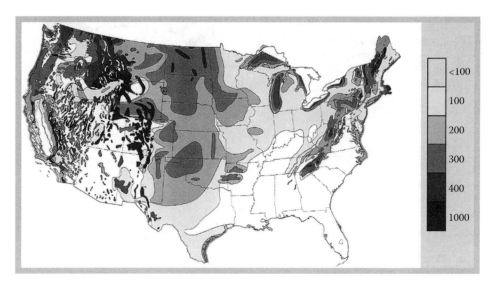

FIGURE 6.38   U.S. average wind power density map in watt per square meter at 10 m elevation above the sea level. (Map courtesy of the U.S. Department of Energy.)

Air density is a function of air pressure, temperature, humidity, elevation, and gravitational acceleration. The expression commonly used to compute air density is

$$\delta = \frac{\mathrm{pr}}{\kappa T} e^{\left(\frac{gh}{\kappa T}\right)} \tag{6.25}$$

where
   pr is the standard atmospheric pressure at sea level (101,325 Pa or N/m^2)
   T is the air temperature (K) (kelvin $= 273.15 + $°C)
   κ is the specific gas constant, for air $\kappa = 287$ (Ws/kg K)
   g is the gravitational acceleration (9.8 m/s^2)
   h is the elevation of the wind above the sea level (m)

Substituting these values into Equation 6.25 yields

$$\delta = \frac{353}{T + 273} \exp\left(\frac{-h}{29.3(T + 273)}\right) \tag{6.26}$$

Notice that the temperature T in Equation 6.26 is in degree Celsius. The equation shows that when temperature decreases, air is denser. Also, air is less dense at high altitudes. For the same velocity, wind with higher air density (heavier air) possesses more KE.

EXAMPLE 6.16

Tehachapi is a desert city in California with an elevation of about 350 m and is known for its extensive wind farms. Compute the power density of wind when air temperature is 30°C and the speed of the wind is 12 m/s.

Solution

To compute the power density of wind, you need to compute air density.

$$\delta = \frac{353}{30 + 273}\exp\left(\frac{-350}{29.3(30 + 273)}\right) = 1.12 \text{ kg/m}^3$$

The power density obtained by using Equation 6.24 is

$$\rho = \tfrac{1}{2}\delta v^3 = \tfrac{1}{2}1.12 \times 12^3 = 967.7 \text{ W/m}^3$$

Compare this power density of wind with the power density of the sun computed in Example 6.1. Can you draw a conclusion?

6.2.2 Wind Turbine

Wind turbine is a generic name given to the wind energy system that converts the KE in wind into electrical energy. The current size of a wind turbine ranges from 50 kW to more than 5 MW. Wind turbines come in a variety of designs and power ratings, but the most common design is the horizontal-axis type shown in Figure 6.39. It consists of the following basic components:

- A tower that keeps the rotating blades at a sufficient height to increase the exposure of the blades to the wind. Large wind turbines, in the megawatt range, have towers as high as 250 m above the base.
- Rotating blades that capture the KE of the wind. They are normally made of fiberglass-reinforced polyester or wood-epoxy material. The length of the rotating blades ranges from

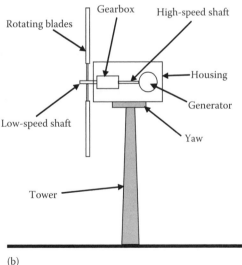

(a)   (b)

FIGURE 6.39 (See color insert following page 300.)   Basic components of a wind-generating system. (a) Horizontal design and (b) mechanical structure.

5 to over 60 m. More advanced blade systems allow the blades to change their pitch angle to maximize their absorption of the wind's KE. Most wind turbines have three rotor blades.

- A yaw mechanism that allows the housing box to rotate and keeps the blades perpendicular to the wind speed. Thus, it increases the exposure of the blades to the wind.
- A gearbox that is used to connect the low-speed rotating blades to the high-speed generator. It also serves as a clutch.
- A generator connected to the high-speed shaft of the gearbox to convert the mechanical energy of the rotating blades into electrical energy.
- A controller that connects the utility system to the wind generator and locks the blades when the wind speed is below the minimum generation limit or when the wind speed exceeds the design limitations of the system.

Figure 6.40 depicts the housing and blade of a wind-generating system. These pictures were taken during the installation process of the turbine in Figure 6.39. This turbine is about 2 MW and is located in the Pacific Northwest of the United States. The enormous size of the housing can be appreciated by comparing it to the size of the workers. Also, the length of the blade can be realized by comparing it to the length of the truck.

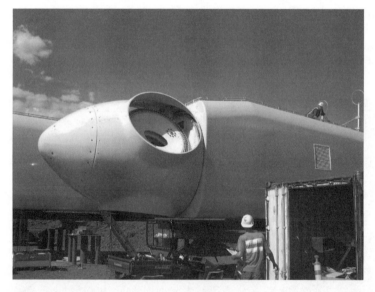

(a)

(b)

FIGURE 6.40 (See color insert following page 300.) (a) Housing and (b) blade of a 2 MW wind-generating system.

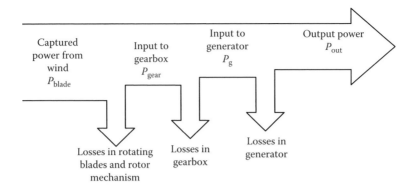

FIGURE 6.41 Power flow of a wind turbine.

EXAMPLE 6.17

A wind turbine at the site discussed in Example 6.16 has three rotating blades; each is 20 m in length. Compute the available power from the wind.

Solution

The wind power density computed in Example 6.16 is 967.7 W/m^2. The area swept by the blade is a circle of 40 m diameter. Hence, the available power from the wind P_{wind} is as follows:

$$P_{wind} = A\rho = \pi r^2 \rho = \pi \times 20^2 \times 967.7 = 1.216 \text{ MW}$$

where r is the length of the blade.

6.2.2.1 Efficiency of Wind Turbine

Not all of the wind energy is converted into electrical energy because of various losses in the system as shown in Figure 6.41. The blades of the wind turbine capture only part of the available wind energy. The ratio of the power captured by the blades P_{blade} to the available power from the wind P_{wind} is known as the power coefficient C_p. The maximum theoretical value of the power coefficient is known as Betz limit, which is 0.5926.

$$C_p = \frac{P_{blade}}{P_{wind}} \tag{6.27}$$

Part of the blade power entering the turbine is wasted as rotational losses (frictional and windage) in the rotating blades and rotor mechanism. The rest of the power P_{gear} enters the gearbox, where part of it is wasted as gearbox rotational loss. The remaining power P_g enters the generator where some gets wasted in the form of electrical losses in the generator's windings and core as well as the generator's own rotational losses. The remaining power P_{out} is the output electric power delivered to the load.

EXAMPLE 6.18

For the wind turbine in Example 6.17, compute the output power assuming that the power coefficient C_p is 0.4, the efficiency of the rotating blades and rotor mechanism η_{blade} is 0.9%, the gearbox efficiency η_{gear} is 95%, and the generator efficiency η_g is 70%.

Solution

The output power of the wind turbine is the wind power multiplied by the total efficiency.

$$P_{out} = \eta_{total} P_{wind} = \left(C_p \eta_{blade} \eta_{gear} \eta_g \right) P_{wind} = (0.4 \times 0.9 \times 0.95 \times 0.7)1.126 = 291 \text{ kW}$$

The output power is about 24% of the wind power. Although the efficiency is low, it is still higher than the efficiency of the PV system.

6.2.2.2 Tip Speed

Figure 6.42 shows a front view of the rotating blade. The linear velocity of the blade is known as the tip velocity v_{tip}. The rotating speed of the blade ω (rad/s) is a function of its tip velocity v_{tip} (m/s) and the length of the blade r (m).

$$v_{tip} = \omega r = 2\pi nr \tag{6.28}$$

where n is the number of revolutions the blade makes in 1 s.

$$n = \frac{v_{tip}}{2\pi r} \tag{6.29}$$

The wind turbine is often designed to have its tip velocity faster than wind speed to allow the turbine to generate electricity at low wind speeds. The ratio of the tip velocity v_{tip} to the wind speed v is known as the tip speed ratio (TSR).

$$\text{TSR} = \frac{v_{tip}}{v} \tag{6.30}$$

The TSR is an important parameter in the design of wind turbines. If the TSR is too small, most of the wind passes through the open areas between the rotor blades and little energy is captured by the blade. If it is too large, the fast-moving blades appear like a solid wall to the wind. When a rotor blade passes through the wind it creates a wake of air turbulence that is often damped out quickly. However, if the TSR is high, the next blade may arrive to the area swept by the earlier blade before the turbulence is damped out. In this case, the efficiency of the blade is substantially reduced because blades capture a smaller amount of energy from turbulent wind. Figure 6.43 shows the

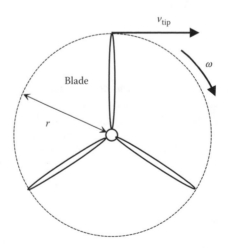

FIGURE 6.42   Tip velocity.

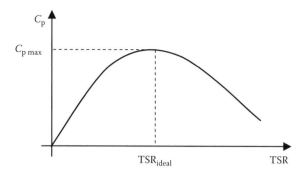

FIGURE 6.43 Power coefficient as a function of TSR.

relationship between the TSR and the efficiency of the blades (power coefficient). This figure shows that the power coefficient is substantially reduced when we deviate from an ideal value of the TSR (TSR$_{ideal}$).

In some advanced types of wind machines, the value of TSR can be adjusted by changing the pitch angle of the blades. At light wind conditions, the pitch angle is set to increase the TSR; therefore, more energy can be extracted from the wind. At high wind speeds, the pitch angle is adjusted to reduce the TSR and maintain the rotor speed of the generator within its design limits. In some systems, the TSR can be reduced to almost zero to lock the blades at excessive wind conditions. Variable TSR wind turbines operate for wider ranges of wind speeds; thus, they produce more energy as they operate for a longer time than fixed pitch turbines.

EXAMPLE 6.19

A wind turbine is designed to produce power when the speed of the generator n_g is at least 905 rpm, which correlates to a wind speed of 5 m/s. The turbine has a fixed TSR of 7 and a sweep diameter of 10 m. Compute the gear ratio.

Solution

Tip speed

$$v_{tip} = TSR(v) = 7 \times 5 = 35 \text{ m/s}$$

Speed of the shaft at the low-speed side of the gearbox

$$n = \frac{v_{tip}}{2\pi r} = \left(\frac{35}{2\pi 5}\right) 60 = 67 \text{ rpm}$$

Gear ratio

$$\frac{n_g}{n} = \frac{905}{67} = 13.5$$

EXAMPLE 6.20

A wind turbine has a power coefficient that can be represented by the equation

$$C_p = 0.4 \sin(TSR) + 0.03 \sin(3TSR - 0.2)$$

Compute the TSR that leads to maximum power coefficient.

Solution

To compute $C_{p\,max}$, we set the derivative of the above equation to zero.

$$\frac{\partial C_p}{\partial TSR} = 0.4\cos{(TSR)} + 0.09\cos{(3TSR - 0.2)} = 0$$

The iterative solution of this equation leads to $TSR_{ideal} = 1.5$, at which the maximum power coefficient is

$$C_{p\,max} = 0.4\sin{(1.5)} + 0.03\sin{(4.5 - 0.2)} = 0.3715$$

6.2.3 WIND FARM PERFORMANCE

When a group of wind turbines is installed at one site, the site is called a wind farm. Wind farms can have clusters of these machines as shown in Figure 6.44. Major wind farms such as the one located

(a)

(b)

FIGURE 6.44 (See color insert following page 300.) Wind farm located in California. (Images courtesy of the U.S. Department of Energy.)

at Tehachapi near Los Angeles, California, can have hundreds of these machines. The amount of energy generated by these farms depends on several factors:

Wind speed and length of wind season: Wind speed determines how much power can be produced by the wind turbine. As seen in Equation 6.24, the wind power density is proportional to the cube of the wind speed. Assuming all other parameters are fixed, a 10% increase in wind speed results in a 33% increase in wind power. Most wind turbines start to generate electricity at wind speeds as low as 4 m/s and reach maximum power output at about 16 m/s.

Diameter of the rotating blades: The power captured by the blades is a function of the area they sweep as shown in Equation 6.23. The area is circular with a radius equal to the length of one blade as shown in Figure 6.42. The power is then proportional to the square of the radius. Hence, a 10% increase in the blade length will result in 21% increase in the captured power.

Efficiency of wind turbine components: As seen in Example 6.18, the low efficiency of the various components of the wind turbine reduces the output power of the system. Improvements in the design of the blades, gearbox, and generator should increase the efficiency of the system.

Pitch control: With pitch control, the TSR can be adjusted to produce power at a wide range of wind speeds. However, controlling the pitch angle adds to the complexity of the system and increases the maintenance cost, so it is more suitable for large turbines.

Yaw control: Most wind turbines are equipped with a yaw mechanism to keep the blades facing into the wind as the wind direction changes. Some turbines are designed to operate on downwind; these turbines do not need yaw mechanisms as the wind aligns these turbines.

Arrangement of the turbines in the farm: Layout of individual wind turbines in a wind farm is very crucial. If a wind turbine is placed in the wind shadow of other turbines, the amount of generated power can be substantially reduced. This is because the blades of the front turbines create wakes of turbulent wind, which can reach the rear turbines. When this happens, the rear turbines capture less energy from the wind because blade efficiency is reduced when wind is turbulent.

Reliability and maintenance: The cost of electricity generated by the wind farm is a function of the capital cost, land use, maintenance, and any other contractual arrangement. The early designs of wind turbines were high-maintenance machines as well as cost-ineffective systems. Newer designs, however, are much better with a reliability rate around 98%.

6.2.4 WIND ENERGY AND ENVIRONMENT

Wind power is an environmentally friendly form of generating electricity with virtually no contribution to air pollution. For this reason, wind energy industries are expanding rapidly worldwide. In 2006, the countries with the most wind power capacity were Germany (21 GW), followed by the United States (12 GW), Spain (11 GW), India (6 GW), and Denmark (3 GW). In 2006 alone, 15 GW of new wind power plants were installed, bringing the world's total wind power generating capacity to about 75 GW. This capacity is enough to power over 40 million average European households. Utilities often measure the wind energy capacity in terms of the penetration factor of the wind energy systems PF_w, which is defined as the ratio of wind power capacity to the power generation capacity from primary resources.

$$PF_w = \frac{P_w}{P_p}$$ (6.31)

where
   P_w is the power capacity of wind turbines
   P_p is the power capacity from primary resources

In some countries, the penetration factor of wind energy is quite high. In Germany, for example, the primary power generation capacity as of 2006 is about 80 GW. This makes their wind energy penetration factor just above 25%. This high percentage is a great indicator of the mature and viable technology of wind energy systems.

Although it is highly popular, wind energy has some critics who often raise several issues such as (1) noise pollution, (2) aesthetics and visual pollution, and (3) bird collision. In addition, wind turbines have two main technical drawbacks: (1) voltage fluctuations and (2) wind variability.

Noise pollution: There are two potential sources of noise from a wind turbine: mechanical noise from the gearbox and generator, and aerodynamic noise from the rotor blades. Aerodynamic noise is the main component of the total noise, which is similar to the swish sound produced by a helicopter. At 400 m downwind, the turbine noise can be as high as 60 dB, which is equivalent to the noise level from a dishwasher or air conditioner. Although the noise from wind farms is far below the threshold of pain, which is 140 dB, it is a major annoyance to some people. Fortunately, newer and better designs of rotor blades have reduced aerodynamic noise dramatically.

Aesthetics: Designers of wind turbines believe their creations look pretty, but, unfortunately, not everyone agrees. Critics see wind farms as defacement of the landscape, forcing developers to install them in remote areas, including offshore as shown in Figure 6.45.

Bird collisions: Instances of birds being hit by the moving blades have been reported. However, newer turbines are designed for low blade speeds, which should reduce the severity of the problem.

Ice accumulation: In cold weather areas, ice can accumulate on the blades. When the turbine rotates, the ice can be dislodged and turned into projectile endangering people and structures in the area. In some models, the blades are heated to prevent the ice from accumulating on the blades.

Sunlight flicker: Some blades reflect sunlight. When these blades rotate, the reflections of the sunlight off these blades flicker. This is an annoyance to some people.

FIGURE 6.45 (See color insert following page 300.) 2 MW offshore wind turbine farm in Denmark. (Image courtesy of the LM Glasfiber Group.)

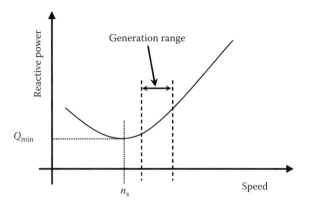

FIGURE 6.46   Reactive power of an induction machine.

Voltage fluctuations: A large number of wind turbines use induction machines as generators. These machines are very rugged and require little maintenance and control. They are also self-synchronized with the power grid without the need for additional synchronization equipment. However, because the induction machine has no field circuit, it demands a significant amount of reactive power from the utility system. When used in the generating mode, the induction machine consumes reactive power from the utility while delivering real power. In some cases, the magnitude of the reactive power imported from the utility exceeds the magnitude of the real power generated. Further, the reactive power consumed by the generator is not constant, but is dependent on the speed of its shaft. Figure 6.46 shows the relationship between the reactive power and the speed of the machine. The machine operates as a generator when its speed is higher than the synchronous speed n_s. This is the linear region identified in the figure. Since the speed of wind changes continuously, the reactive power consumed by all induction machines in the farm also changes continuously as shown in Figure 6.47. The voltage at the wind farm is dependent on the reactive power consumed by all wind

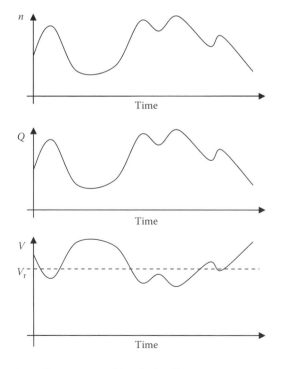

FIGURE 6.47   Wind speed, reactive power, and terminal voltage.

turbines in the farm; the higher the reactive power, the lower the voltage. Thus, the voltage at the site is a mirror image of the reactive power, but at a different scale as shown in Figure 6.47. V_r is the average voltage. If the voltage variation is slow, it is probably unnoticeable. However, if the variations are fast and large in magnitude, annoying voltage flickers can be produced at nearby areas.

Variability of wind: Most of the electric loads peak in the morning and early evening. During these times, utilities often fire their fossil fuel power plants to compensate for the extra demands. Renewable energy can play a great role during these periods by supplying the extra energy. However, because wind is not reliable, energy from wind farms is not always synchronized with the increase in demands.

6.3 FUEL CELL

FCs are electrochemical devices that use chemical reactions to produce electricity. The FC technology was invented over a century ago, but received little attention until the 1950s when NASA used it in its space programs. Sir William Grove in 1839 was the first to develop an FC device. Grove, who was an attorney by education, conducted an experiment in which he immersed parts of two platinum electrodes in sulfuric acid and sealed the remaining part of one electrode in a container of oxygen and the other in a container of hydrogen. He noticed that a constant current flew between the electrodes and the sealed containers eventually included water and gases. This was the first known FC. In 1939, Francis Bacon built a pressurized FC from nickel electrodes, which was reliable enough to attract the attention of NASA who used his FC in its Apollo spacecraft. Since then NASA has used various designs of FCs in several of its space vehicles including the Gemini and the space shuttles.

During the last 30 years, research and development in FC technology has exploded. New materials and technologies have a made FCs safer, more reliable, lighter, and more efficient than the earlier models. The application areas of FCs include ground transportation, marine applications, distributed power, cogeneration, and even consumer products.

6.3.1 HYDROGEN FUEL

Most FCs utilize hydrogen and oxygen to produce electricity. Hydrogen is found in many compounds such as water and fossil fuels. Although it is the most abundant element in the universe, hydrogen is not found alone, but always combined with other elements. To separate hydrogen from its compounds, a reformer process, such as the one shown in Figure 6.48, is often used. Inside the reformer, a fuel such as hydrocarbon (CH_2) is chemically treated to produce hydrogen. The hydrocarbon compound, which contains hydrogen and carbon atoms, can be found in fossil fuels such as natural gas and oil. The by-products of the reformer process, however, are carbon dioxide (CO_2) and carbon monoxide (CO). These undesired gases are responsible for global warming and increased hazards to human health as discussed in Chapter 5. Since CO is more

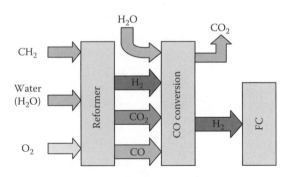

FIGURE 6.48 Generation of hydrogen.

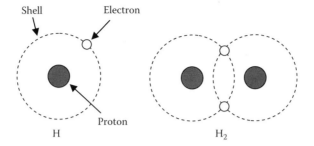

FIGURE 6.49 Hydrogen atom and hydrogen gas.

hazardous than CO_2, CO is further oxidized by the CO-converter. In this chemical process, water is added to the output of the reformer to convert CO into CO_2. Carbon dioxide is vented in air, and hydrogen is used in the FC.

A new generation of FCs use fuels such as methane directly without the need for reformers, which are called direct FCs. In these systems, hydrogen is extracted directly from methane inside the FC.

In its stable gas form, hydrogen is covalent bonded, which means the electrons are shared between two hydrogen atoms as shown in Figure 6.49. The symbol of this hydrogen gas is H_2.

6.3.2 TYPES OF FUEL CELLS

FCs come in various configurations based on the gas used and their electrolyte. Although FCs have various designs and operating characteristics, they all work on a similar principle. Different FCs operate at diverse temperature ranges and have different start-up times. These differences make FCs useful in a wide range of applications from consumer electronics to power plants. Table 6.1 shows the main types of FCs and their electrolytes as well as their gases. The table also shows their typical efficiencies and operating temperatures. These FCs are covered in more detail in the following sections.

TABLE 6.1
Main Types of FCs and Their Operating Characteristics

FC	Electrolyte	Anode Gas	Cathode Gas	Approximate Temperature (°C)	Typical Efficiency (%)
PEM	Solid polymer membrane	Hydrogen	Pure or atmospheric oxygen	80	35–60
AFC	Potassium hydroxide	Hydrogen	Pure oxygen	65–220	50–70
PAFC	Phosphorous	Hydrogen	Atmospheric oxygen	150–210	35–50
SOFC	Ceramic oxide	Hydrogen, methane	Atmospheric oxygen	600–1000	45–60
MCFC	Alkali-Carbonates	Hydrogen, methane	Atmospheric oxygen	600–650	40–55
DMFC	Solid polymer membrane	Methanol solution in water	Atmospheric oxygen	50–120	35–40

6.3.2.1 Proton Exchange Membrane Fuel Cell

The basic components of the proton exchange membrane (PEM) FC shown in Figure 6.50 are the anode, the cathode, and the electrolyte, which is a membrane of solid polymer coated with a metal catalyst such as platinum. Figure 6.51 depicts a system consisting of several FCs in series. Pressurized hydrogen enters the anode, which is a flat plate with channels built into it to disperse hydrogen gas over the surface of the catalyst. A catalyst (such as platinum) causes two hydrogen gas atoms ($2H_2$) to oxidize into four hydrogen ions ($4H^+$) and give up four electrons ($4\varepsilon^-$). An ion is an atom or a group of atoms that lost or gained one or more electrons. The hydrogen ion is just the proton of the atom. This process is known as the anode reaction and can be represented by the chemical equation

$$2H_2 \Rightarrow 4H^+ + 4\varepsilon^- \tag{6.32}$$

The free electrons take the path of least resistance through the external load to the other electrode (cathode). Since the electric current is the flow of electrons, the load is then energized.

The hydrogen ions pass through the membrane from the anode to the cathode. When the electrons enter the cathode, they react with the oxygen from the outside air and the hydrogen ions at the cathode to form water. This is known as the cathode reaction.

$$O_2 + 4H^+ + 4\varepsilon^- \Rightarrow 2H_2O \tag{6.33}$$

The overall chemical reaction of the anode and cathode can be represented by Equation 6.34. The equation shows that the FC combines hydrogen and oxygen to produce water and in the process, energy is released.

$$2H_2 + O_2 \Rightarrow 2H_2O + \text{energy} \tag{6.34}$$

Keep in mind that the reformer and the CO-converter require water. Therefore, it is convenient that the FC produces water, which can be fed back to the reformer and CO-converter. Also, the cathode reaction produces heat that can be used in various applications.

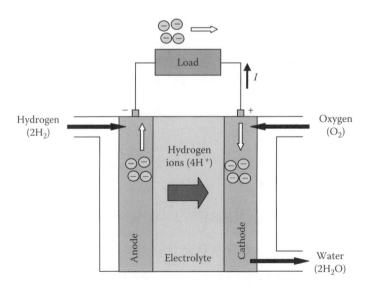

FIGURE 6.50   PEM FC.

FIGURE 6.51 (See color insert following page 300.)   PEM FC module.

The PEM FC operates at relatively low temperatures, about 80°C, and its output power can be regulated fairly easily. It is relatively lightweight, has high energy density, and can start very quickly (within a few milliseconds). For these reasons, the PEM FC system, such as the one shown in Figure 6.51, is suitable for a large number of applications including transportation and distributed generation for residential loads.

The PEM FC uses platinum as a catalyst; this costly metal makes the PEM FC expensive. Furthermore, platinum is extremely sensitive to CO and the elimination of CO is crucial for the longevity of the FC. This adds another expense to the overall cost of the system.

6.3.2.2   Alkaline Fuel Cell

The AFC is the oldest type of FCs that was developed by Francis Bacon in the last part of 1930s. Its electrolyte is liquid alkaline solution of potassium hydroxide (KOH). The AFC operates at elevated temperatures, 65°C–220°C; therefore, it is slower at starting relative to the PEM cell.

The main components of the AFC are shown in Figure 6.52. At the anode, hydrogen reacts with the hydroxyl ions OH^- to produce water and free electrons. The hydroxyl ions are generated at the cathode by combining oxygen, water, and free electrons. Notice that water produced at the anode is migrated back to the cathode.

The anode reaction of the AFC is expressed by

$$2H_2 + 4OH^- \Rightarrow 4H_2O + 4\varepsilon^- \tag{6.35}$$

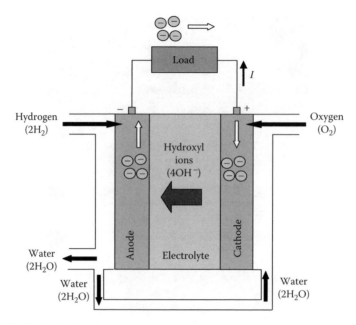

FIGURE 6.52 Alkaline fuel cell.

And the cathode reaction is

$$O_2 + 2H_2O + 4\varepsilon^- \Rightarrow 4OH^- \tag{6.36}$$

The overall cell reaction is

$$2H_2 + O_2 \Rightarrow 2H_2O + \text{energy} \tag{6.37}$$

The AFC is very susceptible to contamination, particularly carbon dioxide (CO_2), which reacts with the electrolyte and quickly degrades the FC performance. In addition, water and methane can also contaminate the FC. Because of its susceptibility to contaminations, the AFC must run on pure hydrogen and oxygen, which increase the cost of its operation. Therefore, the application of AFC is limited to controlled environments such as spacecraft.

6.3.2.3 Phosphoric Acid Fuel Cell

Phosphoric acid (H_3PO_4) is the electrolyte medium for the PAFC shown in Figure 6.53. Its operating temperature is high, 150°C–210°C, but is considered suitable for small and midsize generation. The PAFC is among the first generation of modern FCs and is typically used for stationary power generation, but some are used to power large buses.

The anode reaction here is similar to that of the PEM FC. The hydrogen introduced at the anode is stripped of its electrons. The hydrogen protons (ions) migrate through the electrolyte to the cathode.

$$2H_2 \Rightarrow 4H^+ + 4\varepsilon^- \tag{6.38}$$

The electrons released at the anode travel through an external circuit where they give their energy to the electric load and return to the cathode. At the cathode, the hydrogen ions are combined with the four electrons and oxygen, usually from air, to produce water.

$$O_2 + 4H^+ + 4\varepsilon^- \Rightarrow 2H_2O \tag{6.39}$$

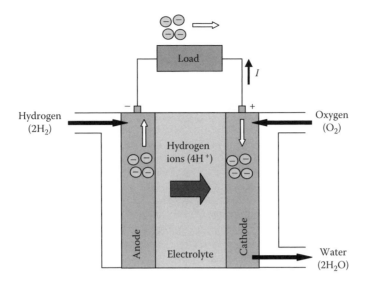

FIGURE 6.53 Phosphoric acid fuel cell.

The overall cell reaction of the PAFC is

$$2H_2 + O_2 \Rightarrow 2H_2O + energy \qquad (6.40)$$

Efficiency of the PAFC is typically higher than 40%. If the steam generated by the FC heat is used in other applications such as cogeneration and air-conditioning, the efficiency of the cell can reach 80%. The electrolyte of the PAFC is not sensitive to CO_2 contamination, so it can use reformed fossil fuel. The relatively simple structure, its less expensive material, and its stable electrolyte make the PAFC more popular than PEM in some applications such as in buildings, hotels, hospitals, and electric utility systems.

6.3.2.4 Solid Oxide Fuel Cell

The electrolyte of the SOFC is a hard ceramic material such as zirconium oxide (Figure 6.54). The cell operates at very high temperatures (600°C–1000°C), which requires a significant time to reach its steady state. Therefore, it is slow at starting and slow at responding to changes in electricity demand. However, the high temperature makes the SOFC less sensitive to impurities in fuels such as sulfur and CO_2. For these reasons, SOFC can use fuels such as natural gas where the high temperature reforms the fuel without the need for external reformers such as the one shown in Figure 6.48. Therefore, the SOFC is suitable for large-scale stationary power generation in the megawatt range.

At the cathode, the oxygen molecules from air are combined with four electrons to produce the negatively charged oxygen ions O^{2-}. The solid ceramic material used in the FC is conductive to these oxygen ions that migrate to the anode and combine with hydrogen to produce water. In the process, the extra four free electrons in the oxygen ions are released and passed through the electric load and then to the cathode. The anode reaction is then

$$2H_2 + 2O^{2-} \Rightarrow 2H_2O + 4\varepsilon^- \qquad (6.41)$$

And the cathode reaction is

$$O_2 + 4\varepsilon^- \Rightarrow 2O^{2-} \qquad (6.42)$$

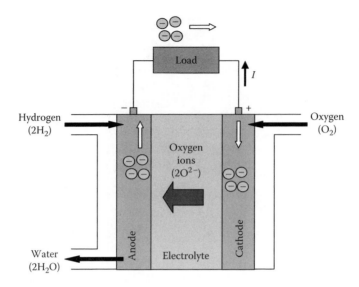

FIGURE 6.54 Solid oxide fuel cell.

Then, the overall cell reaction of the SOFC is

$$2H_2 + O_2 \Rightarrow 2H_2O + \text{energy} \tag{6.43}$$

The operating efficiency of the SOFC is among the highest of the FCs, about 60%. If the steam generated by the FC is used in cogeneration or air-conditioning, the overall efficiency can reach 80%.

6.3.2.5 Molten Carbonate Fuel Cell

The electrolyte of the MCFC is a mixture of lithium carbonate and potassium carbonate, or lithium carbonate and sodium carbonate (Figure 6.55). This FC is also a high-temperature type, operating at 600°C–650°C, and is suitable for large systems in the megawatt range.

When the electrolyte of the MCFC is heated to a temperature around 600°C, the salt mixture melts and becomes conductive to carbonate ions CO_3^{2-}. These negatively charged ions flow from the cathode to the anode where they combine with hydrogen to produce water, carbon dioxide, and free electrons. The anode reaction is

$$2CO_3^{2-} + 2H_2 \Rightarrow 2H_2O + 2CO_2 + 4\varepsilon^- \tag{6.44}$$

The carbon dioxide is routed to the cathode where the chemical reaction is

$$2CO_2 + O_2 + 4\varepsilon^- \Rightarrow 2CO_3^{2-} \tag{6.45}$$

The overall cell reaction of the MCFC is

$$2H_2 + O_2 \Rightarrow 2H_2O + \text{energy} \tag{6.46}$$

Notice that the CO_2 produced at the anode is also consumed at the cathode under ideal conditions. If the FC design can fully manage the carbon dioxide, the cell would emit no CO_2.

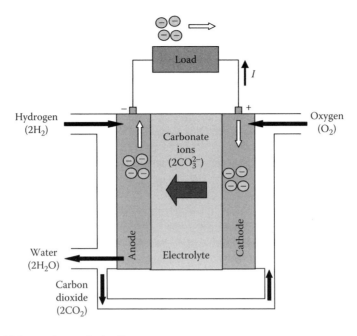

FIGURE 6.55 Molten carbonate fuel cell.

The high operating temperature of the MCFC has the same advantages and disadvantages mentioned for the SOFC. One of the distinctive disadvantages of this FC is the internal corrosion that is caused by the carbonate electrolyte.

6.3.2.6 Direct Methanol Fuel Cell

The DMFC is a low-temperature cell (50°C–120°C) and can use methanol directly without the need for a reformer (Figure 6.56). For these two reasons, it is suitable for consumer electronics

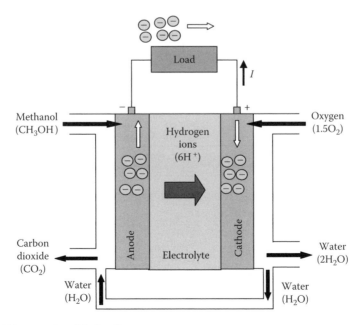

FIGURE 6.56 Direct methanol fuel cell.

applications such as mobile phones, entertainment devices, and portable computers as well as electric vehicles.

The DMFC electrolyte is a polymer, which is similar to that used in PEM FCs. At the anode, liquid methanol (CH_3OH) is oxidized by water, producing carbon dioxide (CO_2), six hydrogen ions (H^+), and six free electrons.

The anode reaction of the DMFC is

$$CH_3OH + H_2O \Rightarrow CO_2 + 6H^+ + 6\varepsilon^- \qquad (6.47)$$

The free electrons travel to the external electric load where they unload their energy. The hydrogen ions pass through the electrolyte to the cathode and react with oxygen from air and the free electrons from the external circuit to form water.

$$\tfrac{3}{2}O_2 + 6H^+ + 6\varepsilon^- \Rightarrow 3H_2O \qquad (6.48)$$

The overall cell reaction of the DMFC is

$$CH_3OH + \tfrac{3}{2}O_2 \Rightarrow CO_2 + 2H_2O + \text{energy} \qquad (6.49)$$

Because methanol is a toxic alcohol, researchers are developing FCs that use different alcohols such as ethanol C_2H_6O. The progress is expected to produce safer FCs with similar performance.

6.3.3 HYDROGEN ECONOMY

Hydrogen economy is a term often used to describe a complete closed-loop system that is based on extracting hydrogen and using it to produce electricity and in the process utilizing the heat of the cell in other applications. Such a system is shown in Figure 6.57. Renewable energy (wind, solar, etc.) is used to generate the electricity needed to reform fuel and produce hydrogen. It could also be used to

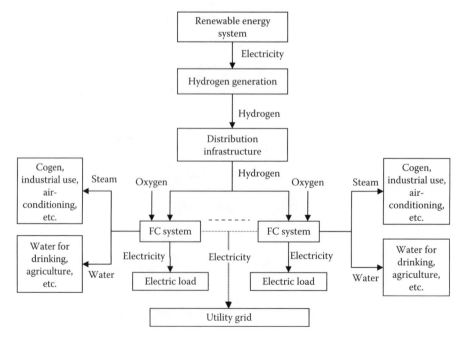

FIGURE 6.57 Hydrogen-based energy system.

extract hydrogen from water by electrolysis process. Once hydrogen is generated, it is distributed to fueling stations much like the gasoline system for automobiles. The FCs use the hydrogen to produce electricity that can be used locally, exported to utility grids, or both. If the FC is used to power automobiles, all the electrical energy is consumed by the load. If the FC is used in stationary sites with access to a utility grid, then part or all of the energy can be exported to the utility.

The by-products of the FC operation are heat and water. The water can be used for drinking, agriculture, or industrial processes. The heat, in the form of steam, can be used to generate electricity, air-condition buildings, and in industrial processes.

One obvious question is why we do not use the electricity generated by the renewable system directly instead of this rather elaborate process. Among the reasons are the following:

- Energy generated by the renewable system depends on favorable weather conditions. During these conditions, the energy produced may not be needed by the local loads or utilities. Instead of shutting down the renewable systems, we can store the energy using the system described in Figure 6.57.
- Renewable energy systems could be located in remote areas (small islands, deserts, etc.) without access to a power grid. In this case, it is better to store the electric energy in the form discussed above.
- For stand-alone systems such as spacecraft and electric ships, PV and FCs can be incorporated into an integrated system. The solar cells power the craft during the day and at the same time generate a supply of hydrogen that is stored and used at night by the FC.

6.3.4 MODELING OF IDEAL FUEL CELLS

To understand the energy conversion of the FC, we need to analyze its thermal and electrical processes. The thermal process gives us an idea on how much energy can be produced by the chemical reaction of the FC, and the electrical process gives us the value of the produced voltage and current. The following analyses are for most types of FCs.

6.3.4.1 Thermal Process of Fuel Cells

To understand the thermal process, we need to review some key thermodynamic equations pertaining to FCs. In thermodynamics, energy is computed for a unit of substance known as mole. One mole of a substance contains 6.002×10^{23} entities (molecules, ions, electrons, or particles). This constant is also known as the Avogadro's number $N_A = 6.002 \times 10^{23}/\text{mol}$.

During the last part of the eighteenth century, the American physicist and chemist Josiah Willard Gibbs developed a method by which the useful energy produced by a chemical reaction can be computed. It is known as Gibbs free energy G, where

$$G = H - Q \tag{6.50}$$

where
   H is the enthalpy of the process
   Q is the entropy of the process

For the FC shown in Figure 6.58, we can view the enthalpy as the energy in the fuel (hydrogen) at the anode and the entropy as the wasted heat during the process of making water from hydrogen and oxygen at the cathode.

At 1 atmospheric pressure and 298°K, hydrogen has $H = 285.83$ kJ/mol and $Q = 48.7$ kJ/mol. Hence, the chemical energy that is converted into electrical energy (Gibbs free energy constant) is

$$G = H - Q = 285.83 - 48.7 = 237.13 \text{ kJ/mol} \tag{6.51}$$

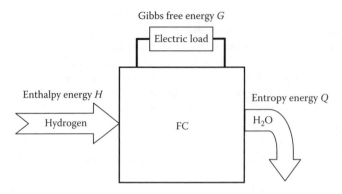

FIGURE 6.58   Gibbs free energy.

EXAMPLE 6.21

Assume ideal conditions and compute the thermal efficiency of the FC.

Solution

The input energy H is

$$H = 285.83 \text{ kJ/mol}$$

The energy that is converted into electricity is

$$G = H - Q = 285.83 - 48.7 = 237.13 \text{ kJ/mol}$$

The thermal efficiency η_t is

$$\eta_t = \frac{\text{output energy}}{\text{input energy}} = \frac{G}{H} = \frac{237.13}{285.83} = 83\%$$

Notice that the thermal efficiency of the FC is very high relative to the thermal efficiency of power plants, which cannot exceed 50%.

6.3.4.2   Electrical Process of Fuel Cells

Faraday has developed a key equation that computes the amount of electric charge q_e in a mole of electrons.

$$q_e = N_A q \qquad (6.52)$$

where
q is the charge of a single electron (1.602×10^{-19} C)
N_A is the Avogadro's number

In FCs, two electrons are released per molecule of hydrogen gas (H_2). Thus, the number of electrons N_e released by 1 mol of H_2 is

$$N_e = 2N_A \qquad (6.53)$$

The total charge of the electrons q_m released by 1 mol of hydrogen is

$$q_m = N_e q \qquad (6.54)$$

In an electric circuit, the charge of an object is defined as the current I passing through an object during a period t.

$$q_m = It \tag{6.55}$$

Since the electric energy E is the power multiplied by time, we can compute the energy as follows:

$$E = VIt = Vq_m \tag{6.56}$$

where the electric energy of the FC is equal to the Gibbs free energy G given in Equation 6.50. Then, the voltage of a single FC under ideal conditions is

$$V = \frac{E}{q_m} = \frac{G}{q_m} \tag{6.57}$$

EXAMPLE 6.22

Assume ideal conditions and compute the output voltage of a PEM FC.

Solution

Using Equation 6.53, we can compute the number of electrons N_e released in 1 mol of H_2.

$$N_e = 2N_A = 1.2004 \times 10^{24}$$

The total charge released by 1 mol of hydrogen can be computed using Equation 6.54.

$$q_m = N_e q = \left(1.2004 \times 10^{24}\right)\left(1.602 \times 10^{-19}\right) = 1.9288 \times 10^5 \text{ C}$$

Finally, the voltage of the ideal FC can be computed using Equation 6.57.

$$V = \frac{E}{q_m} = \frac{G}{q_m} = \frac{237.13 \times 10^3}{1.9288 \times 10^5} = 1.23 \text{ V}$$

6.3.5 MODELING OF ACTUAL FUEL CELLS

From Example 6.22, the voltage of the FC under ideal conditions is about 1.23 V. However, the actual voltage is lower due to the internal losses of the cell, the sluggish reactions at the anodes and cathodes, and the degradation of the FC due to the corrosion of its electrodes or the contamination of the electrolyte.

The losses of the FC can generally be divided into three categories: activation loss, ohmic loss, and mass transport loss. Activation loss occurs because of the way the anode and cathode reactions are carried out, especially at low currents or when the cell is activated. This is known as electrode kinetic, which is the way fuel and oxygen are diffused at the electrodes. Because they are not fully diffused at starting, the performance of the FC is reduced during the activation period.

Ohmic loss occurs mainly because of the resistances of the electrolyte and electrodes. This loss is small at low currents.

Mass transport loss occurs because of conditions at which the input reaction is less than the output reaction. This situation arises when the output current is high and the input reaction cannot meet the needed demand.

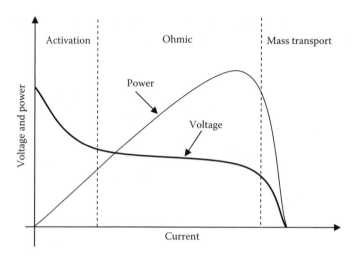

FIGURE 6.59   Polarization and power curves of FC.

6.3.5.1 Polarization Characteristics of Fuel Cells

The polarization curve is the steady-state voltage versus current. This curve is essential in determining the best operating conditions for an FC. A typical polarization curve is shown in Figure 6.59. The thick line represents the cell voltage and the thin line represents its output power. The figure is divided into three regions: activation, ohmic, and mass transport. At no load, the voltage of the cell is close to its ideal value. When the current is slightly increased, the voltage of the cell drops rapidly because of activation loss. Since the current is small, the ohmic loss and the mass transport loss are insignificant. In the ohmic region, the current is high enough to cause internal losses in the electrolyte and electrodes. In this region, the ohmic loss is more dominant than the activation loss or mass transport loss. The voltage in the ohmic region is fairly stable. At high currents, the mass transport loss is very dominant and the FC cannot cope with the high current demand of the load. In this case, the anode reaction is not fast enough to provide the needed output current. As a result, the voltage of the cell collapses rather rapidly. This is the region we should avoid as it causes the collapse of the FC. Keep in mind that all three losses exist at all times, but the dominant ones depend on the operating region of the cell. The power curve in the figure shows the maximum power occurring in the ohmic region.

Example 6.23

A polarization curve of an FC can be represented by the empirical formula

$$V = 0.77 - 0.128 \tan (I - 1.3)$$

where
    V is the voltage of the FC (V)
    I is the current (A)

Plot the polarization and power curves. Also compute the maximum power of the cell and the current at maximum power.

Solution

The plot of the polarization curve of this FC is given in Figure 6.60. The power of the FC is the multiplication of voltage times current.

$$P = VI = [0.77 - 0.128 \tan (I - 1.3)]I$$

To compute the maximum power, we set the derivative of the power equation to zero and compute the current at maximum power.

$$\frac{\partial P}{\partial I} = 0.77 - 0.128 \times \tan{(I-1.3)} - 0.128 \times I\sec^2{(I-1.3)} = 0$$

The numerical solution of this equation yields a current of 2.145 A at which the power is maximum. The maximum power of the cell is

$$P_{\max} = 0.77 \times 2.145 - 0.128 \times 2.145 \tan{(2.145 - 1.3)} = 1.3422 \text{ W}$$

6.3.6 EVALUATION OF FUEL CELLS

Experts expect FCs to eventually dominate the power generation market, especially for transportation, household use, and utility size generation. In automotive applications, using FCs to propel an electric motor is an ever-growing trend. The cars that are powered by FCs and use electric motors as the driving engines do not need the elaborate cooling and lubricating systems that are required by the internal combustion engines, making the cars smaller, lighter, more efficient, quieter, and cheaper to maintain. Several generations of FC automobiles and buses are already roaming city streets. In addition to transportation, FCs are used as backup systems or independent sources of energy. Several sensitive installations such as hospitals, satellites, and military installations use FCs as backup systems.

The efficiency of the FC alone is relatively high; the range is about 30%–80%. If the heat generated by the FC is wasted, the efficiency of the cell is at the low end of the range. When we add the efficiency of the reformer plus CO-converter, the efficiency range of the FC system drops to about 26%–40%. If the FC is used to drive an electric motor (electric car), the efficiency of the drive system is about 25%–40%. In contrast, modern internal combustion engines have efficiency around 15%–22%.

High-temperature FCs produce enough heat, which can be used in industrial processes, in heating buildings, and even in cogeneration applications. These systems are suitable for high-power applications where portability is unnecessary. However, low-temperature FCs are suitable for a wider range of applications that require small units with quick start-up time such as electric cars.

A single FC produces a dc at less than 1.5 V. This is barely enough to power small consumer electronic devices. For higher voltage applications, FCs are stacked in series as shown in Figure 6.51. The dc produced by the FC system cannot be used directly to power ac devices; therefore, a converter is needed to convert the dc into ac.

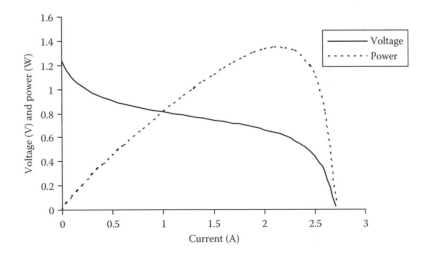

FIGURE 6.60 Polarization and power curves of the FC in Example 6.23.

One major concern about FCs is their relatively short lifetime. Their various components can suffer from pollution and corrosion. Some types of electrolytes can be contaminated by CO, CO_2, or sulfur. Carbonate electrolyte can cause internal corrosion that can rapidly degrade the cell's performance.

6.3.7 FUEL CELLS AND ENVIRONMENT

If we compare FCs with conventional thermal power generation, we can easily fall in love with FCs as they produce no pollution and most of them operate silently. The same assessment is true if we compare FC-powered automobiles with conventional internal combustion engine cars.

Some scientists, however, believe that the widespread use of FCs could have some negative environmental impacts due to unintended release of hydrogen in the atmosphere. This could increase the water vapor in the stratosphere zone, which could cool down the stratosphere causing more destruction to the ozone.

A more thorough analysis of the environmental impact of FCs must include the way we generate hydrogen. The processes by which hydrogen is harnessed and stored initiate most of the environmental concerns.

6.3.7.1 Generation of Hydrogen

Hydrogen is a clean gas that is completely nontoxic. Therefore, one valid question is why not just burn hydrogen in thermal power plants instead of using it in FCs? Burning hydrogen produces tremendous heat that cannot be efficiently harnessed by thermal turbines, thus reducing the system efficiency to less than 10%. But when hydrogen is used in an FC, the overall efficiency is much higher.

Because hydrogen cannot be found in nature as a free element, it must be extracted from hydrogen-rich compounds and safely stored. About 50% of world hydrogen is produced by reforming natural gas, 28% is produced by processing crude oil, 18% by processing coal, and 4% by other means.

Hydrogen reforming: This process is discussed in Section 6.3.1 and the system is shown in Figure 6.48. When a reformer alone is used, carbon monoxide and carbon dioxide are released in the atmosphere; both are polluting gases. When a reformer and CO-converter are used, only carbon dioxide is produced, which is one of the greenhouse gases.

Electrolysis of water: This process sounds very simple since water contains only hydrogen and oxygen. However, the electrolysis of water is an energy intensive process that reduces the overall efficiency of the entire system to less than 10%. Furthermore, electrolysis generates indirect pollution as the electricity used in the process is probably produced by a polluting power plant.

Futuristic methods: Recent research claims that hydrogen can be liberated from a solution of sodium hydride (NaH) in the presence of certain catalysts without using electric energy. Also, researchers believe that the old algae can be genetically altered to produce hydrogen instead of oxygen during their photosynthesis process. Yet another method is based on a phenomenon known in metallurgy where some metal hydrides absorb hydrogen and store it between their molecules when they are cooled. The hydride alloys release the stored hydrogen when heated. The potential of these techniques to produce hydrogen at industrial scales remains to be seen, and the environmental impacts of the processes need to be assessed.

6.3.7.2 Safety of Hydrogen

Hydrogen is currently used in several applications such as direct combustion and rocket propulsion. However, because hydrogen is a very explosive gas, it creates tremendous anxiety among the public. Most people associate hydrogen with the well-known Hindenburg zeppelin fire, which occurred in 1937. This luxury airship used hydrogen to provide the needed lift and was burned while attempting to dock at Lakehurst, New Jersey. The spectacular fire of Hindenburg is the main factor affecting public perception regarding the safety of hydrogen.

The hydrogen used in FC electric cars is stored in pressurized tanks. The tanks are often made from carbon fibers and aluminum cylinders and are mounted on the top of the car. These tanks can survive a direct force from collisions without exploding. In the worst scenario, if the hydrogen catches fire, it will escape upward leaving the people inside the car below the tank relatively safe.

An alternative option to storing hydrogen is to use natural gas and install a reformer and CO-converter inside the car. However, the storage of pressurized natural gas has its own safety concerns. Safer systems such as direct FCs that use methanol and perhaps ethanol are expected to be more viable options.

6.4 SMALL HYDROELECTRIC SYSTEMS

Hydroelectric power is one of the oldest forms of electricity generated, which dates back to 1882 when the first hydroelectric power plant was constructed across the Fox River in Appleton, Wisconsin. As seen in Chapter 4, hydroelectric power plants are enormous in size and capacity. These power plants require enormous water resources, which are available only in a limited number of areas worldwide. However, the world has an immense number of small rivers that can be utilized to generate small amounts of electric energy enough to power neighborhoods. These types of power plants are called small hydroelectric systems.

Small hydroelectric systems are at the low end of power ratings, up to a few megawatt. The popularity of these systems is due to their mature and proven technology, reliable operation, suitability for sensitive ecology, and capability to produce electricity even in small rivers. There are two versions of hydroelectric systems: reservoir type and diversion type. The reservoir type may require a small dam to store water at higher elevations. The diversion type does not require a dam, and relies on the speed of the current to generate electricity.

6.4.1 RESERVOIR-TYPE SMALL HYDROELECTRIC SYSTEM

A schematic of a simple reservoir-type hydroelectric plant is shown in Figure 6.61. The system is similar to the impoundment hydroelectric system discussed in Chapter 4, but the head of the reservoir is much shorter and its capacity is much smaller. The reservoir is either a natural lake at a higher elevation than a downstream river or a lake created by a dam. The system consists mainly of a reservoir, a penstock, a turbine, and a generator. If the water is allowed to flow to a lower elevation through the penstock, the potential energy (PE) of the water is converted into KE where some of it is captured by the turbine. After passing through the turbine, the water exits to the stream

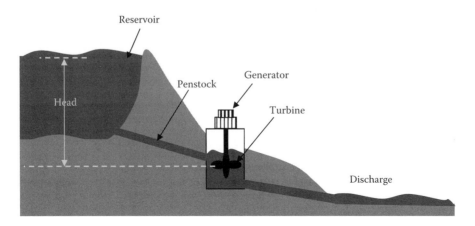

FIGURE 6.61 Small hydroelectric system with reservoir.

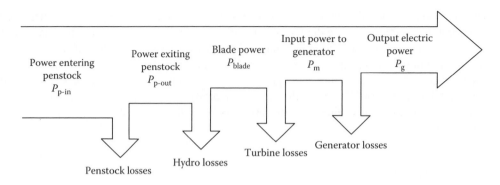

FIGURE 6.62 Power flow in a small hydroelectric system.

at the lower elevation. The turbine rotates due to its acquired KE from the flow of water, thus rotating the generator and electricity is generated.

The power flow inside a reservoir-type small hydroelectric system is shown in Figure 6.62. The input power entering the penstock is $P_{p\text{-}in}$; part of this power is wasted in the penstock due to losses such as water friction. The rest of the power $P_{p\text{-}out}$ exits the penstock and enters the turbine. Part of $P_{p\text{-}out}$ is captured by the turbine blade power P_{blade}. Part of P_{blade} is lost in the form of turbine rotational losses and the rest is converted into mechanical power P_m, which is the input to the electrical generator. Part of P_m is lost due to the various losses inside the generator, and the rest is the output electric power P_g of the small hydroelectric system.

The amount of electric power generated by the reservoir-type small hydroelectric system depends on three key parameters:

1. Water head H, which is the average vertical distance between the water surface of the reservoir and the turbine
2. Flow of water, which is the mass of water passing through the penstock per unit time
3. Efficiency of the system components

The PE of the water behind the reservoir PE_r is the weight of the water W in the reservoir multiplied by the elevation of the water H with respect to the turbine.

$$PE_r = WH \tag{6.58}$$

The weight of the water behind the dam is equal to its mass M multiplied by the acceleration of gravity g.

$$W = Mg \tag{6.59}$$

$$PE_r = MgH \tag{6.60}$$

where PE_r is the PE of the entire reservoir. If we are to calculate the PE of the water entering the penstock $PE_{p\text{-}in}$, we use the water mass m passing through the penstock.

$$PE_{p\text{-}in} = mgH \tag{6.61}$$

where
   $PE_{p\text{-}in}$ is in joules (J)
   m is in kilogram (kg)
   H is in meter (m)
   g is in meter per second (m/s)

The water flow f inside the penstock is defined as the mass of water m passing through the penstock during a time interval t.

$$f = \frac{m}{t} \tag{6.62}$$

Hence, Equation 6.61 can be rewritten in terms of water flow as

$$PE_{\text{p-in}} = fgHt \tag{6.63}$$

The PE given by Equation 6.61 is converted into KE inside the penstock. The output KE of the penstock $KE_{\text{p-out}}$ is defined as

$$KE_{\text{p-out}} = \tfrac{1}{2}mv^2 \tag{6.64}$$

where v is the speed of water exiting the penstock. Due to the losses inside the penstock, such as water friction, the KE exiting the penstock $KE_{\text{p-out}}$ is less than the PE at the entrance of the penstock $PE_{\text{p-in}}$.

$$\eta_{\text{p}} = \frac{KE_{\text{p-out}}}{PE_{\text{p-in}}} \tag{6.65}$$

where η_{p} is the penstock efficiency. The mass m in Equation 6.64 can be substituted by the water volume "vol" times the water density ρ.

$$KE_{\text{p-out}} = \tfrac{1}{2}mv^2 = \tfrac{1}{2}\text{vol}\,\rho v^2 \tag{6.66}$$

The water density is about 1000 kg/m^3. Hence,

$$KE_{\text{p-out}} = 500\,\text{vol}\,v^2 \tag{6.67}$$

The water volume passing through the penstock during a time interval t can be computed as

$$\text{vol} = Avt \tag{6.68}$$

where A is the cross-sectional area of the penstock. Substituting Equation 6.68 into 6.67 yields

$$KE_{\text{p-out}} = 500\,Av^3t \tag{6.69}$$

The blades of the turbine cannot capture all of the kinetic energy $KE_{\text{p-out}}$ exiting the penstock. The ratio of the energy captured by the blades KE_{blade} to $KE_{\text{p-out}}$ is known as the power coefficient C_{p}.

$$C_{\text{p}} = \frac{KE_{\text{blade}}}{KE_{\text{p-out}}} \tag{6.70}$$

The energy captured by the blades KE_{blade} is not all converted into the mechanical energy entering the generator KE_{m} due to the various losses in the turbine. The ratio of the two energies is known as turbine efficiency η_{t}.

$$\eta_t = \frac{KE_m}{KE_{blade}} \tag{6.71}$$

The output electric energy of the generator E_g is equal to its input kinetic energy KE_m minus the losses of the generator; hence, the generator efficiency η_g is defined as

$$\eta_g = \frac{E_g}{KE_m} \tag{6.72}$$

Using Equations 6.63, 6.65, 6.70 through 6.72, we can write the equation of the output electrical energy as a function of the water flow and head.

$$E_g = fgHt(C_p \eta_p \eta_t \eta_g) \tag{6.73}$$

EXAMPLE 6.24

A small hydroelectric site with a reservoir has a head of 5 m. The penstock passes water at the rate of 100 kg/s. The efficiency of the penstock is 95%, the power coefficient is 47%, the turbine efficiency is 85%, and the efficiency of the generator is 90%. Assume that the owner of this small hydroelectric system sells the generated energy to the local utility at \$0.15/kW h. Compute his income in 1 month.

Solution

Using Equation 6.73, we can compute the output electric energy for 1 month.

$$E_g = fgHt(C_p \eta_p \eta_t \eta_g) = 100 \times 9.8 \times 5(30 \times 24)(0.47 \times 0.95 \times 0.85 \times 0.9) = 1.206 \text{ MW h}$$

The income in 1 month is as follows:

$$\text{Income} = E_g \times 0.15 = \$181$$

EXAMPLE 6.25

For the system in Example 6.24, compute the speed of the water exiting the penstock.

Solution

Using Equation 6.61, we can compute the PE of the water at the entrance of the penstock.

$$PE_{p\text{-in}} = mgH = 100 \times 9.8 \times 5 = 4.75 \text{ kJ}$$

From Equation 6.65, the KE exiting the penstock is

$$KE_{p\text{-out}} = \eta_p PE_{p\text{-in}} = 0.95 \times 4.75 = 4.75 \text{ kJ}$$

Now using Equation 6.64, the water speed can be computed as

$$v = \sqrt{\frac{2KE_{p\text{-out}}}{m}} = \sqrt{\frac{2 \times 4.51 \times 10^3}{100}} = 9.49 \text{ m/s}$$

Example **6.26**

A man owns a property with a small river. He wants to build a dam to create a reservoir for a small hydroelectric system. The site can accommodate a penstock of 3 m in diameter. In order for him to generate 2 MW of electricity, compute the height of the dam. Assume that the penstock efficiency is 97%, the power coefficient is 40%, the turbine efficiency is 90%, and the generator efficiency is 95%.

Solution

We can solve this problem by using the power flow shown in Figure 6.62 but in the backward direction. Using Equation 6.72, we can compute the mechanical power of the turbine.

$$\eta_g = \frac{E_g}{KE_m} = \frac{P_g}{P_m}$$

$$P_m = \frac{P_g}{\eta_g} = \frac{2}{0.95} = 2.105 \text{ MW}$$

Compute the blade power using Equation 6.70.

$$P_{blade} = \frac{P_m}{\eta_t} = \frac{2.105}{0.9} = 2.33 \text{ MW}$$

Compute the output power of the penstock using Equation 6.71.

$$P_{p\text{-out}} = \frac{P_{blade}}{C_p} = \frac{2.33}{0.4} = 5.83 \text{ MW}$$

Use Equation 6.69 to compute the velocity of the water that yields 5.83 MW of power.

$$KE_{p\text{-out}} = 500 \, A v^3 t$$

$$P_{p\text{-out}} = 500 \, A v^3$$

Hence,

$$v = \sqrt[3]{\frac{P_{p\text{-out}}}{500 \, A}} = \sqrt[3]{\frac{5.83 \times 10^6}{500 \, (\pi \times 1.5^2)}} = 11.5 \text{ m/s}$$

Now use Equation 6.65 to compute the head of water behind the dam.

$$\eta_p = \frac{KE_{p\text{-out}}}{PE_{p\text{-in}}} = \frac{v^2}{2gH}$$

$$H = \frac{v^2}{2g\eta_p} = \frac{(11.5)^2}{2 \times 9.8 \times 0.97} = 6.95 \text{ m}$$

Hence, the height of the dam should be at least 6.95 m.

6.4.2 Diversion-Type Small Hydroelectric System

The diversion-type small hydroelectric system does not require a dam, and therefore is considered more environmentally sensitive. The schematic of this system is shown in Figure 6.63. The river for this small hydroelectric system must have strong enough current for a realistic power generation.

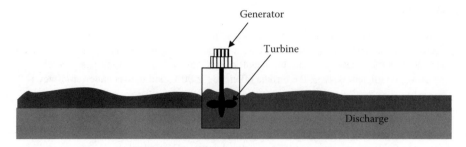

FIGURE 6.63   Diversion-type small hydroelectric system.

Let us assume that the volume of water "vol" entering the turbine during a time t is

$$\text{vol} = A_s vt \qquad (6.74)$$

where A_s is the sweep area of the turbine's blades in one revolution. The KE entering the turbine KE_t is

$$\text{KE}_t = \tfrac{1}{2}mv^2 = \tfrac{1}{2}\text{vol}\,\rho v^2 = \tfrac{1}{2}A_s\rho v^3 t \qquad (6.75)$$

At a water density of about 1000 kg/m^3, the power entering the turbine P_t is

$$P_t = 500\,A_s v^3 \qquad (6.76)$$

The blades of the turbine cannot capture all of the power entering the turbine. The ratio of the power captured to the power entering is known as the power coefficient C_p.

$$C_p = \frac{P_{\text{blade}}}{P_t} \qquad (6.77)$$

The mechanical output power of the turbine P_m is equal to the blade power P_{blade} minus the turbine losses. Hence,

$$\eta_t = \frac{P_m}{P_{\text{blade}}} \qquad (6.78)$$

where η_t is the turbine efficiency. The output electric power of the generator P_g is equal to P_m minus the losses of the generator; hence, the generator efficiency η_g is defined as

$$\eta_g = \frac{P_g}{P_m} \qquad (6.79)$$

EXAMPLE 6.27

A diversion-type small hydroelectric system is installed across a small river with a current speed of 5 m/s. The diameter of the swept area of the turbine is 1.2 m. Assume that the power coefficient is 50%, the turbine efficiency is 90%, and the efficiency of the generator is 90%. Compute the output power of the plant and the energy generated in 1 year. If the price of the energy is $0.05/kW h, compute the income from this small hydroelectric plant in 1 year.

Solution

Use Equation 6.76 to compute the input power to the turbine.

$$P_t = 500\,A_s v^3 = 500\,(\pi \times 0.6^2)5^3 = 70.69 \text{ kW}$$

The output electric power is P_t multiplied by the efficiencies.

$$P_g = P_t(C_p \eta_t \eta_g) = 70.69 \, (0.5 \times 0.9 \times 0.9) = 25.45 \text{ kW}$$

The energy generated in 1 year is

$$E_g = 25.45 \, (24 \times 365) = 222.942 \text{ MW h}$$

Income from the plant is

$$\text{Income} = 0.05 \times 222{,}942 = \$11{,}147.00$$

6.5 GEOTHERMAL ENERGY

Geothermal energy, which is the stored heat in the earth's crust, is an enormous, underused energy resource. At just a few meters under the earth's surface, the geothermal temperature in winter is about 10°C–20°C higher than the ambient temperature and about 10°C–20°C lower in the summer. At a greater depth, the magma (molten rock) has extremely high temperature that can produce a tremendous amount of steam suitable for generating large amounts of electricity. As a matter of fact, the thermal energy in the uppermost 10 km of the earth's crust is much more than the energy from all oil and gas resources known to us so far. The thermal energy of the magma can be stored for thousands of years as the thermal isolation of the magma by the surrounding material cools down the magma at an extremely slow rate.

The depth at which the magma is located and its surrounding material determine the way we can harness geothermal energy. In shallow depths, heat pumps can be used to heat houses in the winter and cool them down in the summer. At greater depths, closer to the molten rocks, enough steam can be produced to generate electricity. These two systems are discussed in Sections 6.5.1 and 6.5.2.

6.5.1 HEAT PUMP

A typical heat pump system is shown in Figure 6.64. It consists of a geo-exchanger, pump, and a heat distribution system. The geo-exchanger is a series of pipes buried in the ground near the building to be conditioned. The pipes are filled with water, or a mixture of water and antifreeze, to absorb the earth's heat in the winter, or dissipate heat to the earth in the summer. Pumps are used to circulate the water mix between the geo-exchanger and the rest of the system. In winter, the building

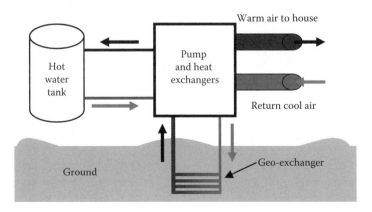

FIGURE 6.64 Heat pump system.

temperature is lower than the geothermal temperature, so the geo-exchanger increases the temperature of the mixture. The pump moves the heated mixture to the building and by using a system of heat exchangers, the air and water of the building are warmed up. In summer, when the earth's temperature is lower than the air temperature, the process is reversed.

6.5.2 GEOTHERMAL POWER PLANT

In nature, steam can be formed in various remarkable ways such as the geyser in Figure 6.65. The geyser throws forth intermittent jets of heated water and steam. Steam is formed when water from rain seeps into the fractured earth's rocks for miles until it gets near the molten rocks where the water is turned into steam and pushes its way upward toward any crack on the surface of the earth. This steam could include other minerals or gases that are mixed with water.

Steam could also be trapped in geothermal reservoirs, which are large pools of steam or very hot water in porous rocks above the magma. When rain or groundwater reaches the molten rocks, the water absorbs the thermal energy and ascends toward a reservoir as shown in Figure 6.66. The steam in some reservoirs is at very high temperatures and can be used directly to generate electricity.

A typical geothermal power plant is shown in Figure 6.67. The steam, which may include hot mist, is drawn from the geothermal reservoir and passes through a mist eliminator where the hot mist is condensed to water and used to provide heat to buildings and industrial processes. The steam enters a thermal turbine that converts the steam energy into mechanical energy much

FIGURE 6.65 (See color insert following page 300.)   Steam generated from rain even in a cold environment.

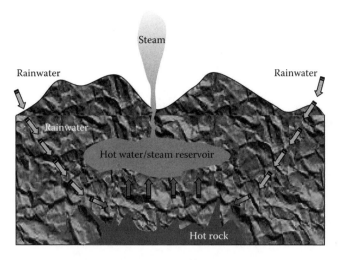

FIGURE 6.66 Geothermal reservoir.

like the thermal power plant discussed in Chapter 4. This mechanical energy is converted by the generator G into electrical energy. The steam exiting the turbine is cooled down in a cooling tower where external cold water is poured in to create the heat sink needed to complete the thermal cycle. The hot water can be used to heat buildings, and then injected back into the geothermal reservoir.

The first experimental geothermal power plant was built in Lardarello, Italy, by Prince Gionori Conti around 1905. The first commercial power plant based on Conti's work was commissioned in 1913 at Larderello. In 1958, Wairakai's geothermal power plant was built in New Zealand, followed by Pathe's plant in Mexico in 1959 and the Geysers plant in California in 1962. By the end of 2006, 21 countries had operational geothermal power plants with a total capacity of 8.2 GW. Iceland produces about 50% of its electrical energy from geothermal power plants. The geothermal power

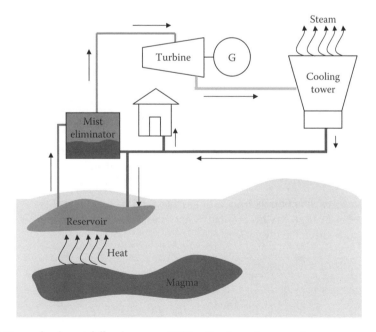

FIGURE 6.67 (See color insert following page 300.) Geothermal power plant.

FIGURE 6.68 (See color insert following page 300.)   The Geysers in northern California, first geothermal power plant in the United States.

plants in the United States have a total generating capacity of 2.7 GW. The Geysers power plant in northern California, shown in Figure 6.68, is still the largest geothermal power plant with a total generation capacity of 1.7 GW. This is enough power for the San Francisco metropolitan area. Other geothermal power plants in the United States are in Nevada, Utah, and Hawaii.

6.5.3 TYPES OF GEOTHERMAL POWER PLANTS

The naturally occurring steam in geothermal reservoirs varies in temperature and pressure. In addition, the depth of the magma and its surrounding material vary from one site to another. These variations make it hard to design one geothermal power plant for all conditions; therefore, currently we have three basic designs:

Dry steam power plants: This system is used when the steam temperature is very high (300°C) and the steam is readily available. This is the system explained in Figure 6.67.

Flash steam power plants: When the reservoir temperature is above 200°C, the reservoir fluid is drawn into an expansion tank that lowers the pressure of the fluid. This causes some of the fluid to rapidly vaporize (flash) into steam. The steam is then used to generate electricity.

Binary-cycle power plants: At a moderate temperature (below 200°C), the energy in the reservoir water is extracted by exchanging its heat with another fluid (called binary) that has a much lower boiling point. The heat from the geothermal water causes the secondary fluid to flash to steam, which is then used to drive the turbines.

6.5.4 EVALUATION OF GEOTHERMAL ENERGY

The earth's magma can provide clean and reliable energy that can last for thousands of years. In some areas, however, geothermal electricity is not yet as cost competitive as the other more traditional methods. Besides cost, geothermal power systems have several challenges such as the ones listed below:

- Geothermal power plants can only be constructed in geothermal sites where their magmas are close enough to the surface to heat reservoirs accessible by the current drilling technology. That is why they are placed in areas with volcanic activity such as the west coasts of the United States and Japan. Similar sites are scarce worldwide.
- The cost of drilling deep wells to reach underground reservoirs surrounded by hard rocks could be very expensive.
- The geothermal fluid of the reservoir may include other minerals and gases besides water that can be pollutants. Geothermal fields could also produce a small amount of carbon dioxide.
- The geothermal fluid can produce objectionable odors due to the presence of compounds such as hydrogen sulfide.

6.6 TIDAL ENERGY

Tides are the daily swells and sags of ocean waters relative to coastlines due to the gravitational pull of the moon and the sun. Although the moon is much smaller in mass than the sun, it exerts a larger gravitational force because of its relative proximity to earth. This force of attraction causes the oceans to rise along a perpendicular axis between the moon and earth. Because of the earth's rotation, the rise of water moves opposite to the direction of the earth's rotation creating the rhythmic rise and sag of coastal water. These tidal waves are slow in frequency (about one cycle every 12 h), but contain tremendous amounts of KE, which is probably one of the major untapped energy resources of earth.

6.6.1 TIDAL ENERGY SYSTEMS

There are a few major tidal energy generating power plants in operation. The main one is the *Le Rance* River in northern France built in 1966 and has a capacity of 240 MW. Other plants are the 20 MW plant in Nova Scotia, Canada, and the 400 kW plant near Murmansk, Russia.

The technology required to convert tidal energy into electricity is very similar to the technology used in either wind energy or hydroelectric power plants. The common designs are the free-flow system (also called tidal stream or tidal mill) and the dam system (also known as barrage or basin system).

6.6.1.1 Free-Flow Tidal System

The free-flow system, such as the ones shown in Figure 6.69, converts the KE of the tidal current into electrical energy much the same way as the wind is converted into electrical energy by wind turbines. The tidal energy turbine has its blades immersed in oceans or rivers at the path of strong tidal currents. The current rotates the blades that are attached to an electrical generator mounted above the water level.

Tidal mills produce much more power than wind turbines because the density of the water is 800–900 times the density of air. Recall Equation 6.23 that computes the power of the wind in the sweep area of the blades. The same equation can be used to compute the power of the tidal current P_{tidal}.

$$P_{tidal} = \tfrac{1}{2}A\delta v^3 \qquad (6.80)$$

where
  A is the sweep area of the turbine blades (m^2)
  v is the velocity of water (m/s)
  δ is the water density (kg/m^3)

Because the water density is high (about 1025 kg/m^3), the tidal current has much higher power density than wind.

(a)

(b)

FIGURE 6.69 (See color insert following page 300.) Free-flow tidal energy system. (a) Free-flow tidal turbine and (b) conceptual design of a farm. (Images courtesy of Marine Current Turbines Limited.)

EXAMPLE 6.28

For a tidal mill with a blade length of 3 m, compute the power captured by the blades when the tidal current is 10 knots and the power coefficient of the tidal mill is 0.45.

Solution

Convert the nautical speed to metric.

$$v = 10 \times 1.852 = 18.52 \text{ km/h}$$
$$v = \frac{18.52 \times 1000}{3600} = 5.14 \text{ m/s}$$

Use Equation 6.80 to compute the tidal power.

$$P_{\text{tidal}} = \tfrac{1}{2}A\delta v^3 = \tfrac{1}{2}\left(\pi \times 3^2\right)1025(5.14)^3 = 1.97 \text{ MW}$$

The power coefficient C_p determines the amount of power captured by the blades.

$$P_{\text{blade}} = C_p P_{\text{tidal}} = 0.45 \times 1.97 = 886.5 \text{ kW}$$

EXAMPLE 6.29

Assume a wind mill has the same diameter as the tidal mill in Example 6.28. Also assume that the wind speed is equivalent to the tidal speed and the power coefficient is 0.45. Compute the power captured by the blades of the wind turbine. Also compute the ratio of the powers from tidal current and wind.

Solution

Assume the air density is 1.2 kg/m^3; use Equation 6.23 to compute the wind power

$$P_{\text{wind}} = \tfrac{1}{2}A\delta v^3 = \tfrac{1}{2}\left(\pi r^2\right)1.2(5.14)^3 = 2.3 \text{ kW}$$

The power coefficient determines the amount of power captured by the blades.

$$P_{\text{blade}} = C_p P_{\text{wind}} = 0.45 \times 2.3 = 1.035 \text{ kW}$$

The ratio of tidal power to wind power is

$$\frac{P_{\text{tidal}}}{P_{\text{wind}}} = \frac{886.5}{1.035} = 856.5$$

Notice that this ratio is the same as the ratio of water density to air density. Hence, the tidal mill can capture much more power than the wind mill.

As seen in the previous examples, when tidal turbines are placed in areas with strong currents, they can produce large amount of power. Strong currents can be found in numerous places around the world such as in the northwest and northeast coasts of the United States and Canada, Australia, southern tip of the Korean peninsula, the western coast of Taiwan, and the western and northern coasts of Europe. The city of Saltstraumen in Norway claims to have the strongest tidal current in the world, about 22 knots.

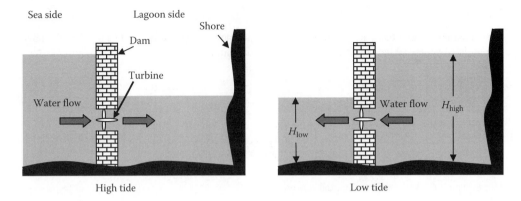

FIGURE 6.70 Dam-type tidal energy system.

6.6.1.2 Barrage System

The barrage energy system, which is also known as a dam-type tidal system, is shown in Figure 6.70. It is most suited for inlets where a channel connects an enclosed lagoon to the open sea. At the mouth of the channel, a dam is constructed to regulate the flow of the tidal water in either direction. A turbine is installed inside a conduit connecting the two sides of the dam. At high tides, the water moves from the sea to the lagoon through the turbine. The turbine and its generator convert the KE of the water into electrical energy. When the tide is low, the water stored in the lagoon at high tides goes back to the sea, and in the process electricity is generated.

The difference in hydraulic heads of the tide determines the amount of energy that can be captured. We can use Equation 6.60 to compute the PE of a body of water with a higher head than the rest of the ocean.

$$\text{PE} = mg\Delta H \tag{6.81}$$

where

m is the mass of water moving from the high head side to the low head side
g is the acceleration of gravity
ΔH is the average of the difference in heads between the waters on the two sides of the dam

$$\Delta H = \frac{H_{\text{high}} - H_{\text{low}}}{2} \tag{6.82}$$

where

H_{high} is the head of the high water side of the dam
H_{low} is the head of the low water side of the dam as shown in Figure 6.70

Because the difference in head is directly proportional to the PE of water, barrage tidal energy should be located in areas with high amplitude of tides such as in the United States, Russia, Canada, United Kingdom, and Australia. In the United States, a large tidal difference of up to 17 m occurs in places such as Maine, Alaska, and the Bay of Fundy on the Atlantic Ocean in Canada. These areas can produce large amounts of electrical energy from tides.

Example 6.30

A dam-type tidal energy system is constructed between a lagoon and the open ocean. The base of the lagoon is approximately semicircle with a radius of 1 km. At high tide, the head on the high water side of the dam is 25 m and the head on the low side is 15 m. Compute the PE in the tidal water.

Solution

The average head difference is

$$\Delta H = \frac{H_{\text{high}} - H_{\text{low}}}{2} = \frac{25 - 15}{2} = 5 \text{ m}$$

The volume of water passing through the turbine during high tide.

$$\text{Vol} = \frac{1}{2}\pi R^2 \Delta H = 0.5 \times \pi \times 1000^2 \times 5 = 7.85 \times 10^6 \text{ m}^3$$

The mass of water is the volume times the water density. Assume the water density is 1000 kg/m³.

$$m = \text{vol} \times \rho = 7.85 \times 10^6 \times 1000 = 7.85 \times 10^9 \text{ kg}$$

The PE of the water passing through the turbine at high tide is

$$\text{PE} = mg\Delta H = 7.85 \times 10^9 \times 9.81 \times 5 = 3.85 \times 10^{11} \text{ J}$$

As seen in this example, there is a tremendous amount of power that can be harvested from tides.

EXAMPLE 6.31

For the system in Example 6.30, compute the electrical energy generated by the tidal system assuming that the power coefficient of the blades is 35%, the efficiency of the turbine is 90%, and the efficiency of the generator is 95%.

Solution

The output electric energy from the system is

$$E_{\text{out}} = \text{PE} \times C_p \eta_t \eta_g = 7.85 \times 10^{11} \times 0.35 \times 0.9 \times 0.95 = 2.35 \times 10^{11} \text{ J}$$

6.6.2 EVALUATION OF TIDAL ENERGY

Tide is a renewable source of energy, and tidal energy has the potential of becoming one of the major electric power generating systems. Because tidal current is predictable with high accuracy for years in advance, it is easy to calculate the electric power generated by a tidal energy system for a few months in the future. This makes the owner able to schedule and sell the power at a higher rate than systems with less predictability such as wind energy. However, tidal systems are not free from problems or limitations. Among the problems are

- Electricity can be generated by water flowing into and out of a lagoon or by tidal currents. It is an intermittent source of energy, which depends on the tidal cycle. Therefore, it can generate electricity only for about 6–12 h daily.
- Cost of building a tidal energy system is high and the return on investment is longer than other methods.
- Changing tidal flows by building dams could result in negative effects on aquatic life, as well as water navigation. It is believed that dams can potentially stimulate the growth of the red tide organism that causes sickness in shellfish.

6.7 BIOMASS ENERGY

Biomass consists of garbage, agricultural waste, and tree products. Garbage, in particular, is a major concern for modern societies. In the United States, each household produces about 800 kg of garbage every year, followed by Norway (520 kg) and Netherlands (490 kg). On the low end of garbage production in the western world are Austria (180 kg) and Portugal (180 kg). Garbage is often collected and dumped in landfills, which are considered by many as major health hazards because they produce unpleasant odors, contaminate underground water by the leachate generated when water is mixed with garbage, and they produce ethanol and methanol that increase fire hazards. In addition, housing developments are often expanded to areas closer to landfills where landfills and garbage trucks are considered sight pollution and safety hazards.

For these reasons, it is becoming harder to build new landfills, and cities all over the United States are limiting the capacities of their landfills or eliminating them altogether. Hence, the burning of garbage to produce electricity (biomass energy) sounds like a reasonable idea.

When biomass is burned in incinerators, the biomass volume is reduced by as much as 90%, and in the process, steam can be produced to generate electricity. One of these systems is shown in Figure 6.71. The biomass material is fed to a furnace where it is burned and the heat produced is used to generate steam by heating water pipes. The steam is used to generate electricity by a turbine-generator system much like the regular thermal power plant. The steam exiting the turbine is cooled to complete the thermal cycle. The ash produced in the furnace is collected and sent to landfills. The volume of the ash is about 10% of the original volume of the biomass material. The gases generated by the combustion enter the filtering stage where more ash is extracted, collected by trucks, and sent to landfills. The remaining gases enter the stack and are released into the air.

Biomass power plants generate electricity and at the same time reduce the amount of material sent to landfills. However, biomass incineration is not free from pollution. Heavy metal and dioxins are formed during the various stages of the incinerations. Dioxin is highly carcinogenic and can cause cancer and genetic defects.

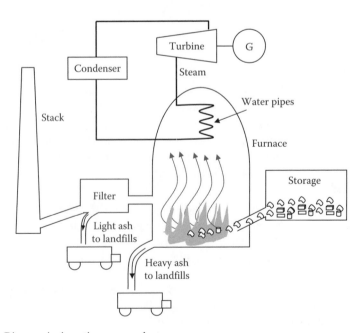

FIGURE 6.71   Biomass incineration power plant.

EXERCISES

1. A solar panel consists of 4 parallel columns of PV cells. Each column has 10 PV cells in series. Each cell produces 2 W at 0.5 V. Compute the voltage and current of the panel.
2. State three factors that determine the amount of power generated by wind machines.
3. What is the function of the reformer and CO-converter used in FC technology?
4. Compute the amount of sulfur dioxide produced when a coal-fired power plant generates 50 MW of electricity.
5. Solar power density for a given area has a standard deviation of 3 h and a maximum power of 200 W at noon. Compute the solar energy in 1 day.
6. An area located near the equator has the following parameters:

$$\alpha_{dt} = 0.82; \quad \alpha_p = 0.92; \quad \beta_{wa} = 0.06$$

The solar power density measured at 11:00 AM is 890 W/m^2. Compute the solar power density at 4:00 PM.
7. Why do we need to dope the silicon in solar cells?
8. Which material is added to make the silicon n-type, and which material makes the silicon p-type?
9. What is depletion zone?
10. Explain how light energy is converted into electric energy.
11. Why do we need to connect solar cells in series?
12. Why do we need to connect solar cells in parallel?
13. What is a solar array?
14. A solar cell with a reverse saturation current of 1 nA is operating at 35°C. The solar current at 35°C is 1.1 A. The cell is connected to a 5 Ω resistive load. Compute the output power of the cell.
15. A solar cell with a reverse saturation current of 1 nA is operating at 35°C. The solar current at 35°C is 1.1 A. Compute the maximum output power of the cell.
16. For the solar cell in the above problem, compute the load resistance at the maximum output power.
17. A PV module is composed of 100 ideal solar cells connected in series. At 25°C, the solar current of each cell is 1.2 A and the reverse saturation current is 10 nA. Find the module power when the module voltage is 45 V.
18. An 80 cm^2 solar cell is operating at 30°C where the output current is 1 A, the load voltage is 0.5 V, and the saturation current of the diode is 1 nA. The series resistance of the cell is 100 mΩ and the parallel resistance is 500 Ω. At a given time, the solar power density is 300 W/m^2. Compute the irradiance efficiency.
19. A solar cell is operating at 30°C where the output current is 1.1 A and the load voltage is 0.5 V. The series resistance of the cell is 20 mΩ and the parallel resistance is 2 kΩ. Compute the electrical efficiency of the PV cell. Assume that the irradiance efficiency is 22%. Compute the overall efficiency of the PV cell.
20. An investor wishes to install a wind farm in the Snoqualmie pass area located in Washington state. The pass is about 920 m above the sea level. The average low temperature of the air is −4°C, and the average height is 18°C. Compute the power density of the wind in winter and summer assuming that the average wind speed is 15 m/s.
21. For the site in the above problem, compute the length of the blades to capture 200 kW of wind energy during the summer.
22. A wind turbine with a gearbox ratio of 200 produces electric energy when the generator speed is at least 910 rpm. The length of each blade is 5 m. The turbine has a variable TSR. At a wind speed of 10 m/s, compute the minimum TSR of the wind turbine.
23. Generate an idea to reduce the voltage flickers associated with wind energy.
24. What are the main advantages of direct FCs?

25. What are the main components used in FC electric vehicles? Explain the operation of the vehicle.
26. What are the different types of FCs?
27. Which FC type is suitable for high power?
28. Which FC type is suitable for mobile energy?
29. What are the advantages of using an FC?
30. Is it safe to use hydrogen in FCs?
31. Why not just burn hydrogen in thermal power plants instead of using it in FCs?
32. The polarization curve of an FC can be represented by the empirical formula

$$V = 0.9 - 0.128 \tan (I - 1.2)$$

    where
       V is the voltage of the FC (V)
       I is the current (A)

    Plot the polarization and power curves. Also compute the voltage at the maximum power of the cell.
33. For the cell in the previous example, compute the drop in voltage for a 10% increase in current in the activations region. Repeat the solution for the ohmic and mass transport regions.
34. A reservoir-type small hydroelectric system has a penstock that is 2 m in diameter. The speed of water at the exit end of the penstock is 10 m/s. Assume that the power coefficient is 40%, the turbine efficiency is 80%, and the generator efficiency is 92%. Compute the output electric power of the generator.
35. Assume that the penstock efficiency in the previous problem is 95%. Compute the water head.
36. A person wants to build a reservoir-type small hydroelectric system on his/her property. The site can accommodate a penstock of 2 m in diameter. In order for him to generate 1 MW of electricity, compute the height of the dam. Assume that the penstock efficiency is 95%, the power coefficient is 50%, the turbine efficiency is 90%, and the generator efficiency is 96%.
37. Person wishes to build a diversion-type small hydroelectric system on his/her property. To measure the speed of the water, a ping-pong ball was dropped in the river and the ball was found to travel 10 m in 4 s. A turbine with a sweep diameter of 0.8 m was selected. Assume that the power coefficient is 50%, the turbine efficiency is 90%, and the efficiency of the generator is 96%. Compute the output power of the plant and the energy generated in 1 year. If the price of the energy sold to the neighboring utility is $0.06/kW h, compute the annual income from this small hydroelectric plant.
38. What are the benefits of using geothermal energy?
39. Why is geothermal energy a renewable resource?
40. What are the environmental effects of geothermal power plants?
41. What makes a geothermal site good for electric power generation?
42. A tidal mill has a blade length of 2 m; compute the blade power when the tidal current is 5 knots. Assume the power coefficient of the tidal mill is 0.5.
43. A dam-type tidal energy system is constructed between a lagoon and the open ocean. The base of the lagoon is approximately semicircle with a diameter of 0.5 km. At high tide, the average head difference between the lagoon and the ocean waters is about 10 m. The lagoon is fully filled in 3 h. Compute the average electrical power generated by the tidal system assuming the power coefficient of the blades is 25%, the efficiency of the turbine is 90%, and the efficiency of the generator is 95%.
44. What is biomass?
45. How is electricity generated from a biomass power plant?
46. Is biomass a renewable source of energy? Why?

7 Alternating Current Circuits

Nikola Tesla proposed the alternating current (ac) system as a better alternative to the direct current (dc) system. His idea was initially met with fierce resistance from Thomas Edison as explained in Chapter 1. Eventually the ac was selected for power systems worldwide essentially because of three main reasons:

1. Voltage of the ac system can be adjusted by transformers, which are relatively simple devices.
2. Voltage of long transmission lines can increase to high levels, thus reducing the current of the line, transmission line losses, and cross section of the conductors of the transmission lines.
3. ac Systems produce rotating magnetic fields that spin electric motors. Keep in mind that electric motors represent the majority of electrical loads.

NOMENCLATURE

The following nomenclatures are used in this chapter:

Instantaneous current	i	Instantaneous voltage	v
Average current	I_{ave}	Average voltage	V_{ave}
Maximum (peak) current	I_{max}	Maximum (peak) voltage	V_{max}
Current magnitude in root mean square	I	Voltage magnitude in root mean square	V
Phasor current in complex form	$\bar{I}$	Phasor voltage in complex form	$\bar{V}$
Impedance in complex form	$\bar{Z}$	Apparent power in complex form	$\bar{S}$
Magnitude of impedance	Z	Magnitude of apparent power	S
Instantaneous power	ρ	Real power	P
Reactive power	Q		

7.1 ALTERNATING CURRENT WAVEFORM

An ideal ac circuit has its voltage and current in sinusoidal forms. The voltage waveform (shown in Figure 7.1) is expressed mathematically by

$$v = V_{max} \sin \omega t \tag{7.1}$$

where
  v is the instantaneous voltage
  V_{max} is the peak (maximum) value of the voltage
  ω is the angular frequency
  t is the time

The unit of ω is radian/second (rad/s) and is expressed by

$$\omega = 2\pi f \tag{7.2}$$

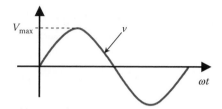

FIGURE 7.1   Sinusoidal waveform of ac voltage.

where f is the frequency of the ac waveform. In places such as North, Central, and South America; South Korea; Taiwan; the Philippines; Saudi Arabia; and western Japan, the frequency of power supply is 60 cycle/s or 60 Hz. For the rest of the world, including eastern Japan, it is 50 Hz.

During the nineteenth century, 60 Hz was chosen by Westinghouse in the United States. Meanwhile, in Germany, the giant power equipment manufacturers AEG and Siemens chose 50 Hz. These companies had a virtual monopoly in Europe, and their 50 Hz standard spread to the rest of Europe and most of the world.

7.2   ROOT MEAN SQUARE

The waveform in Figure 7.1 has its magnitude changing with time. For 60 Hz systems, the waveform is repeated 60 times every second. Quantifying the value of this sinusoidal voltage is rather tricky. If we quantify the sinusoidal waveform by its maximum value, it would be unreliable since the peak is often skewed by the transients and harmonics that exist in power networks. Using the average value is not a good idea either, as the average value is always zero because of the symmetry of the waveform around the time axis.

To address this problem, engineers developed an effective method called root mean square (rms) to quantify ac waveforms. The concept of this method is shown in Figure 7.2, where the rms value is computed in three steps:

Step 1: Compute the square of the waveform.
Step 2: Compute the average value of step 1.
Step 3: Compute the square root of step 2.

The idea behind step 1 (squaring the waveform) is to create another waveform that is always positive so that its average value in step 2 is nonzero. The square root in step 3 is imposed to

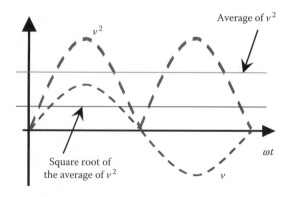

FIGURE 7.2 (See color insert following page 300.)   Concept of rms.

somehow compensate for the initial squaring in step 1. Today, all ac waveforms in power circuits are quantified by their rms values.

The rms value can be computed by following the three steps mentioned earlier. Let us consider the waveform in Equation 7.1. The first step is to square this waveform.

$$v^2 = V_{max}^2 \sin^2 \omega t = \frac{V_{max}^2}{2}(1 - \cos 2\omega t) \tag{7.3}$$

The second step is to find the average value ave of Equation 7.3.

$$\text{ave} = \frac{1}{2\pi} \int_0^{2\pi} \frac{V_{max}^2}{2}(1 - \cos 2\omega t)\, d\omega t = \frac{V_{max}^2}{2} \tag{7.4}$$

The final step is to find the rms voltage V by computing the square root of Equation 7.4.

$$V = \sqrt{\frac{V_{max}^2}{2}} = \frac{V_{max}}{\sqrt{2}} \tag{7.5}$$

The rms voltage for households in North America is 120 V. For Europe and most of the Middle East, the rms voltage is 220–240 V. There are a few exceptions such as Japan where the voltage is 100 V.

As mentioned in Chapter 2, it appears that 120 V was chosen somewhat arbitrarily. Actually, Thomas Edison came up with a high-resistance lamp filament that operated well at 120 V. Since then, the 120 V has been used in the United States. Other nations chose a higher voltage, 220–240 V, to reduce the current in the electric wires, and therefore use wires with smaller cross sections.

EXAMPLE 7.1

The outlet voltage measured by an rms voltmeter is 120 V. Compute the maximum value of the voltage waveform.

Solution

$$V_{max} = \sqrt{2}V = \sqrt{2} \times 120 = 169.7 \text{ V}$$

Notice that the voltage of all household equipment is given in rms, not the maximum value.

EXAMPLE 7.2

Compute the rms voltage of the waveform in Figure 7.3.

Solution

The general equation of the rms value of v is

$$V = \sqrt{\frac{1}{\tau} \int_0^\tau v^2 \, dt}$$

where τ is the period. Since the waveform in Figure 7.3 is discrete, the rms voltage can be expressed by

$$V = \sqrt{\frac{1}{\tau} \sum_k (v_k^2 t_k)}$$

$$V = \sqrt{\frac{1}{16} \sum 2^2 \times 2 + 4^2 \times 2 + 6^2 \times 4 + 2^2 \times 4} = 3.535 \text{ V}$$

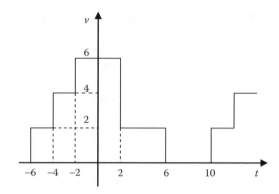

FIGURE 7.3 Voltage waveform.

7.3 PHASE SHIFT

The ideal waveforms for currents and voltages in ac circuits are sinusoidal. For a purely resistive load R, the current waveform is in phase with the voltage waveform, as given in Equation 7.6 and shown in Figure 7.4.

$$v = V_{max} \sin(\omega t) = i_R R$$
$$i_R = \frac{V_{max}}{R} \sin(\omega t) \tag{7.6}$$

If the circuit has a purely inductive load (no resistance or capacitance), the voltage–current relationship is

$$v = V_{max} \sin(\omega t) = L \frac{di_L}{dt} \tag{7.7}$$

where L is the inductance of the load. The current in this purely inductive load i_L can be obtained from Equation 7.7.

$$i_L = \frac{1}{L} \int v \, dt = \frac{V_{max}}{L} \int \sin(\omega t) \, dt = -\frac{V_{max}}{\omega L} \cos(\omega t) = -\frac{V_{max}}{X_L} \cos(\omega t) \tag{7.8}$$

where $X_L = \omega L$ is the magnitude of the inductive reactance of the load. As seen in Equation 7.8, the current i_L of a purely inductive load is a negative cosine waveform, which is shown in Figure 7.5.

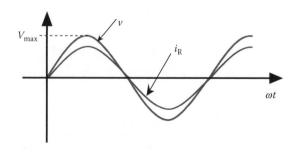

FIGURE 7.4 Voltage and current of purely resistive load.

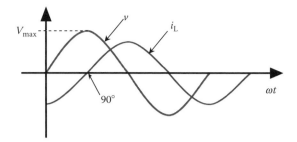

FIGURE 7.5   Voltage and current of purely inductive load.

The positive peak of the current occurs 90° after the voltage reaches its own positive peak. In this case, the current is said to be lagging the voltage by 90°.

For a purely capacitive load, the voltage–current relationship is given by

$$v = V_{max} \sin(\omega t) = \frac{1}{C} \int i_C \, dt$$

$$i_C = C \frac{dv}{dt}$$

$$(7.9)$$

where C is the load capacitance. The current, i_C, can be computed as

$$i_C = C \frac{d}{dt}[V_{max} \sin(\omega t)] = \omega C V_{max} \cos(\omega t) = \frac{V_{max}}{X_C} \cos(\omega t) \qquad (7.10)$$

where $X_C = 1/\omega C$ is the magnitude of the capacitive reactance of the load. As seen in Equation 7.10, the current of a purely capacitive load is a cosine waveform. Figure 7.6 shows the voltage across the capacitor and the current i_C. Note that the positive peak of the current waveform occurred 90° before the voltage peak. Hence, the current is said to be leading the voltage by 90°.

If the load is composed of a mix of elements such as resistances, capacitances, and inductances, the phase shift angle θ of the current can be any value between −90° and +90°. Figure 7.7 shows the current and voltage waveforms of an inductive load that is composed of a resistance and an inductance.

The current waveform in Figure 7.7 can be expressed mathematically by

$$i = I_{max} \sin(\omega t - \theta) \qquad (7.11)$$

7.4   CONCEPT OF PHASORS

Phasors are handy graphical representations for ac circuit analysis. They give quick information on the magnitude and phase shift of any waveform. Figure 7.8a shows a sinusoidal waveform without any phase shift. The waveform can be represented by the phasor in Figure 7.8b. The length of the phasor is proportional to the rms value of the waveform, and its phase angle with respect

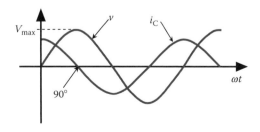

FIGURE 7.6   Voltage and current of purely capacitive load.

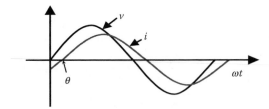

FIGURE 7.7 Lagging current.

to the *x*-axis represents the phase shift of the sinusoidal waveform. Since the waveform has no phase shift, the phasor of the voltage is aligned with the *x*-axis.

Consider the circuit in Figure 7.9 for a purely resistive load. Since the resistance does not cause any phase shift in the current as shown in Figure 7.4, the current phasor is in phase with the voltage as shown in Figure 7.9b.

For a purely inductive load, the current lags the voltage by 90° as given in Equation 7.8 and shown in Figure 7.5. The phasor diagram in this case is depicted in Figure 7.10b. The lagging angles are always in the clockwise direction.

For a purely capacitive load, the current leads the voltage by 90° as given in Equation 7.10 and shown in Figure 7.6. The phasor diagram in this case is given in Figure 7.11b. The leading angles are always in the counterclockwise direction.

Now let us consider the waveforms in Figure 7.7. As seen in the figure, the current is lagging the voltage by an angle θ. These waveforms can be represented by the phasor diagram in Figure 7.12. The voltage is considered as the reference with no phase shift, so it is aligned with the *x*-axis. Since the current lags the voltage by θ, the phasor for the current lags the reference by θ in the clockwise direction.

7.5 COMPLEX NUMBER ANALYSIS

It is inconvenient to numerically analyze electric circuits by the graphical method of phasor diagrams. An alternative method is to use the complex number analysis where any phasor is represented by a magnitude and an angle. For example, the voltage and current in Figure 7.7 or Figure 7.12 can be represented by the following complex variable equations

$$\overline{V} = V \angle 0° \tag{7.12}$$

$$\overline{I} = I \angle -\theta° \tag{7.13}$$

where
  $\overline{V}$ is the phasor representation of the voltage. It has a bar on the top indicating that the variable is a complex number
  V is the magnitude of the voltage in rms. The symbol $\angle$ is followed by the angle of the phasor, which is zero for voltage and $-\theta$ for the lagging current

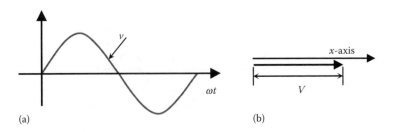

(a)   (b)

FIGURE 7.8 Phasor representation of a sinusoidal waveform. (a) Waveform and (b) phasor representation.

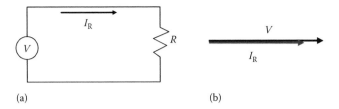

FIGURE 7.9 Phasor representation of the current and voltage of a purely resistive load. (a) Circuit representation and (b) phasor diagram.

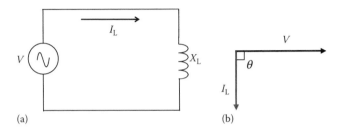

FIGURE 7.10 Phasor representation of the current and voltage of a purely inductive load. (a) Circuit representation and (b) phasor diagram.

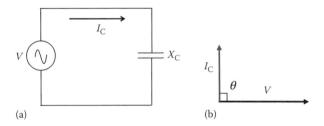

FIGURE 7.11 Phasor representation of the current and voltage of a purely capacitive load. (a) Circuit representation and (b) phasor diagram.

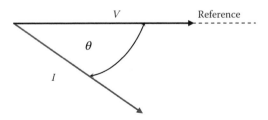

FIGURE 7.12 Phasor representation of the current and voltage in Figure 7.7.

The complex forms in Equations 7.12 and 7.13 are called polar forms (also known as trigonometric forms). The mathematics of the polar forms is very simple for multiplication and division. Let us assume that we have two complex numbers

$$\overline{A} = A\angle\theta_1 \tag{7.14}$$

$$\overline{B} = B\angle\theta_2 \tag{7.15}$$

The multiplication and division of these two complex numbers are

$$\overline{A}\,\overline{B} = A\angle\theta_1 \; B\angle\theta_2 = AB\angle(\theta_1 + \theta_2) \tag{7.16}$$

$$\frac{\overline{A}}{\overline{B}} = \frac{A\angle\theta_1}{B\angle\theta_2} = \frac{A}{B}\angle(\theta_1 - \theta_2) \tag{7.17}$$

The addition and subtraction of complex numbers in polar forms are harder to implement. However, if we switch the polar form to the rectangular form (also known as Cartesian form), we can perform these operations very quickly. To convert from polar to rectangular forms, consider the phasor $\overline{A}$ in Figure 7.13. The reference in the figure represents the x-axis. The component of $\overline{A}$ projected on the reference is $X = A\cos\theta$, and the vertical projection is $Y = A\sin\theta$. These two components can be represented by Equation 7.18.

$$\overline{A} = A\angle\theta = A[\cos\theta + j\sin\theta] = A\cos\theta + jA\sin\theta = X + jY \tag{7.18}$$

The first component along the x-axis is known as the real component. The second component, preceded by an operator j, is known as the imaginary component. Keep in mind that there is nothing imaginary about the quantity $A\sin\theta$; we simply use the label to distinguish between the directions of the real and imaginary vectors. The operator j is used to represent the counterclockwise rotation of a vector by 90°; any quantity preceded by j is shifted in the counterclockwise direction by 90°. The basic relations of the operator j are summarized in Equation 7.19.

$$
\begin{aligned}
j &= 1\angle 90° \\
j^2 &= 1\angle 180° = -1 \\
j^3 &= 1\angle 270° = 1\angle{-90°} = -j \\
j^4 &= 1\angle 0° = 1 \\
j^5 &= 1\angle 90° = j \\
&\text{etc.}
\end{aligned}
\tag{7.19}
$$

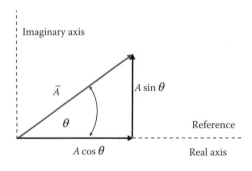

FIGURE 7.13 X and Y components of phasor $\overline{A}$.

The mathematical expressions in Equation 7.19 can also be achieved by defining the operator $j = \sqrt{-1}$. By using Figure 7.13, the conversion of $\overline{A}$ from the rectangular form to the polar form is

$$\overline{A} = X + jY = A\angle\theta$$
$$A = \sqrt{X^2 + Y^2} \qquad (7.20)$$
$$\theta = \tan^{-1}\frac{Y}{X}$$

The addition or subtraction of complex numbers in rectangular forms is done by adding or subtracting the real and imaginary components independently.

$$\overline{A} + \overline{B} = A(\cos\theta_1 + j\sin\theta_1) + B(\cos\theta_2 + j\sin\theta_2)$$
$$= (A\cos\theta_1 + B\cos\theta_2) + j(A\sin\theta_1 + B\sin\theta_2) \qquad (7.21)$$

$$\overline{A} - \overline{B} = A(\cos\theta_1 + j\sin\theta_1) - B(\cos\theta_2 + j\sin\theta_2)$$
$$= (A\cos\theta_1 - B\cos\theta_2) + j(A\sin\theta_1 - B\sin\theta_2) \qquad (7.22)$$

One of the convenient operations used in complex number analyses is the conjugate of a vector. Two complex numbers are conjugates if their

- Real components are equal
- Imaginary components are equal
- Imaginary components have different signs

If $\overline{A}$ is a complex vector, and $\overline{A}*$ is its conjugate, then

$$\overline{A} = X + jY = \sqrt{X^2 + Y^2}\ \angle\theta$$
$$\overline{A}* = X - jY = \sqrt{X^2 + Y^2}\ \angle-\theta \qquad (7.23)$$

The conjugate is useful in complex operations such as inverting a phasor. Let us assume that the phasor $\overline{A}$ is in rectangular form. To compute $1/\overline{A}$, follow the process below.

$$\frac{1}{\overline{A}} = \frac{1}{X + jY}\frac{\overline{A}*}{\overline{A}*}$$
$$\frac{1}{\overline{A}} = \frac{1}{X + jY}\frac{X - jY}{X - jY} = \frac{X - jY}{X^2 + Y^2} \qquad (7.24)$$
$$\frac{1}{\overline{A}} = \frac{X}{X^2 + Y^2} - j\frac{Y}{X^2 + Y^2}$$

EXAMPLE 7.3

Assume $\overline{A} = 10\angle60°$ and $\overline{B} = 5\angle40°$; compute the following:

(a) $\overline{C} = \overline{A}\,\overline{B}$

(b) $\overline{C} = \dfrac{\overline{A}}{\overline{B}}$

(c) $\overline{C} = \overline{A} + \overline{B}$

(d) $\overline{C} = \overline{A} - \overline{B}$

Solution

(a) $\overline{C} = \overline{A}\,\overline{B} = 10\angle 60°\ 5\angle 40° = (10 \times 5)\angle(60° + 40°) = 50\angle 100°$

(b) $\overline{C} = \dfrac{\overline{A}}{\overline{B}} = \dfrac{10\angle 60°}{5\angle 40°} = \dfrac{10}{5}\angle(60° - 40°) = 2\angle 20°$

(c) $\overline{C} = \overline{A} + \overline{B} = 10\angle 60° + 5\angle 40° = 10(\cos 60° + j\sin 60°) + 5(\cos 40° + j\sin 40°)$

$\overline{C} = (5 + j8.66) + (3.83 + j3.21) = 8.83 + j11.87$

$\overline{C} = \sqrt{(8.83)^2 + (11.87)^2}\,\angle\tan^{-1}\left(\dfrac{11.87}{8.83}\right) = 14.79\angle 53.35°$

(d) $\overline{C} = \overline{A} - \overline{B} = 10\angle 60° - 5\angle 40° = 10(\cos 60° + j\sin 60°) - 5(\cos 40° + j\sin 40°)$

$\overline{C} = (5 + j8.66) - (3.83 + j3.21) = 1.17 + j5.45$

$\overline{C} = \sqrt{(1.17)^2 + (5.45)^2}\,\angle\tan^{-1}\left(\dfrac{5.45}{1.17}\right) = 5.57\angle 77.88°$

7.6   COMPLEX IMPEDANCE

The impedance in ac circuits is composed of any combination of resistances, inductances, and capacitances. These elements are also complex quantities and can be computed as given in Equations 7.25 through 7.27.

$$\overline{R} = \frac{\overline{V}}{\overline{I}_R} = \frac{V\angle 0}{I_R\angle 0} = R\angle 0 \tag{7.25}$$

$$\overline{X}_L = \frac{\overline{V}}{\overline{I}_L} = \frac{V\angle 0}{I_L\angle -90°} = \frac{V}{I_L}\angle 90° = X_L\angle 90° \tag{7.26}$$

$$\overline{X}_C = \frac{\overline{V}}{\overline{I}_C} = \frac{V\angle 0}{I_C\angle 90°} = \frac{V}{I_C}\angle -90° = X_C\angle -90° \tag{7.27}$$

where the magnitudes $X_L = \omega L$ and $X_C = \dfrac{1}{\omega C}$ as given in Equations 7.8 and 7.10.

7.6.1   SERIES IMPEDANCE

When the load is composed of different elements, the total ohmic value of the load is called impedance. In Figure 7.14, a resistance is connected in series with an inductive reactance. Each of these components is a complex parameter with a magnitude and phase angle, as given in Equations 7.25 and 7.26. The impedance $\overline{Z}$ of this load is

$$\overline{Z} = \overline{R} + \overline{X}_L = R\angle 0 + X_L\angle 90° = R + jX_L \tag{7.28}$$

The phasor diagram of the impedance is shown in Figure 7.14b. Note that the impedance angle θ is positive and is equal in magnitude to the phase angle of the current in the impedance.

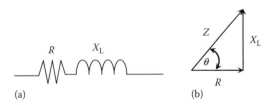

FIGURE 7.14   Series impedance of an inductive load. (a) Components of impedance and (b) phasor representation of the impedance.

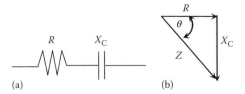

FIGURE 7.15   Series impedance of a capacitive load. (a) Components of impedance and (b) phasor representation of the impedance.

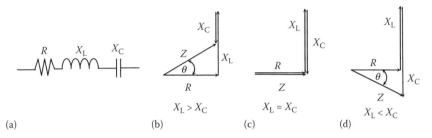

FIGURE 7.16   Series impedance of a complex load. (a) Components of impedance and (b) phasor representation of the impedance.

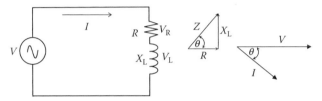

FIGURE 7.17   Circuit connection and phasor diagrams.

Similarly, the capacitive load in Figure 7.15 is composed of a resistor and a capacitor connected in series. The impedance of this load can be expressed by Equation 7.29.

$$\bar{Z} = \bar{R} + \bar{X}_C = R\angle0 + X_C\angle-90° = R - jX_C \tag{7.29}$$

The phasor diagram of the capacitive impedance is shown in Figure 7.15b. Note that the capacitive reactance X_C lags the resistance by 90°.

Now let us assume a general case where a resistor is in series with a capacitor and an inductor. The circuit is shown in Figure 7.16. The impedance of the load can be expressed by

$$\bar{Z} = \bar{R} + \bar{X}_L + \bar{X}_C = R\angle0 + X_L\angle90° + X_C\angle-90° = R + j(X_L - X_C) \tag{7.30}$$

Depending on the magnitude of the inductive reactance X_L with respect to the capacitive reactance X_C, the phasor diagram of the circuit in Figure 7.16a can be any one of the three phasor diagrams in Figure 7.16b–d. In Figure 7.16b, $X_L > X_C$ and the impedance angle is positive. Hence, the circuit acts as if it contains the resistance and an equivalent inductive reactance of a magnitude equal to $(X_L - X_C)$. In Figure 7.16c, when $X_L = X_C$, the impedance angle of the circuit is zero, and the circuit impedance is equivalent to the resistance only. In Figure 7.16d, when $X_L < X_C$, the impedance angle is negative, and the circuit acts as if it contains the resistance and an equivalent capacitive reactance of a magnitude equal to $(X_C - X_L)$.

EXAMPLE 7.4

A load is composed of a 4 Ω resistance connected in series with a 3 Ω inductive reactance. The load is connected across a 120 V source as shown in Figure 7.17. Compute the following:

(a) Current of the circuit
(b) Voltage across the resistance
(c) Voltage across the inductive reactance

Solution

(a) Current of the circuit is

$$\bar{I} = \frac{\bar{V}}{\bar{Z}}$$

where

$$\bar{Z} = R\angle 0° + X_L\angle 90° = R + jX_L = 4 + j3 = 5\angle 36.87°$$

hence

$$\bar{I} = \frac{120\angle 0}{5\angle 36.87°} = 24\angle -36.87° \text{A}$$

(b) Voltage across the resistance V_R is

$$\bar{V}_R = \bar{I}\bar{R} = (24\angle -36.87°)(4\angle 0) = 96\angle -36.87° \text{ V}$$

(c) Voltage across the inductive reactance V_L is

$$\bar{V}_L = \bar{I}\bar{X}_L = (24\angle -36.87°)(3\angle 90°) = 72\angle 53.13° \text{ V}$$

Keep in mind that the sum of the voltage across the resistance and across the inductive reactance yields the value of the source voltage. This, however, must be done using complex mathematics. Try it!

Also note that the impedance angle and the angle of the current are the same in magnitude, but opposite in sign, as shown in Figure 7.17.

EXAMPLE 7.5

A 120 V adjustable frequency ac source is connected in series with a resistor, an inductor, and a capacitor. At 60 Hz, the resistance is 5 Ω, inductive reactance is 3 Ω, and the capacitive reactance is 4 Ω. Compute the following:

(a) Load impedance at 60 Hz
(b) Frequency at which the total impedance is equal to the load resistance only

Solution

(a) As given in Equation 7.30, the load impedance is

$$\bar{Z} = R + j(X_L - X_C) = 5 + j(3 - 4) = 5 - j = 5.1\angle -11.31 \text{ Ω}$$

(b) First, let us compute the inductance and capacitance of the load

$$X_L = 2\pi f L$$

$$L = \frac{3}{2\pi 60} = 7.96 \text{ mH}$$

Similarly

$$X_C = \frac{1}{2\pi f C}$$

$$C = \frac{1}{2\pi 60 \times 4} = 663.15 \text{ μF}$$

If $\overline{Z} = R$, then $X_L = X_C$. This can occur when the frequency of the supply voltage is changed until $X_L = X_C$. The frequency in this case is called *resonance frequency* or f_o.

$$X_L = X_C$$

$$2\pi f_o L = \frac{1}{2\pi f_o C}$$

$$f_o = \frac{1}{2\pi\sqrt{LC}} = \frac{10^3}{2\pi\sqrt{7.96 \times 0.66315}} = 69.27 \text{ Hz}$$

7.6.2 PARALLEL IMPEDANCE

In parallel connections, such as the one shown in Figure 7.18, the current of the source is divided among the parallel elements of the circuit according to the values of their impedances. The phasor sum of all branch currents in a parallel circuit is equal to the current of the source. This can be represented mathematically as given in Equation 7.31.

$$\overline{I} = \overline{I}_R + \overline{I}_L + \overline{I}_C = \frac{\overline{V}}{R} + \frac{\overline{V}}{\overline{X}_L} + \frac{\overline{V}}{\overline{X}_C} = \overline{V}\left[\frac{1}{R} + \frac{1}{\overline{X}_L} + \frac{1}{\overline{X}_C}\right] = \frac{\overline{V}}{\overline{Z}} \tag{7.31}$$

Hence, the total impedance $\overline{Z}$ is defined as

$$\frac{1}{\overline{Z}} = \frac{1}{\overline{R}} + \frac{1}{\overline{X}_L} + \frac{1}{\overline{X}_C}$$

$$\tag{7.32}$$

or

$$\overline{Y} = \overline{G} + \overline{B}_L + \overline{B}_C$$

where

$$\overline{Y} = \frac{1}{\overline{Z}}$$

$$\overline{G} = \frac{1}{\overline{R}} = \frac{1}{R\angle 0°} = \frac{1}{R}\angle 0°$$

$$\overline{B}_L = \frac{1}{\overline{X}_L} = \frac{1}{X_L\angle 90°} = \frac{1}{X_L}\angle -90° \tag{7.33}$$

$$\overline{B}_C = \frac{1}{\overline{X}_C} = \frac{1}{X_C\angle -90°} = \frac{1}{X_C}\angle 90°$$

where
$\overline{G}$ is called conductance, and its angle is zero
$\overline{B}_L$ is called inductive susceptance
$\overline{B}_C$ is called capacitive susceptance

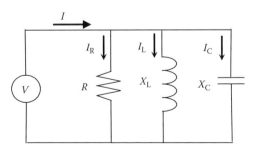

FIGURE 7.18 Parallel connection.

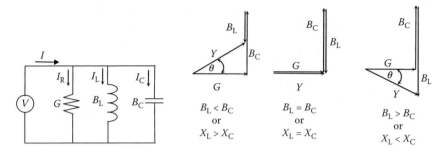

FIGURE 7.19 Parallel circuit.

Note that the angle of B_L is $-90°$ as opposed to the angle of inductive reactance X_L, which is $+90°$. Similarly, the angle of B_C is $+90°$ as opposed to the angle of capacitive reactance X_C, which is $-90°$. The sum of all conductances and susceptances is called admittance $\overline{Y}$.

$$\overline{Y} = \overline{G} + \overline{B}_L + \overline{B}_C = G + j(B_C - B_L) \tag{7.34}$$

The unit of G, B_L, B_C, and Y is mho, which is the reverse spelling of the word "ohm." The impedance diagram of the parallel circuit is shown in Figure 7.19. Compare these phasor diagrams with the ones shown in Figure 7.16 for series impedance.

EXAMPLE 7.6

A 120 V adjustable frequency ac source is connected in parallel with a resistor, an inductor, and a capacitor as shown in Figure 7.18. At 60 Hz, the resistance is 5 Ω, inductive reactance is 10 Ω, and the capacitive reactance is 2 Ω. Compute the following:

(a) Load impedance at 60 Hz
(b) Frequency at which the total impedance is equivalent to the load resistance only

Solution

(a) As given in Equation 7.34, the load admittance Y is

$$\overline{Y} = G + j(B_C + B_L) = \frac{1}{5} + j\left(\frac{1}{2} - \frac{1}{10}\right) = 0.2 + j0.4 \text{ mho}$$

The load impedance Z is

$$\overline{Z} = \frac{1}{\overline{Y}} = \frac{1}{0.2 + j0.4} = 1.0 - j2.0 \ \Omega$$

(b) First, let us compute the inductance and capacitance of the load.

$$X_L = 2\pi f L$$

$$L = \frac{10}{2\pi 60} = 26.5 \text{ mH}$$

Similarly,

$$X_C = \frac{1}{2\pi f C}$$

$$C = \frac{1}{2\pi 60 \times 2} = 1.326 \text{ mF}$$

If $\overline{Z} = R$, then $B_L = B_C$. This can occur when the frequency of the supply voltage is changed.

$$f_o = \frac{1}{2\pi\sqrt{LC}} = \frac{10^3}{2\pi\sqrt{26.5 \times 1.326}} = 26.85 \text{ Hz}$$

7.7 ELECTRIC POWER

Instantaneous power ρ is defined as the multiplication of the instantaneous voltage v by the instantaneous current i.

$$\rho = v\,i \tag{7.35}$$

Let us assume that the waveforms of the voltage and current are sinusoidal as described by Equations 7.1 and 7.11. Hence, the instantaneous power is

$$
\begin{aligned}
\rho &= v\,i = [V_{max} \sin(\omega t)]\,[I_{max} \sin(\omega t - \theta)] \\
\rho &= \frac{V_{max}I_{max}}{2}[\cos(\theta) - \cos(2\omega t - \theta)] \\
\rho &= \frac{V_{max}}{\sqrt{2}}\frac{I_{max}}{\sqrt{2}}[\cos(\theta) - \cos(2\omega t - \theta)] \\
\rho &= V\,I[\cos(\theta) - \cos(2\omega t - \theta)]
\end{aligned}
\tag{7.36}
$$

where V and I are the rms values of the sinusoidal waveforms v and I, respectively. θ is the impedance angle. Now, let us rewrite Equation 7.36 as follows:

$$
\begin{aligned}
\rho &= V\,I\cos(\theta) - V\,I\cos(2\omega t - \theta) \\
\rho &= P + h
\end{aligned}
\tag{7.37}
$$

The first term P in Equation 7.37 is time invariant since V, I, and θ are all time independent. The second term h is a time-varying sinusoid with a frequency equal to twice the frequency of the supply voltage and is shifted by θ. P and h are plotted in Figure 7.20.

For a purely resistive load, θ is zero.

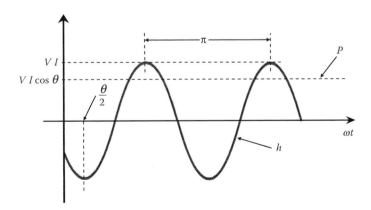

FIGURE 7.20 Two terms of the instantaneous power in Equation 7.37.

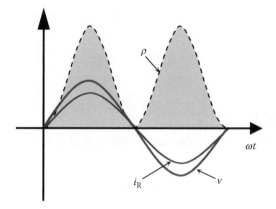

FIGURE 7.21 Instantaneous power of a purely resistive load.

Hence

$$\rho = vi = VI\cos(\theta) - VI\cos(2\omega t - \theta)$$
$$\rho = VI[1 - \cos(2\omega t)]$$

$$(7.38)$$

The power waveform for this case is shown in Figure 7.21. Note that the instantaneous power is always positive.

For a purely inductive load, the impedance angle $\theta = 90°$, and its instantaneous power is given by Equation 7.39 and is shown in Figure 7.22.

$$\rho = vi = VI[\cos(\theta) - \cos(2\omega t - \theta)]$$
$$\rho = -VI\sin(2\omega t)$$

$$(7.39)$$

Note that the average value of the instantaneous power is zero, which indicates that the inductor consumes power during the first one fourth of the voltage cycle and then returns the power back during the second one fourth cycle. The inductor, in this case, does not store any energy.

For a purely capacitive load, the impedance angle $\theta = -90°$, and its instantaneous power is given by Equation 7.40 and shown in Figure 7.23.

$$\rho = vi = VI[\cos(\theta) - \cos(2\omega t - \theta)]$$
$$\rho = VI\sin(2\omega t)$$

$$(7.40)$$

Note that the average value of the instantaneous power for a capacitor is zero, which is similar to that for the inductance. In this case, the capacitor also exchanges energy with the source. In one cycle, the total energy consumed by the capacitor is zero.

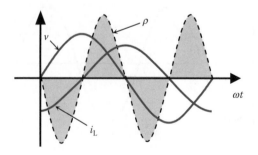

FIGURE 7.22 Instantaneous power of a purely inductive load.

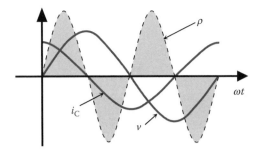

FIGURE 7.23 Instantaneous power of a purely capacitive load.

For the general case of loads with any phase shift, Equation 7.37 can be used. For the case of lagging current, the waveforms are shown in Figure 7.24. Note that the average of the instantaneous power is nonzero.

7.7.1 REAL POWER

The power that produces energy is the average value of ρ given in Equation 7.37. Since the first term P in the equation is time invariant, its average value is the term itself. For the second term, h, its average value is zero since it is sinusoidal (i.e., symmetrical across the time axis). P is called *active* power or *real* power, and its units are watt, kilowatt, megawatt, etc.

$$P = \frac{1}{2\pi} \int_0^{2\pi} \rho \, d\omega t = \frac{1}{2\pi} \int_0^{2\pi} (P + h) \, d\omega t = \frac{VI}{2\pi} \int_0^{2\pi} [\cos\theta - \cos(2\omega t - \theta)] \, d\omega t \tag{7.41}$$

$$P = VI\cos(\theta)$$

7.7.2 REACTIVE POWER

Reactive power is the power exchanged between the source and the inductive or capacitive elements of the circuit. Recall that inductive and capacitive elements have their instantaneous powers in sinusoidal forms, as shown in Figures 7.22 and 7.23. This means the power is drawn from the source and then delivered back to the source.

To identify the reactive power mathematically, let us rewrite Equation 7.37 by expanding the term h.

$$\rho = VI\cos(\theta) - VI[\cos(\theta)\cos(2\omega t) + \sin(\theta)\sin(2\omega t)]$$
$$\rho = VI\cos(\theta)[1 - \cos(2\omega t)] - VI\sin(\theta)\sin(2\omega t) \tag{7.42}$$

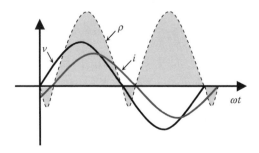

FIGURE 7.24 Instantaneous power of a complex load.

Define

$$P = V I \cos(\theta)$$
$$Q \equiv V I \sin(\theta)$$

(7.43)

Hence

$$\rho = P[1 - \cos(2\omega t)] - Q \sin(2\omega t)$$

(7.44)

The term Q is called *imaginary* power or *reactive* power. It is time invariant and its unit is voltampere reactive (VAr), kilovoltampere reactive (kVAr), etc.

Another method to compute reactive power is using Ohm's law. Assume that an inductive load is composed of a resistance in series with an inductive reactance. Since the magnitude of the voltage across the load is

$$V = I Z$$

(7.45)

and

$$\sin(\theta) = \frac{X_L}{Z}$$

(7.46)

then

$$Q_L = V I \sin(\theta) = (I Z) I \frac{X_L}{Z}$$
$$Q_L = I^2 X_L$$

(7.47)

Similarly, for a capacitive load, it can be shown that the reactive power is

$$Q_C = I^2 X_C$$

(7.48)

7.7.3 COMPLEX POWER

Complex power is the phasor sum of the real and reactive powers. Let us first define complex power S as

$$\overline{S} \equiv \overline{V} \overline{I}*$$

(7.49)

where $\overline{I}*$ is the conjugate of the current $\overline{I}$. If $\overline{I} = I \angle -\theta$, then $\overline{I}* = I \angle \theta$. Hence, for lagging current the complex power is

$$\overline{S} = \overline{V} \overline{I}* = V \angle 0 \; I \angle \theta = V I \angle \theta = V I \cos(\theta) + j V I \sin(\theta)$$

(7.50)

Substituting the values of P and Q in Equation 7.43 into Equation 7.50 yields

$$\overline{S} = P + jQ$$

(7.51)

The complex power S is also known as *apparent* power. Its units are voltampere, kilovoltampere, etc. The phasor quantity $\overline{S}$ in Equation 7.51 can be represented by the phasor diagram in Figure 7.25.

7.7.4 SUMMARY OF AC PHASORS

For the inductive load in Figure 7.17, the phasor diagrams of volt–current, impedance, and power are shown in Figure 7.26. In the volt–current diagram, the voltage is taken as the reference and the current lags the voltage by θ. In the impedance diagram, the resistance is always the reference and the inductive reactance leads the resistance by $90°$. For the power diagram, the real power is always the reference, and the inductive reactive power leads the real power by $90°$.

For a capacitive load, the phasor diagrams are shown in Figure 7.27. Note that the reactive power of the capacitive load is lagging the real power by $90°$, while the reactive power of the inductive load is leading.

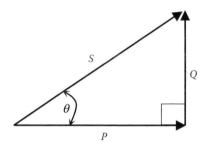

FIGURE 7.25   Phasor diagram of the complex power.

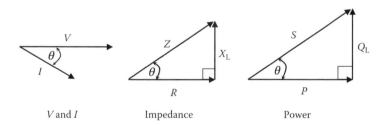

FIGURE 7.26   Phasor diagrams of an inductive load.

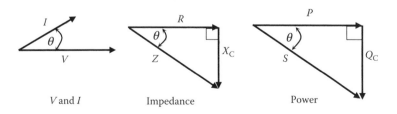

FIGURE 7.27   Phasor diagrams of a capacitive load.

EXAMPLE 7.7

The voltage and current waveforms of an electric load are

$$v = 150 \sin(377t + 0.2)$$

$$i = 25 \sin(377t - 0.5)$$

Compute the following:
  (a) Frequency of the supply voltage
  (b) Phasor voltage
  (c) Phasor current
  (d) Real power of the load
  (e) Reactive power of the load

Solution

(a) $\omega = 2\pi f = 377$ rad/s

$$f = \frac{\omega}{2\pi} = \frac{377}{2\pi} = 60 \text{ Hz}$$

(b) The phasor voltage is represented by the magnitude in rms and the phase angle in degrees. Note that the angles in the waveform equations are in radians.

$$\overline{V} = \frac{V_{max}}{\sqrt{2}} \angle \theta_V = \frac{150}{\sqrt{2}} \angle \left(0.2 \frac{180}{\pi}\right) = 106.06 \angle 11.46° \text{ V}$$

(c) Similarly, the phasor current is

$$I = \frac{I_{max}}{\sqrt{2}} \angle \theta_I = \frac{25}{\sqrt{2}} \angle \left(-0.5 \frac{180}{\pi}\right) = 17.67 \angle -28.65° \text{ A}$$

(d) The angle between the voltage and current is

$$\theta = \theta_V - \theta_I = 11.46 + 28.65 = 40.11°$$

The real power of the load is

$$P = VI \cos(\theta) = 106.06 \times 17.67 \times \cos(40.11) = 1.433 \text{ kW}$$

(e) The reactive power of the load is

$$Q = VI \sin(\theta) = 106.06 \times 17.67 \times \sin(40.11) = 1.207 \text{ kVAr}$$

7.7.5 POWER FACTOR

The apparent power S and the real power P are linearly related by the term $\cos(\theta)$.

$$P = VI \cos(\theta) = S \cos(\theta)$$
$$\frac{P}{S} = \cos(\theta) \tag{7.52}$$

where
  $\cos(\theta)$ is known as the power factor pf
  θ is known as the power factor angle

The power factor can be computed by various methods. Consider Figures 7.26 and 7.27. The pf in these figures is

$$\text{pf} = \cos(\theta) = \frac{P}{S} = \frac{R}{Z} \tag{7.53}$$

The power factor can be either lagging or leading depending on the angle of the current with respect to the voltage. When the current lags the voltage, the power factor is lagging. When the current leads the voltage, the power factor is leading.

7.7.6 PROBLEMS RELATED TO REACTIVE POWER

Reactive power does not produce work, and its presence in the system can create several problems. Consider the simple system in Figure 7.28 where the source is assumed to be at a distance from a purely reactive load whose inductive reactance is X_L. The load is connected to the source by a cable (wire or transmission line). The wire has a resistance R_{wire} and an inductive reactance X_{wire}. The voltage of the source is V_S and the voltage across the load is V_{load}.

The currents in the wire are

$$\bar{I} = \frac{\bar{V}_S}{R_{wire} + j(X_{wire} + X_L)} \tag{7.54}$$

and the reactive power of the load is

$$Q_L = I^2 X_L \tag{7.55}$$

Keep in mind that only real power produces energy and almost all utilities charge residential customers for real power only. It is highly unlikely that you pay for any reactive power that you consume in your house. Hence, the inductive load in Figure 7.28 does not generate any revenue to the utility; besides it creates several problems such as the following:

- It increases the losses of the transmission line.
- It reduces the spare capacity of the line.
- It reduces the voltage across the load.

For the first problem, because of the inductive load, the line current that feeds the inductive load results in real power loss in the wire P_{loss}, where

$$P_{loss} = I^2 R_{wire} \tag{7.56}$$

For the second problem, each transmission line has a capacity defined by the maximum current of the line, which is determined by the wire's cross section and its material. The spare capacity is defined as the line capacity minus the actual current in the line. Because the reactive current is

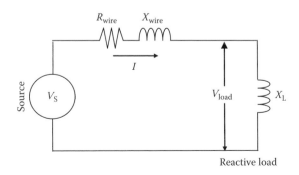

FIGURE 7.28 Currents due to an inductive load.

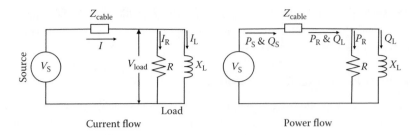

FIGURE 7.29 Current and power flow.

transmitted through the line, the spare capacity is reduced. This reduction in the spare capacity limits the ability of the utility to use the line to deliver real power to additional customers.

The third problem can be explained by considering the system in Figure 7.28, where the voltage on the load side $\overline{V}_{\text{load}}$ can be expressed by

$$\overline{V}_{\text{load}} = \overline{I}\, jX_{\text{L}} = \overline{V}_{\text{S}} \frac{jX_{\text{L}}}{R_{\text{wire}} + j(X_{\text{wire}} + X_{\text{L}})} \tag{7.57}$$

Hence, the magnitude of the load voltage V_{load} is

$$V_{\text{load}} = \frac{V_{\text{S}}}{\sqrt{\left(\frac{R_{\text{wire}}}{X_{\text{L}}}\right)^2 + \left(1 + \frac{X_{\text{wire}}}{X_{\text{L}}}\right)^2}} \tag{7.58}$$

Note that the load voltage decreases when the inductive reactance X_{L} decreases (i.e., a more reactive load is added). Only when $X_{\text{L}} = \infty$ (i.e., no reactive load), the load voltage is equal to the source voltage.

EXAMPLE 7.8

The circuit in Figure 7.29 has a source voltage of 110 V at 60 Hz. The load, which is heavily inductive, consists of 20 Ω resistance in parallel with 10 Ω inductive reactance. The cable resistance is 1 Ω, and its inductive reactance is 5 Ω. Compute the following:

(a) Load impedance
(b) Line current
(c) Load voltage
(d) Real and imaginary components of the load current
(e) Real and reactive powers of the load
(f) Real and reactive losses of the cable
(g) Real and reactive power delivered by the source

Solution

(a) Load admittance Y is

$$\overline{Y} = G - jB_{\text{L}} = \frac{1}{20} - j\frac{1}{10} = 0.05 - j0.1 \text{ mho}$$

Load impedance Z is

$$\overline{Z} = \frac{1}{\overline{Y}} = \frac{1}{0.05 - j0.1} = 4 + j8 \text{ ohm}$$

(b) To compute the line current, we need to compute the total impedance of the circuit.

$$\overline{Z}_{total} = \overline{Z} + \overline{Z}_{cable} = (4 + j8) + (1 + j5) = 5 + j13 \text{ ohm}$$

$$\overline{I} = \frac{\overline{V}_S}{\overline{Z}_{total}} = \frac{110\angle 0°}{5 + j13} = 2.835 - j7.37 = 7.9\angle -68.96° \text{ A}$$

(c) Load voltage can be computed by

$$\overline{V}_{load} = \overline{I}\,\overline{Z} = (7.9\angle -68.96°)(4 + j8) = 70.63\angle -5.52° \text{ V}$$

Note that the load voltage in this case is about 64% of the source voltage. This low voltage is mainly due to the voltage drop across the cable.

(d) The real and imaginary components of the load current are

$$\overline{I}_R = \frac{\overline{V}_{load}}{R} = \frac{70.63\angle -5.52°}{20} = 3.531\angle -5.52° \text{ A}$$

$$\overline{I}_L = \frac{\overline{V}_{load}}{jX_L} = \frac{70.63\angle -5.52°}{10\angle 90°} = 7.06\angle -84.48° \text{ A}$$

Note that the reactive current is higher than the real current. This is because X_L is smaller in magnitude than R.

(e) The real and reactive powers of the load are

$$P_R = I_R^2\, R = (3.531)^2 \times 20 = 249.36 \text{ W}$$
$$Q_L = I_L^2\, X_L = (7.06)^2 \times 10 = 498.437 \text{ VAr}$$

Note that the reactive power is higher than the real power. This is because I_L is larger in magnitude than I_R.

(f) The real power loss of the cable is

$$P_{loss} = I^2\, R_{cable} = 7.9^2 \times 1 = 62.41 \text{ W}$$

The reactive power loss of the cable is

$$Q_{loss} = I^2\, X_{cable} = 7.9^2 \times 5 = 312.05 \text{ VAr}$$

(g) The real power delivered by the source is the load's real power plus the real power loss of the line.

$$P_s = P_R + P_{loss} = 249.36 + 62.41 = 311.77 \text{ W}$$

Similarly, the reactive power delivered by the source is

$$Q_s = Q_L + Q_{loss} = 498.437 + 312.05 = 810.487 \text{ VAr}$$

Example 7.9

Consider the system in Example 7.8, but assume that the load is purely resistive with no inductive elements. Compute the following:

(a) Load impedance
(b) Line current
(c) Load voltage

(d) Real and imaginary components of the load current
(e) Real and reactive powers of the load
(f) Real and reactive losses of the cable
(g) Real and reactive power delivered by the source

Solution

(a) Load impedance Z is just the resistance.

$$\bar{Z} = R = 20 \ \Omega$$

(b) To compute the line current, we need to compute the total impedance of the circuit.

$$\bar{Z}_{total} = \bar{Z} + \bar{Z}_{cable} = 20 + (1.0 + j5) = 21 + j5 \ \Omega$$

$$\bar{I} = \frac{\bar{V}_S}{\bar{Z}_{total}} = \frac{110 \angle 0°}{21 + j5} = 5.09 \angle -13.39° \ A$$

(c) Load voltage is

$$\bar{V}_{load} = \bar{I}\bar{Z} = 5.09 \angle -13.39° \times 20 = 101.8 \angle -13.39° \ V$$

Note that the load voltage in this case is about 92% of the source voltage. This is a much higher magnitude than the case in Example 7.8.

(d) The real component of the load current is the same as the line current since the imaginary current of the load is zero.

$$\bar{I}_R = \bar{I} = 5.09 \angle -13.39° \ A$$

(e) The real power of the load is

$$P_R = I_R^2 \ R = 5.09^2 \times 20 = 518.16 \ W$$
$$Q_L = 0 \ VAr$$

(f) The real power loss of the cable is

$$P_{loss} = I^2 R_{cable} = 5.09^2 \times 1 = 25.91 \ W$$

The reactive power loss of the cable is

$$Q_{loss} = I^2 X_{cable} = 5.09^2 \times 5 = 129.54 \ VAr$$

(g) The real power delivered by the source is the load real power plus the line losses.

$$P_s = P_R + P_{loss} = 518.16 + 25.91 = 544.07 \ W$$

The reactive power delivered by the source is

$$Q_s = Q_L + Q_{loss} = 0 + 129.54 = 129.54 \ VAr$$

7.7.7 POWER FACTOR CORRECTION

As seen in the previous section, the inductive load increases cable losses, reduces the voltage across the load, and reduces the spare capacity of the transmission lines. For these reasons, most utilities

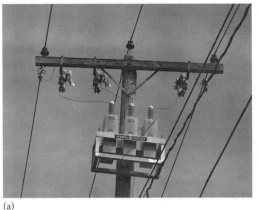 (a)
 (b)

FIGURE 7.30 (See color insert following page 300.) Capacitors for power factor correction, (a) on a distribution pole and (b) in a distribution substation.

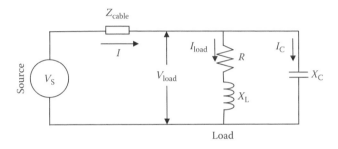

FIGURE 7.31 Circuit consists of resistive and inductive loads.

install capacitors near the service areas to compensate for the reactive power consumed by the load. Figure 7.30 shows two compensation methods: Figure 7.30a shows three rectangular shaped capacitors mounted on a distribution pole close to a residential area, and Figure 7.30b shows a much larger compensation at a substation where a large number of capacitors are mounted on racks.

The circuit in Figure 7.31 shows a source feeding a load through a cable. Across the load, a compensation capacitor is installed. Since the reactive power of the purely inductive load is positive and that of the purely capacitive load is negative, the reactive power of the capacitor cancels the reactive power of the inductor. This process is called *power factor correction*, or *reactive power compensation*. If the two reactive powers are summed up to zero, the source delivers only real power to the load. Thus, the problems associated with the reactive current on the transmission line are eliminated.

EXAMPLE 7.10

The circuit in Figure 7.31 has a voltage source of 110 V, 60 Hz. The source is connected to an inductive load through a cable. The load impedance is $\bar{Z} = 3 + j4$ Ω, the cable resistance is 1 Ω, and the cable inductive reactance is 2 Ω.
Assume the capacitor is not connected to the circuit and compute the following:

(a) Power factor of the load
(b) Line current
(c) Load voltage
(d) Real and reactive powers of the load
(e) Losses of the cable

If a 6 Ω capacitive reactance is connected in parallel with the load, compute the following:

(f) Power factor of the load plus the capacitor
(g) Line current
(h) Load voltage
(i) Real and reactive powers of the load plus the capacitor

Solution

Without the capacitor

(a) Power factor angle of the load is equal in magnitude to the angle of the load impedance.

$$\overline{Z} = R + jX_L = 3.0 + j4.0 = 5\angle 53.13° \ \Omega$$

The power factor is

$$pf = \cos(53.13) = 0.6 \text{ lagging}$$

Note that the power factor must be identified with either *lagging* or *leading*. Lagging is when the current lags the voltage, which is the case of the inductive load.

(b) To compute the line current, we need to compute the total impedance of the circuit

$$\overline{Z}_{total} = \overline{Z} + \overline{Z}_{cable} = (3.0 + j4.0) + (1.0 + j2.0) = 4.0 + j6.0 \ \Omega$$

$$\overline{I} = \frac{\overline{V}_S}{\overline{Z}_{total}} = \frac{110\angle 0°}{4.0 + j6.0} = 8.461 - j12.69 = 15.254\angle -56.31° \text{ A}$$

(c) Load voltage can be computed as

$$\overline{V}_{load} = \overline{I}\,\overline{Z} = (15.254\angle -56.31°)(3 + j4) = 76.27\angle -3.18° \text{ V}$$

Note that the load voltage in this case is just about 69% of the source voltage.

(d) Real and reactive powers of the load can be computed as

$$P_R = I^2 R = 15.254^2 \times 3 = 698.05 \text{ W}$$
$$Q_L = I^2 X_L = 15.254^2 \times 4 = 930.74 \text{ VAr}$$

(e) Losses of the cable are

$$P_{loss} = I^2 R_{cable} = 15.254^2 \times 1 = 232.68 \text{ W}$$

$$Q_{loss} = I^2 X_{cable} = 15.254^2 \times 2 = 465.36 \text{ VAr}$$

With the capacitor

(f) Load impedance Z_1 is the load impedance in parallel with the capacitive reactance.

$$\overline{Y}_1 = \frac{1}{\overline{Z}} + \frac{1}{\overline{X}_C} = \frac{1}{3.0 + j4.0} + \frac{1}{-j6.0} = 0.12 + j0.007 \text{ mho}$$

$$\overline{Z}_1 = \frac{1}{\overline{Y}_{total}} = 8.32\angle -3.18° \ \Omega$$

The power factor is

$$pf = \cos(3.18) = 0.9985 \text{ leading}$$

Note that the power factor is near unity.

(g) The total impedance of the circuit is

$$\overline{Z}_{\text{total}} = \overline{Z}_1 + \overline{Z}_{\text{cable}} = 8.32\angle-3.18° + (1.0 + j2.0) = 9.307 + j1.538 \ \Omega$$

$$\overline{I} = \frac{\overline{V}_S}{\overline{Z}_{\text{total}}} = \frac{110\angle0°}{9.307 + j1.538} = 11.505 - j1.901 = 11.66\angle-9.38° \ \text{A}$$

Note that the line current with the capacitor installed is 11.66 A, which is smaller than the line current computed in part b (15.254 A). This is about 23.5% reduction in the line current, which increases the spare capacity of the cable.

(h) Load voltage is

$$\overline{V}_{\text{load}} = \overline{I}\overline{Z}_1 = 11.66\angle-9.38° \ (8.32\angle-3.18°) = 97.01\angle-12.56° \ \text{V}$$

Note that the load voltage in this case is much higher than the case without the capacitor (in part c).

(i) Before we compute the powers, we need to compute the load current I_{load} and the capacitor current I_C

$$\overline{I}_{\text{load}} = \frac{\overline{V}_{\text{load}}}{\overline{Z}} = \frac{97.01\angle-12.56°}{5.0\angle53.13°} = 19.4\angle-65.69° \ \text{A}$$

$$\overline{I}_C = \frac{\overline{V}_{\text{load}}}{\overline{X}_C} = \frac{97.01\angle-12.56°}{6.0\angle-90°} = 16.17\angle77.44° \ \text{A}$$

The real and reactive powers of the load are

$$P_R = I_{\text{load}}^2 R = 19.4^2 \times 3 = 1.129 \ \text{kW}$$

$$Q_L = I_{\text{load}}^2 X_L = 19.4^2 \times 4 = 1.505 \ \text{kVAr}$$

$$Q_C = I_C^2 X_C = 16.17^2 \times 6 = 1.569 \ \text{kVAr}$$

The total real power P_{total} of the load with the capacitor installed is the real power of the load alone. This is because the capacitor and the inductor consume no real power.

$$P_{\text{total}} = P_R = 1.129 \ \text{kW}$$

The total reactive power Q_{total} is the phasor sum of the inductive and capacitive powers

$$\overline{Q}_{\text{total}} = \overline{Q}_L + \overline{Q}_C = j1.505 - j1.569 = -j64 \ \text{VAr}$$

Note that the total reactive power on the load side without the capacitor is 930.74 VAr.

EXAMPLE 7.11

The voltage source of the circuit in Figure 7.32 is 240 V, 60 Hz. The load consists of a 4 Ω resistance in series with a 3 Ω inductive reactance. A capacitor is installed to improve the power factor at the source side to unity. Compute the value of X_C.

Solution

We can solve this problem by several methods. One of them is to compute the reactive power of the load, then compute the value of X_C that produces capacitive reactive power equal in magnitude to the inductive power. Another method is to compute the value of X_C that would set the imaginary component of the total impedance to zero.

First method

The current of the load is

$$\bar{I}_{\text{load}} = \frac{\bar{V}_{\text{S}}}{\bar{Z}} = \frac{240\angle 0°}{4.0 + j3.0} = 38.4 - j28.8 = 48\angle -36.87° \text{ A}$$

The reactive power of the load is

$$Q_{\text{L}} = I_{\text{load}}^2 X_{\text{L}} = (48)^2 \times 3 = 6.912 \text{ kVAr}$$

The reactive power of the capacitor must be equal (in magnitude) to the reactive power of the load.

$$Q_{\text{C}} = I_{\text{C}}^2 X_{\text{C}} = \frac{V_{\text{S}}^2}{X_{\text{C}}} = \frac{240^2}{X_{\text{C}}} = 6.912 \text{ kVAr}$$

Hence,

$$X_{\text{C}} = \frac{240^2}{6912} = 8.33 \text{ } \Omega$$

Second method

Set the imaginary component of the total impedance equal to zero.

$$\text{Im}\left(\frac{1}{\bar{Z}} + \frac{1}{\bar{X}_{\text{C}}}\right) = 0$$

$$\text{Im}\left(\frac{1}{4.0 + j3.0} + \frac{1}{-jX_{\text{C}}}\right) = 0$$

$$0.12 - \frac{1}{X_{\text{C}}} = 0$$

$$X_{\text{C}} = 8.33 \text{ } \Omega$$

7.8 ELECTRIC ENERGY

The energy E consumed by a load is the power delivered to the load P over a period of time. The units of energy are watt hours (W h), kW h, MW h, etc. For constant power during a given period τ, the energy is the shaded area under the power line in Figure 7.33.

$$E = P\tau \qquad (7.59)$$

For time varying power, such as the one shown in Figure 7.34, the energy during a time period τ is the integral of the power over the period τ.

$$E = \int_0^\tau P \, dt \qquad (7.60)$$

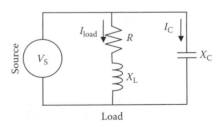

Load

FIGURE 7.32 Circuit consists of resistive and inductive loads.

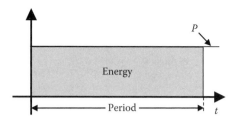

FIGURE 7.33   Energy of constant power.

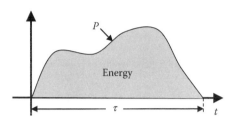

FIGURE 7.34   Energy of variable power.

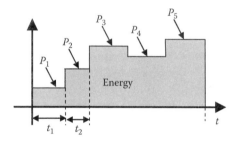

FIGURE 7.35   Energy of discrete power.

For discrete power, as in Figure 7.35, the energy is the summation of the power for each time segment.

$$E = \sum_{i} P_i t_i \qquad (7.61)$$

EXAMPLE 7.12

An electric load is connected across a 120 V source. The load impedance is changing over a period of 24 h as follows:

Time Period	Load Impedance (Ω)	Power Factor Angle (Degrees)
8:00–10:30 AM	10	30
11:00 AM–1:00 PM	20	0
3:00–5:00 PM	15	60
5:00–8:00 PM	5	45

Compute the energy consumed by the load during this 24 h period.

Solution

Time Period	Load Impedance Z (Ω)	Pf Angle θ (Degrees)	Load Current $\frac{V_s}{Z}$ (A)	Load Power $V_s I \cos \theta$ (W)	Time Period (h)	Energy (W h)
8:00–10:30 AM	10	30	12	1247.08	2.5	3117.7
11:00 AM–1:00 PM	20	0	6	720.0	2.0	1440.0
3:00–5:00 PM	15	60	8	480.0	2.0	960.0
5:00–8:00 PM	5	45	24	2036.47	3.0	6109.41
				Total energy		11627.11

EXAMPLE 7.13

A load is connected across a 120 V source. The load power is represented by

$$P = 25 + 100 \sin (20t) \text{ kW}$$

where t is the time (in hours). Compute the energy consumed by the load after 1 h and after 2 h.

Solution

$$E = \int_0^\tau P \, dt = \int_0^\tau (25 + 100 \sin (20t)) \, dt = \left[25t - 100 \frac{\cos (20t)}{20} \right]_0^\tau$$

$$E = 25\tau - 5 \cos (20\tau) + 5$$

After 1 h, $\tau = 1$

$$E = 25(1) - 5 \cos (20) + 5 = 25.3 \text{ kW h}$$

After 2 h, $\tau = 2$

$$E = 25(2) - 5 \cos (40) + 5 = 51.17 \text{ kW h}$$

EXERCISES

1. Given the following vectors:

$$\bar{A} = 12 + j12$$
$$\bar{B} = -6 + j10$$

   Compute $\bar{A} + \bar{B}$, $\bar{A} - \bar{B}$, $\bar{A}\,\bar{B}$, and $\bar{A}/\bar{B}$.

2. Voltage and current equations of an electric load are

$$v = V_{max} \sin \omega t$$
$$i = I_{max} \cos (\omega t - 30°)$$

   Compute the phase shift of the current.

3. Voltage and current of an electric load can be expressed by the following equations:

$$v = 340 \sin (628.318t + 0.5236)$$
$$i = 100 \sin (628.318t + 0.87266)$$

Calculate the following:
(a) rms voltage
(b) Frequency of the current
(c) Phase shift between current and voltage in degrees
(d) Average voltage
(e) Load impedance

4. Current and voltage waveforms of an electric circuit are

$$i(t) = 25 \sin\left(377t + \frac{\pi}{3}\right) \text{ A}; \quad v(t) = 169.7 \sin\left(377t - \frac{\pi}{6}\right) \text{ V}$$

Compute the following:
(a) rms voltage
(b) rms current
(c) Frequency of the supply voltage
(d) Phase angle of the current with respect to the voltage (indicate leading or lagging)
(e) Real power consumed by the circuit
(f) Reactive power consumed by the circuit
(g) Impedance of the circuit

5. An ac source is feeding a load that consists of a resistance and inductive reactance connected in series. The voltage and current of the source are

$$v = 400 \sin (377t + 0.5236) \text{ V}$$

$$i = 100 \sin (377t + 0.87266) \text{ A}$$

Calculate the following:
(a) Resistance and inductive reactance of the load impedance
(b) rms voltage across the resistance
(c) Frequency of the supply voltage
(d) Power factor at the source side, state leading or lagging
(e) Real power delivered to the load
(f) Reactive power delivered to the load

6. An electric load consists of a 4 Ω resistance, a 6 Ω inductive reactance, and an 8 Ω capacitive reactance connected in series. The total impedance of the load is connected across a voltage source of 120 V.
Compute the following:
(a) Power factor of the load
(b) Source current
(c) Real power of the circuit
(d) Reactive power of the circuit

7. A load impedance consists of a 25 Ω resistance in series with a 38 Ω inductive reactance. The load is connected across a 240 V source. Compute the real and reactive powers consumed by the load.

8. A voltage source $\overline{V} = 120\angle 30°$ is connected across a circuit consisting of $R = 3.5\ \Omega$ in series with $L = 0.0833$ H. If the circuit is energized in the United States, compute the following:
   (a) Load current
   (b) Load power factor
   (c) Real power consumed by the load
   (d) Reactive power consumed by the load
9. A voltage source $\overline{V} = 120\angle 30°$ is connected to a circuit consisting of $R = 3.5\ \Omega$ in series with $L = 0.0833$ H. If the circuit is energized in Europe, compute the following:
   (a) Load current
   (b) Load power factor
   (c) Real power consumed by the load
   (d) Reactive power consumed by the load
10. Inductive load consists of $R = 3\ \Omega$ in series with $X_L = 4\ \Omega$. The load is connected in parallel with another load of unknown impedance. The voltage source of the system is 120 V. The total real and reactive powers delivered by the source are 3 kW and 2 kVAr, respectively. Compute the impedance of the unknown load.

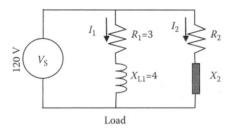

Load

11. The rms voltage and current of an inductive load are 110 V and 10 A, respectively. The frequency of the voltage waveform is 60 Hz. The instantaneous power consumed by the load has no average value. Calculate the following:
    (a) Real (active) power consumed by the load
    (b) Reactive power consumed by the load
    (c) Power factor
    (d) Frequency of the reactive power
12. An electric load consists of a 5 Ω resistance, a 20 Ω capacitive reactance, and a 10 Ω inductive reactance, all connected in parallel. A voltage source of 100 V is applied across the load. Compute the following:
    (a) Real power of the load
    (b) Reactive power of the load
    (c) Current of the load
13. An inductive load consists of a 1 Ω resistance and a 1.0 mH inductance connected in series. The load is connected across a 120 V, 60 Hz source.
    (a) Compute the power factor of the load.
    (b) A capacitor is connected in parallel with the entire load so that no reactive power is delivered by the source. Compute the value of the capacitance.
14. The power consumption of an electric load is represented by

$$P(t) = 100(1 - e^{\frac{-t}{10}})\ \text{kW}$$

where t is the time of the day in hours.
    Compute the energy consumed by the load in a period of 24 h.

8 Three-Phase Systems

High-power equipment such as generators, transformers, and transmission lines are built as three-phase equipment. The three-phase system has many advantages over the single-phase system; the most important ones are

1. Three-phase system produces a rotating magnetic field inside the alternating current (ac) motors, and therefore causes the motors to rotate without the need for extra controls. Since electric motors constitute the majority of the electric energy consumed worldwide, having a rotating magnetic field is a very important advantage of three-phase systems. The theory of the rotating fields is given in Chapter 12.
2. Three-phase generator produces more power than a single-phase generator of equivalent volume.
3. Three-phase transmission line transmits three times the power of a single-phase line.
4. Three-phase system is more reliable—when one phase is lost, the other two phases can still deliver some power to the loads.

8.1 GENERATION OF THREE-PHASE VOLTAGES

According to Faraday's law, when a conductor cuts magnetic field lines, a voltage is induced across the conductor. The synchronous generator uses this phenomenon to produce a three-phase voltage. A simple schematic of the generator is shown in Figure 8.1 and a more detailed analysis of the machine is given in Chapter 12. The generator consists of an outer frame called stator, and a rotating magnet called rotor. At the inner perimeter of the stator, coils are placed inside slots. Each of the two slots separated by 180° houses a single coil (a–a', b–b', or c–c'). The coil is built by placing a wire inside a slot in one direction (e.g., a), and winding it back inside the opposite slot (a'). Figure 8.1 shows three coils; each coil is separated by 120° from the other coils.

Now let us assume that the magnet is spinning clockwise inside the machine by an external prime mover. The magnetic field then cuts all coils and, therefore, induces a voltage across each of them. If each coil is connected to a load impedance, a current would flow into the load and the generator produces electrical energy that is consumed by the load. The dot inside the coil indicates a current direction toward the reader, and the cross indicates a current in the opposite direction.

The voltage across any of the three coils can be expressed by Faraday's law.

$$e = nBl\omega \qquad (8.1)$$

where
  e is the voltage induced across the coil
  n is the number of turns in the coil
  B is the flux density of the magnetic field
  l is the length of the slot
  ω is the angular speed of the rotor

The voltage induced on a conductor is proportional to the perpendicular component of the magnetic field with respect to the conductor. At the rotor position in Figure 8.1, coil a–a' has the

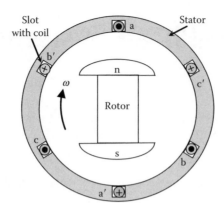

FIGURE 8.1   Three-phase generator.

maximum perpendicular field and, hence, it has the maximum induced voltage. Coil b–b′ will have its maximum voltage when the rotor moves clockwise by 120°, and coil c–c′ will have its maximum voltage when the rotor is at 240° position. If the rotation is continuous, the voltage across each coil is sinusoidal, all coils have the same magnitude of the maximum voltage, and the induced voltages are shifted by 120° from each other as shown in Figure 8.2. These characteristics form what is known as a balanced three-phase system.

To express the continuous waveforms of a balanced system mathematically, we need to select one of the waveforms as a reference and express all other waveforms in relative relations to it. Assuming that $v_{aa'}$ is our reference voltage, the waveforms in Figure 8.2 can be expressed by Equation 8.2.

$$v_{aa'} = V_{max} \cos(\omega t)$$
$$v_{bb'} = V_{max} \cos(\omega t - 120°) \tag{8.2}$$
$$v_{cc'} = V_{max} \cos(\omega t - 240°) = V_{max} \cos(\omega t + 120°)$$

where $v_{aa'}$, $v_{bb'}$, and $v_{cc'}$ are the instantaneous voltages of the three coils; these voltages are known as *phase voltages*. The phase in this case is the coil; so phase a is coil aa′, phase b is coil bb′, and phase c is coil cc′. Note that $v_{bb'}$ in Figure 8.2 lags $v_{aa'}$ by 120°, and $v_{cc'}$ leads $v_{aa'}$ by 120°. The maximum values of the phase voltages V_{max} are equal. As given in Chapter 7, the waveforms in Equation 8.2 can be written in polar form.

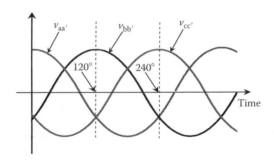

FIGURE 8.2   Waveforms of the three phases.

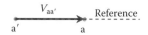

(a) $\overline{V}_{aa'}$

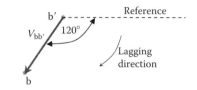

(b) $\overline{V}_{bb'}$

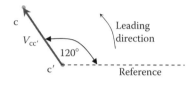

(c) $\overline{V}_{cc'}$

FIGURE 8.3 (See color insert following page 300.) Phasor diagram of balanced three phases. (a) Phasor diagram of phase $\overline{V}_{aa'}$, (b) phasor diagram of phase $\overline{V}_{bb'}$, and (c) phasor diagram of phase $\overline{V}_{cc'}$.

$$\overline{V}_{aa'} = \frac{V_{max}}{\sqrt{2}} \angle 0° = V \angle 0°$$

$$\overline{V}_{bb'} = \frac{V_{max}}{\sqrt{2}} \angle{-120°} = V \angle{-120°}$$

(8.3)

$$\overline{V}_{cc'} = \frac{V_{max}}{\sqrt{2}} \angle 120° = V \angle 120°$$

where
  $\overline{V}$ is the phasor voltage in complex number form
  V is the magnitude of the rms voltage

As explained in Chapter 7, the phasor diagram of $\overline{V}_{aa'}$ (the reference voltage) can be drawn as shown in Figure 8.3a. The direction of the arrow is from a′ to a. The length of the arrow is equal to the rms value of $\overline{V}_{aa'}$. Similarly, the phasor diagrams of $\overline{V}_{bb'}$ and $\overline{V}_{cc'}$ are shown in Figure 8.3b and 8.3c, respectively.

8.2 CONNECTIONS OF THREE-PHASE CIRCUITS

The three-phase generator shown in Figure 8.1 has three independent coils and each coil represents a phase. Since each coil has two terminals, the generator has six terminals. To transmit the generated power from the power plant to the load centers, six wires seem to be needed. Since the transmission lines are often very long (hundreds of miles), the cost of the six wires is very high. To reduce the cost of the transmission system, the three coils are often connected in a wye or delta configuration. In each of these configurations, only three wires are needed to transmit the power. These three wires are considered to be one circuit.

FIGURE 8.4   Transmission line with double circuits.

Figure 8.4 shows a 500 kV transmission line that consists of two circuits. Each circuit is in three-phase configuration mounted on one side of the tower. Figure 8.5 shows a distribution pole with one circuit. The voltage level of this circuit is 15 kV. The figure also shows transformers mounted on the pole to step down the voltage from 15 kV to the household level.

FIGURE 8.5 (See color insert following page 300.)   Single-circuit distribution line.

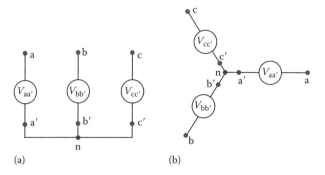

FIGURE 8.6   Connection of the three coils in wye. (a) Circuit connection and (b) convenient representation.

8.2.1 WYE-CONNECTED BALANCED SOURCE

The wye connection is also referred to as "Y" or "star." For the generator in Figure 8.1, assume that the induced voltages in the three coils are represented by three voltage sources. To connect the coils in wye configuration, terminals a′, b′, and c′ of the three coils are bonded into a common point called neutral or n as shown in Figure 8.6a. Since the coils are mechanically separated by 120° with respect to each other, we can also draw the three-phase wye configuration in the convenient diagram shown in Figure 8.6b.

The phasor diagram of the three-phase voltages in wye configuration can be obtained by grouping the three phasors in Figure 8.3 as shown in Figure 8.7a. Since the common point is n, the phasors can be labeled $\overline{V}_{an}$, $\overline{V}_{bn}$, and $\overline{V}_{cn}$ as shown in Figure 8.7b. Keep in mind that $\overline{V}_{an}$, $\overline{V}_{bn}$, and $\overline{V}_{cn}$ are the phase voltages.

For balanced systems, the magnitudes of the rms voltages for all phases are equal, that is

$$V_{an} = V_{bn} = V_{cn} \tag{8.4}$$

By this wye connection, the three terminals of the generator (a, b, and c) in Figure 8.6 are the only ones connected to the three-wire transmission line as shown in Figure 8.8. The neutral point is often connected to the ground and no additional wire is needed. The phase voltage of this system is the voltage between any line and the ground (or neutral). Hence,

$$V_{ph} = V_{an} = V_{bn} = V_{cn} \tag{8.5}$$

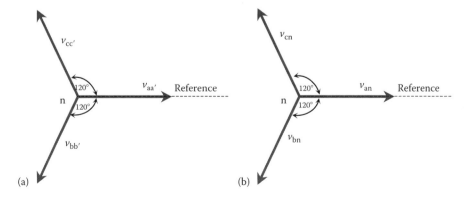

FIGURE 8.7   Phasor diagram of the three phases.

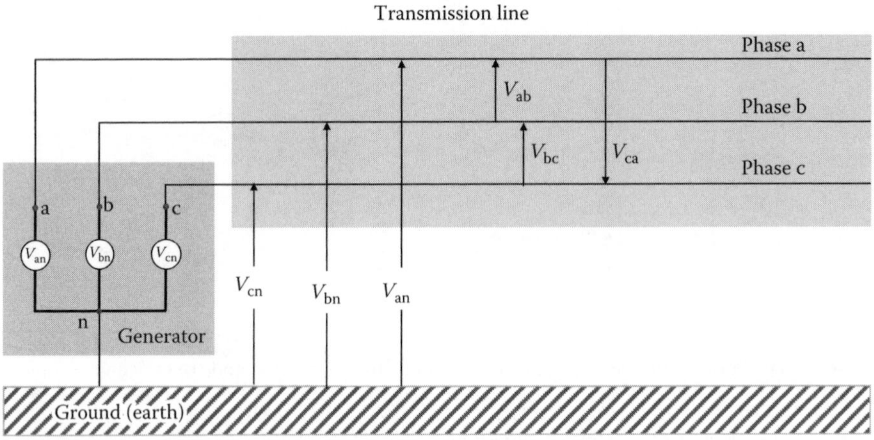

Transmission line

FIGURE 8.8 Three-phase wye generator connected to a three-phase transmission line.

where V_{ph} is the magnitude of the phase voltage. Equation 8.3 can then be rewritten as

$$\overline{V}_{an} = V_{ph}\angle 0°$$
$$\overline{V}_{bn} = V_{ph}\angle -120° \tag{8.6}$$
$$\overline{V}_{cn} = V_{ph}\angle 120°$$

The voltage between any two lines is known as the line-to-line voltage. The line-to-line voltage between a and b is $\overline{V}_{ab}$; it is the potential of phase a minus the potential of phase b

$$\overline{V}_{ab} = \overline{V}_{an} - \overline{V}_{bn} \tag{8.7}$$

Substituting the values of $\overline{V}_{an}$ and $\overline{V}_{bn}$ of Equation 8.6 into Equation 8.7 yields

$$\overline{V}_{ab} = V_{ph}\angle 0° - V_{ph}\angle -120° = \sqrt{3}V_{ph}\angle 30° \tag{8.8}$$

Two important observations can be made from Equation 8.8:

1. Magnitude of the line-to-line voltage is larger than the magnitude of the phase voltage by a factor of $\sqrt{3}$.
2. Line-to-line voltage $\overline{V}_{ab}$ leads the phase voltage $\overline{V}_{an}$ by 30°.

Similarly, the line-to-line voltages $\overline{V}_{bc}$ and $\overline{V}_{ca}$ can be computed as given in Equation 8.9.

$$\overline{V}_{bc} = \overline{V}_{bn} - \overline{V}_{cn} = V_{ph}\angle -120° - V_{ph}\angle 120° = \sqrt{3}V_{ph}\angle -90°$$
$$\overline{V}_{ca} = \overline{V}_{cn} - \overline{V}_{an} = V_{ph}\angle 120° - V_{ph}\angle 0° = \sqrt{3}V_{ph}\angle 150° \tag{8.9}$$

Note that phasor $\overline{V}_{bc}$ leads $\overline{V}_{bn}$ by 30°, and phasor $\overline{V}_{ca}$ leads $\overline{V}_{cn}$ by 30°. The phasor diagram of all phase and line-to-line voltages is shown in Figure 8.9. Keep in mind that the direction of any phasor is determined by the subscript of the phasor, for example the direction of $\overline{V}_{ca}$ is from point a to point c.

From Equations 8.8 and 8.9, we can write the line-to-line voltages as

$$\overline{V}_{bc} = \overline{V}_{ab}\angle -120°$$
$$\overline{V}_{ca} = \overline{V}_{ab}\angle 120° \tag{8.10}$$

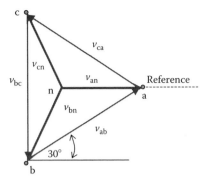

FIGURE 8.9 Phasor diagram of the phase and line-to-line voltages.

Hence, the line-to-line voltages are also balanced; they are equal in magnitude and are shifted by 120° from each other. In a generic term, we can relate the magnitude of the line-to-line voltage V_{ll} to the phase voltage V_{ph} by

$$V_{ll} = \sqrt{3}V_{ph} \tag{8.11}$$

EXAMPLE 8.1

A balanced three-phase system has its phase voltage $\overline{V}_{an} = 120\angle40°$ V. Compute the line-to-line voltage $\overline{V}_{bc}$.

Solution

Note that $\overline{V}_{an}$ is not in phase with the reference. Since $\overline{V}_{bc}$ lags $\overline{V}_{an}$ by 120°, and $\overline{V}_{cn}$ leads $\overline{V}_{an}$ by 120°,

$$\overline{V}_{bn} = \overline{V}_{an}\angle-120° = (120\angle40°)\angle-120° = 120\angle-80° \text{ V}$$

$$\overline{V}_{cn} = \overline{V}_{an}\angle120° = (120\angle40°)\angle120° = 120\angle160° \text{ V}$$

The line-to-line voltages can be computed as follows:

$$\overline{V}_{bc} = \overline{V}_{bn} - \overline{V}_{cn} = (120\angle-80°) - (120\angle160°) = 133.59 - j159.22 = 207.84\angle-50° \text{ V}$$

Note that if we use the basic relationships between the phase and line-to-line voltages given in Equations 8.8 through 8.10, we can solve this problem without the need for complex number computations. First, we know that the magnitude of the line-to-line voltage is $\sqrt{3}V_{ph}$; then

$$V_{ab} = \sqrt{3}V_{ph} = \sqrt{3} \times 120 = 207.84 \text{ V}$$

Second, we know that $\overline{V}_{ab}$ is leading $\overline{V}_{an}$ by 30° as given in Equation 8.8; then

$$\overline{V}_{ab} = 207.84\angle(40° + 30°) = 207.84\angle70° \text{ V}$$

From Equation 8.10, $\overline{V}_{bc}$ lags $\overline{V}_{ab}$ by 120°

$$\overline{V}_{bc} = \overline{V}_{ab}\angle-120° = 207.84\angle(70° - 120°) = 207.84\angle-50° \text{ V}$$

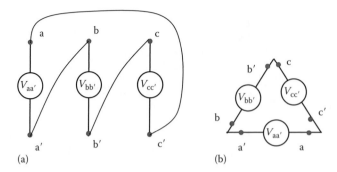

FIGURE 8.10 Delta connection.

8.2.2 Delta-Connected Balanced Source

The delta (Δ) configuration can be made by cascading the coils of the generator in Figure 8.1; terminal a′ is connected to b, b′ is connected to c, and c′ is connected to a as shown in Figure 8.10a. The voltage sources of the delta connection can also be drawn in the convenient diagram shown in Figure 8.10b. Since point a′ is the same as b, b′ is the same as c, and c′ is the same as a. Hence,

$$\begin{aligned} \overline{V}_{aa'} &= \overline{V}_{ab} \\ \overline{V}_{bb'} &= \overline{V}_{bc} \\ \overline{V}_{cc'} &= \overline{V}_{ca} \end{aligned} \tag{8.12}$$

The wiring connection and the phasor diagram of the delta configuration are shown in Figure 8.11. In the wiring connection, a′, b′, and c′ are removed as they are the same points as b, c, and a, respectively. Because the waveforms of the three phases are continuous, the reference phasor can be arbitrarily selected. The phasor diagram in Figure 8.11 has $\overline{V}_{ab}$ as a reference.

As we stated earlier, $\overline{V}_{aa'}$ is the voltage of coil aa′, which is defined as the phase voltage of the generator. Also, $\overline{V}_{ab}$ has been defined earlier as the line-to-line voltage. Since $\overline{V}_{aa'} = \overline{V}_{ab}$ in a delta-connected generator, the phase voltage is equal to the line-to-line voltage. A delta-connected generator is attached to a three-phase transmission line as shown in Figure 8.12. Note that the delta connection has only three wires with no neutral terminal.

8.2.3 Wye-Connected Balanced Load

Residential loads in the United States are normally single-phase loads at 120 and 240 V. Small loads that draw small currents such as lights, televisions, and radios are powered at 120 V. Heavier loads such as furnaces, stoves, and dryers are powered at 240 V. Industrial and commercial loads are mostly three-phase loads at various line-to-line voltage levels (208 V, 480 V, 5 kV, and 15 kV). As a general rule, the larger the power demand, the higher the voltage.

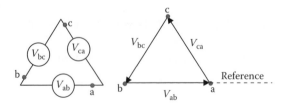

FIGURE 8.11 Delta connection and its phasor diagram.

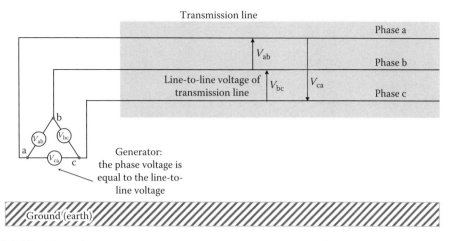

FIGURE 8.12 Three-phase generator connected to a three-phase transmission line.

Although household equipment are single-phase loads, clustered residential areas are powered by three phases. Each group of houses is powered by one phase while other groups are powered by the other phases. This is done to balance the loads among the three phases.

The three-phase loads are connected in various configurations with the most common ones being the wye and the delta. All residential loads and most commercial and industrial loads are connected in wyes to ensure that the voltage across the load is constant regardless of any fluctuation in the load currents. Figure 8.13 shows a wye-connected load where Z_{an}, Z_{bn}, and Z_{cn} are called the phase impedances. For balanced systems, the phase impedance of each phase is equal to that of the other phases, that is

$$\overline{Z}_{an} = \overline{Z}_{bn} = \overline{Z}_{cn} = \overline{Z} \tag{8.13}$$

For residential loads, each of Z_{an}, Z_{bn}, or Z_{cn} represents a group of houses. Although the demands in any neighborhood differ from house to house due to the various energy consumption habits of the customers, three-phase loads are almost balanced over a wide residential area.

Figure 8.14 shows a wye-connected load powered by a wye-connected source. Four lines connect the load to the source. The neutral points of the load and that of the source are connected either directly by wire or indirectly through the ground.

Assume that the load is balanced and the voltage $\overline{V}_{an}$ is phase shifted from an arbitrary reference by an angle θ. The load currents, which flow from the source to the load, can then be computed as

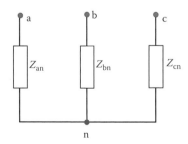

FIGURE 8.13 Three loads connected in wye.

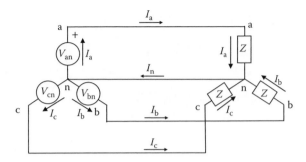

FIGURE 8.14   Loads connected in wye.

$$\bar{I}_a = \frac{\overline{V}_{an}}{\overline{Z}_{an}} = \frac{\overline{V}_{an}}{\overline{Z}} = \frac{V_{ph}\angle\theta}{Z\angle\phi} = \frac{V_{ph}}{Z}\angle(\theta - \phi)$$

$$\bar{I}_b = \frac{\overline{V}_{bn}}{\overline{Z}_{bn}} = \frac{\overline{V}_{bn}}{\overline{Z}} = \frac{V_{ph}\angle(\theta - 120°)}{Z\angle\phi} = \frac{V_{ph}}{Z}\angle(\theta - \phi - 120°) \qquad (8.14)$$

$$\bar{I}_c = \frac{\overline{V}_{cn}}{\overline{Z}_{cn}} = \frac{\overline{V}_{cn}}{\overline{Z}} = \frac{V_{ph}\angle(\theta + 120°)}{Z\angle\phi} = \frac{V_{ph}}{Z}\angle(\theta - \phi + 120°)$$

Four observations can be made by examining Equation 8.14:

1. Magnitudes of $\bar{I}_a$, $\bar{I}_b$, and $\bar{I}_c$ are equal.
2. Current $\bar{I}_a$ leads $\bar{I}_b$ by 120°, and $\bar{I}_b$ leads $\bar{I}_c$ by 120°.
3. Currents of the source (known as the phase currents of the source) are equal to the currents flowing in the lines connecting the source and the load (known as the line currents).
4. Line currents are equal to the currents flowing into the phase impedances (also known as the phase current of the load).

The currents and voltages of the circuit in Figure 8.14 are shown in the phasor diagram in Figure 8.15. Using Kirchhoff's current rule, the sum of the three currents $\bar{I}_a$, $\bar{I}_b$, and $\bar{I}_c$ entering the neutral node n of the load is equal to the current exiting the node $\bar{I}_n$. Hence

$$\bar{I}_n = \bar{I}_a + \bar{I}_b + \bar{I}_c \qquad (8.15)$$

If the load is balanced, the sum of all load currents in Equation 8.14 is zero.

$$\bar{I}_n = \bar{I}_a + \bar{I}_a\angle-120° + \bar{I}_a\angle120° = \bar{I}_a(1 + 1\angle-120° + 1\angle120°) = 0 \qquad (8.16)$$

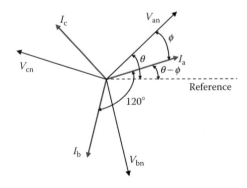

FIGURE 8.15   Phasor diagram of loads connected in wye.

FIGURE 8.16   Single-phase representation of the circuit in Figure 8.14.

Since $\bar{I}_n = 0$, there is no need to have a wire connecting the neutral points of the wye-connected generator and the wye-connected load.

For the three-phase system in Figure 8.14, there are three independent loops with the neutral wire as a common branch. If the system is balanced, we need to analyze only one phase; the voltages and currents of the other two phases can be obtained by the relationships in Equations 8.10 and 8.14. Figure 8.16 shows a single-phase representation of a balanced three-phase system.

EXAMPLE 8.2

The phase voltage of a balanced three-phase source is $\bar{V}_{an} = 120\angle{-40°}$ V. The source is energizing a balanced three-phase load connected in wye. The phase impedance of the load is $\bar{Z} = 20\angle{30°}$ Ω. Compute the following:

(a) Load currents of each phase
(b) Neutral current
(c) Line-to-line voltages at the load side

Solution

Use the single-phase representation of the balanced three-phase system shown in Figure 8.16.

(a)
$$\bar{I}_a = \frac{\bar{V}_{an}}{\bar{Z}} = \frac{120\angle{-40°}}{20\angle{30°}} = 6\angle{-70°} \text{ A}$$

The currents of the other phases can be computed by the relationships of the balanced system.

$$\bar{I}_b = \bar{I}_a\angle{-120°} = 6\angle{-190°} \text{ A}$$

$$\bar{I}_c = \bar{I}_a\angle{120°} = 6\angle{50°} \text{ A}$$

(b) $\bar{I}_n = \bar{I}_a + \bar{I}_b + \bar{I}_c = 6\angle{-70°} + 6\angle{-190°} + 6\angle{50°} = 0$

(c) $\bar{V}_{ab} = \sqrt{3}\,\bar{V}_{an}\angle{30°} = \sqrt{3}(120\angle{-40°})\angle{30°} = 207.84\angle{-10°}$ V

$$\bar{V}_{bc} = \bar{V}_{ab}\angle{-120°} = (207.84\angle{-10°})\angle{-120°} = 207.84\angle{-130°} \text{ V}$$

$$\bar{V}_{ca} = \bar{V}_{ab}\angle{120°} = 207.84\angle{110°} \text{ V}$$

The main conclusions of the wye-connected load are the following:

1. Magnitude of the line-to-line voltage is greater than the phase voltage by a factor of $\sqrt{3}$.
2. Line-to-line voltage $\bar{V}_{ab}$ leads the phase voltage of the load $\bar{V}_{an}$ by 30°. Similar conclusions can be made for the other two phases.
3. Line currents are equal to the phase currents of the load.
4. No current flows through the neutral wire, so there is no need for a wire between the neutral points in Figure 8.14.

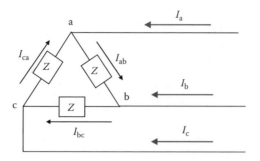

FIGURE 8.17   Three-phase load connected in delta.

8.2.4 DELTA-CONNECTED BALANCED LOAD

A balanced three-phase load connected in delta is shown in Figure 8.17. The load impedance of each phase is connected between two lines. Hence, the voltage across any load impedance is the line-to-line voltage. Although each load impedance is connected between two lines, it is still called phase impedance. There is no line-to-line impedance.

The line currents feeding the load are $\bar{I}_a$, $\bar{I}_b$, and $\bar{I}_c$. The currents of the load impedances are $\bar{I}_{ab}$, $\bar{I}_{bc}$, and $\bar{I}_{ca}$, which are known as the phase currents (they are the currents of the phase impedances). The relationship between the line and phase currents can be obtained by examining Kirchhoff's nodal equations at points a, b, and c.

$$\bar{I}_{ab} = \bar{I}_a + \bar{I}_{ca}$$
$$\bar{I}_{bc} = \bar{I}_b + \bar{I}_{ab} \qquad (8.17)$$
$$\bar{I}_{ca} = \bar{I}_c + \bar{I}_{bc}$$

The phase currents can be computed by dividing the voltage across the phase impedance by the phase impedance. However, the voltage across the phase impedance is equal to the line-to-line voltage. Thus,

$$\bar{I}_{ab} = \frac{\bar{V}_{ab}}{\bar{Z}}$$
$$\bar{I}_{bc} = \frac{\bar{V}_{bc}}{\bar{Z}} \qquad (8.18)$$
$$\bar{I}_{ca} = \frac{\bar{V}_{ca}}{\bar{Z}}$$

Since the impedances of all phases are equal and the line-to-line voltages are balanced, the phase currents must also be balanced.

EXAMPLE 8.3

The phase voltage of a balanced three-phase source is $\bar{V}_{an} = 120\angle{-40°}$ V. The source is energizing a three-phase load connected in delta. The phase impedance of the load is $\bar{Z} = 10\angle{-30°}$ Ω. Compute the phase currents of the load.

Solution

The first step is to compute the voltages across the load, which are the same as the line-to-line voltages of the source.

$$\bar{V}_{ab} = \sqrt{3}\,\bar{V}_{an}\angle{30°} = \sqrt{3}(120\angle{-40°})\angle{30°} = 207.84\angle{-10°} \text{ V}$$

$$\overline{V}_{bc} = \overline{V}_{ab}\angle{-120°} = 207.84\angle{-130°} \text{ V}$$

$$\overline{V}_{ca} = \overline{V}_{ab}\angle{120°} = 207.84\angle{110°} \text{ V}$$

Now compute the phase currents.

$$\overline{I}_{ab} = \frac{\overline{V}_{ab}}{\overline{Z}} = \frac{207.84\angle{-10°}}{10\angle{-30°}} = 20.784\angle{20°} \text{ A}$$

$$\overline{I}_{bc} = \frac{\overline{V}_{bc}}{\overline{Z}} = \frac{207.84\angle{-130°}}{10\angle{-30°}} = 20.784\angle{-100°} \text{ A}$$

$$\overline{I}_{ca} = \frac{\overline{V}_{ca}}{\overline{Z}} = \frac{207.84\angle{110°}}{10\angle{-30°}} = 20.784\angle{140°} \text{ A}$$

Note that the phase currents are equal in magnitude and are separated by 120° from one another.

As seen in Example 8.3, the phase currents are balanced. Let us rewrite Equation 8.17 as given below.

$$\overline{I}_a = \overline{I}_{ab} - \overline{I}_{ca}$$
$$\overline{I}_b = \overline{I}_{bc} - \overline{I}_{ab} \tag{8.19}$$
$$\overline{I}_c = \overline{I}_{ca} - \overline{I}_{bc}$$

Let us assume that the phase current $\overline{I}_{ab}$ is the chosen reference. Then,

$$\overline{I}_{ab} = I_{ph}\angle{0}$$
$$\overline{I}_{bc} = \overline{I}_{ab}\angle{-120°} = I_{ph}\angle{-120°} \tag{8.20}$$
$$\overline{I}_{ca} = \overline{I}_{ab}\angle{120°} = I_{ph}\angle{120°}$$

where I_{ph} is the magnitude of the phase current. Now substitute the values of the phase currents in Equation 8.20 into Equation 8.19.

$$\overline{I}_a = I_{ph}\angle{0} - I_{ph}\angle{120°} = \sqrt{3}\,I_{ph}\angle{-30°} = \sqrt{3}\,\overline{I}_{ab}\angle{-30°}$$
$$\overline{I}_b = I_{ph}\angle{-120°} - I_{ph}\angle{0} = \sqrt{3}\,I_{ph}\angle{-150°} = \sqrt{3}\,\overline{I}_{bc}\angle{-30°} \tag{8.21}$$
$$\overline{I}_c = I_{ph}\angle{120°} - I_{ph}\angle{-120°} = \sqrt{3}\,I_{ph}\angle{90°} = \sqrt{3}\,\overline{I}_{ca}\angle{-30°}$$

As seen in Equation 8.21, the line current $\overline{I}_a$ is greater than the phase current $\overline{I}_{ab}$ by a factor of $\sqrt{3}$ and lags the phase current by 30°. Similar relationships can be obtained for the other two phases.

EXAMPLE 8.4

Calculate the line currents in Example 8.3. Also sketch the phasor diagram showing the line-to-line voltages, phase currents, and line currents.

Solution

$$\overline{I}_a = \sqrt{3}\,\overline{I}_{ab}\angle{-30°} = \sqrt{3}(20.784\angle{20°})\angle{-30°} = 36\angle{-10°} \text{ A}$$

$$\overline{I}_b = \sqrt{3}\,\overline{I}_{bc}\angle{-30°} = \sqrt{3}(20.784\angle{-100°})\angle{-30°} = 36\angle{-130°} \text{ A}$$

$$\overline{I}_c = \sqrt{3}\,\overline{I}_{ca}\angle{-30°} = \sqrt{3}(20.784\angle{140°})\angle{-30°} = 36\angle{110°} \text{ A}$$

The phasor diagram is sketched in Figure 8.18.

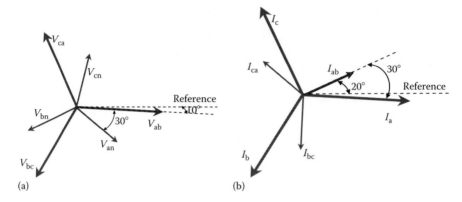

FIGURE 8.18 Phasor diagram of Example 8.4. (a) Phasor diagram of voltages and (b) phasor diagram of currents.

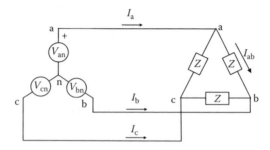

FIGURE 8.19 Circuit with wye source and delta load.

The main conclusions of the delta-connected load are the following:

1. Magnitude of the line-to-line voltage is equal to the voltage across the load.
2. Line current $\bar{I}_a$ lags the phase current $\bar{I}_{ab}$ by 30°. A similar conclusion can be made for the other two phases.
3. Line current is greater than the phase current by a factor of $\sqrt{3}$.
4. Delta connection has no neutral terminal.

8.2.5 CIRCUITS WITH MIXED CONNECTIONS

The source and the load are not always connected in the same fashion; the load could be connected in delta and the source in wye, or vice versa. In either case, careful attention must be given to the line and phase quantities. Examine Example 8.5 carefully.

EXAMPLE 8.5

Figure 8.19 shows a balanced delta-connected load energized by a balanced wye-connected source. The phase voltage of the source is $\bar{V}_{an} = 120\angle 0°$ V, and the line current is $\bar{I}_c = 10\angle 75°$ A. Compute the load impedance.

Solution

To compute the impedance, we need to compute the voltage across the load and the current of the load. The voltage across the load is

$$\bar{V}_{ab} = \sqrt{3}\,\bar{V}_{an}\angle 30° = \sqrt{3}(120\angle 0°)\angle 30° = 207.84\angle 30° \text{ V}$$

The current of the load can be computed by using the relationship

$$\bar{I}_a = \sqrt{3}\,\bar{I}_{ab}\angle{-30°}$$

where

$$\bar{I}_a = \bar{I}_c\angle{-120°} = (10\angle75°)\angle{-120°} = 10\angle{-45°}\text{ A}$$

Hence,

$$\bar{I}_{ab} = \frac{10\angle{-45°}}{\sqrt{3}\angle{-30°}} = 5.774\angle{-15°}\text{ A}$$

and

$$\bar{Z} = \frac{\bar{V}_{ab}}{\bar{I}_{ab}} = \frac{207.84\angle30°}{5.774\angle{-15°}} = 36\angle45°\text{ }\Omega$$

The loads in three-phase circuits can be connected in parallel even if their configurations are not the same. An example is shown in Figure 8.20 where the source is wye-connected, one of the loads is delta-connected, and the other is wye-connected. This circuit can be easily analyzed by using the superposition theorem. For example, to solve for the line current $\bar{I}_a$, you first compute the contribution of each load to the line current independently. Then sum up the two contributions to find the line current due to both loads. Hence, the circuit in Figure 8.20 can be divided into the two subcircuits in Figures 8.21 and 8.22. In Figure 8.21, the contribution of the wye-connected load to the total line current is $\bar{I}_{a1}$. In Figure 8.22, the contribution of the delta-connected load to the total line current is $\bar{I}_{a2}$. The line current due to both loads is just the sum of $\bar{I}_{a1}$ and $\bar{I}_{a2}$.

$$\bar{I}_a = \bar{I}_{a1} + \bar{I}_{a2} \tag{8.22}$$

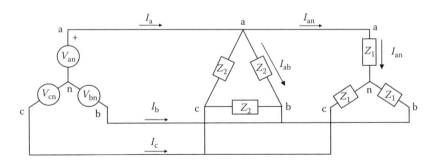

FIGURE 8.20 Circuit with wye and delta loads.

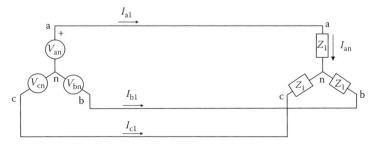

FIGURE 8.21 Contribution of the wye load.

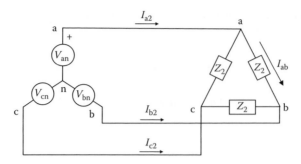

FIGURE 8.22 Contribution of the delta load.

EXAMPLE 8.6

Assume that the source voltage in Figure 8.20 is $\overline{V}_{an} = 120\angle0°$ V, the phase impedance of the wye load is $\overline{Z}_1 = 10\angle20°$ Ω, and the phase impedance of the delta load is $\overline{Z}_2 = 15\angle-30°$ Ω. Compute the line currents of the source.

Solution

First find the contribution of the wye load to the line current.

$$\overline{I}_{a1} = \overline{I}_{an} = \frac{\overline{V}_{an}}{\overline{Z}_1} = \frac{120\angle0}{10\angle20°} = 12\angle-20° \text{ A}$$

Now let us compute the contribution of the delta load $\overline{I}_{a2}$.

$$\overline{I}_{ab} = \frac{\overline{V}_{ab}}{\overline{Z}_2} = \frac{\sqrt{3}\,\overline{V}_{an}\angle30°}{15\angle-30°} = \frac{\sqrt{3}(120\angle0)\angle30°}{15\angle-30°} = 13.85\angle60° \text{ A}$$

$$\overline{I}_{a2} = \sqrt{3}\,\overline{I}_{ab}\angle-30° = \sqrt{3}(13.85\angle60°)\angle-30° = 24\angle30° \text{ A}$$

The total line current at the source side is

$$\overline{I}_a = \overline{I}_{a1} + \overline{I}_{a2} = 12\angle-20° + 24\angle30° = 33\angle13.84° \text{ A}$$

8.2.6 WYE–DELTA TRANSFORMATION

When the source and the load are both in wye or delta, the analysis of the circuit is simple. However, when the connections are different, the process is more involved as you must be attentive to the phase shifts and the changes in magnitude between the line and phase quantities. An alternative method is to find an equivalent load connection that matches that of the source. For example, if the source is wye-connected and the load is delta-connected, it would be easier to analyze the circuit if the delta load is replaced by an equivalent wye load. This is known as the wye–delta transformation.

Let us consider the delta load in Figure 8.23 where the phase impedance of the load is labeled $\overline{Z}_\Delta$. Our objective is to find its equivalent wye load shown in Figure 8.24. The phase impedance of the equivalent wye is $\overline{Z}_Y$. But how do we find the value of $\overline{Z}_Y$ that makes the two circuits equivalent? The simplest way is to treat both circuits as an impedance box with three terminals. The two circuits are equivalent if the voltage applied between any two terminals in both circuits produces the same terminal currents. This means the impedance between any two terminals in the delta circuit is equal to the impedance between the same terminals in the wye circuit. For example, the impedance measured between terminals a and b, $\overline{Z}_{ab}$, must be the same for both circuits.

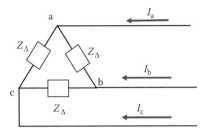

FIGURE 8.23   Delta load.

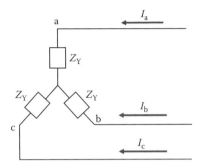

FIGURE 8.24   Wye load equivalent to the delta load in Figure 8.23.

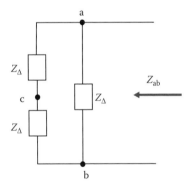

FIGURE 8.25   Terminal impedance of delta load.

$\overline{Z}_{ab}$ of the delta-connected load can be quickly computed by rearranging the impedances in Figure 8.23 as shown in Figure 8.25. It is easy to see that the impedance measured between terminals a and b is a parallel combination of $\overline{Z}_\Delta$ and $2\overline{Z}_\Delta$ as given in Equation 8.23. Keep in mind that the third terminal c is unconnected.

$$\overline{Z}_{ab} = \frac{\overline{Z}_\Delta(2\overline{Z}_\Delta)}{3\overline{Z}_\Delta} = \frac{2}{3}\overline{Z}_\Delta \tag{8.23}$$

Now let us do the same with the equivalent wye load. Rearrange the wye load as shown in Figure 8.26. Since the third terminal c is floating, the impedance between terminals a and b is a series combination of two $\overline{Z}_Y$ as given in Equation 8.24.

$$\overline{Z}_{ab} = 2\overline{Z}_Y \tag{8.24}$$

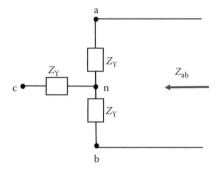

FIGURE 8.26   Terminal impedance of wye load.

To make the wye load equivalent to the original delta load, the terminal impedances of both circuits must be equal, that is Equations 8.23 and 8.24 are equal. Hence,

$$\overline{Z}_Y = \frac{\overline{Z}_\Delta}{3} \tag{8.25}$$

The relationship in Equation 8.25 is valid for transforming either wye into delta or delta into wye.

EXAMPLE 8.7

Repeat Example 8.6 using the wye–delta transformation.

Solution

Convert the delta load in Figure 8.20 to wye.

$$\overline{Z}_Y = \frac{\overline{Z}_\Delta}{3} = \frac{\overline{Z}_2}{3} = \frac{15\angle-30°}{3} = 5\angle-30° \ \Omega$$

After the delta is converted into wye, we have two wye-connected loads in parallel as shown in Figure 8.27. These two loads can be replaced by one equivalent wye load as shown in Figure 8.28. The new value of the equivalent load impedance is

$$\overline{Z}_t = \frac{\overline{Z}_Y \overline{Z}_1}{\overline{Z}_Y + \overline{Z}_1} = \frac{(5\angle-30°)(10\angle20°)}{(5\angle-30°) + (10\angle20°)} = 3.6343\angle-13.84° \ \Omega$$

Now, the line current can be directly computed as

$$\overline{I}_a = \frac{\overline{V}_{an}}{\overline{Z}_t} = \frac{120\angle0}{3.6343\angle-13.84°} = 33\angle13.84° \ A$$

This process is obviously much simpler than that in Example 8.6.

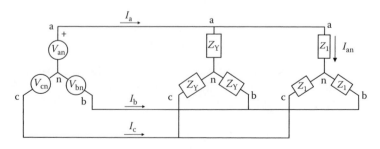

FIGURE 8.27   Equivalent circuit of the system in Figure 8.20.

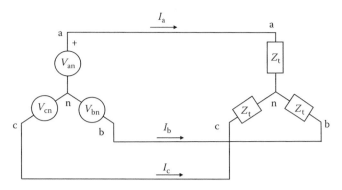

FIGURE 8.28 Equivalent wye load.

8.3 POWER CALCULATIONS OF BALANCED THREE-PHASE CIRCUITS

The power in a three-phase circuit is the sum of the powers of each phase. The power per phase is

$$
\begin{aligned}
P_{ph} &= V_{ph}I_{ph}\cos\theta \\
Q_{ph} &= V_{ph}I_{ph}\sin\theta
\end{aligned}
\tag{8.26}
$$

where
P_{ph} is the real power consumed by the phase impedance
Q_{ph} is the reactive power consumed by the phase impedance
V_{ph} is the phase voltage (the voltage across the phase impedance)
I_{ph} is the phase current (the current in the phase impedance)
θ is the power factor angle (the angle between the phase voltage and the phase current) θ is also the impedance angle

Since the system is balanced, the three-phase power (P) is just three times the power of a single phase.

$$
P = 3P_{ph} = 3V_{ph}I_{ph}\cos\theta
\tag{8.27}
$$

A similar equation can be obtained for reactive power.

$$
Q = 3Q_{ph} = 3V_{ph}I_{ph}\sin\theta
\tag{8.28}
$$

8.3.1 THREE-PHASE POWER OF BALANCED WYE LOADS

When the load is connected in wye, the line current I_l is the same as the phase current I_{ph}, and the line-to-line voltage V_{ll} is greater than the phase voltage V_{ph} by a factor of $\sqrt{3}$. Hence,

$$
P = 3P_{ph} = 3V_{ph}I_{ph}\cos\theta = 3\frac{V_{ll}}{\sqrt{3}}I_l\cos\theta
\tag{8.29}
$$

Rewriting Equation 8.29 yields

$$
P = \sqrt{3}\,V_{ll}I_l\cos\theta
\tag{8.30}
$$

Similarly, reactive power is

$$
Q = \sqrt{3}\,V_{ll}I_l\sin\theta
\tag{8.31}
$$

Keep in mind that θ is the angle between the phase voltage and the phase current of the load.

8.3.2 THREE-PHASE POWER OF BALANCED DELTA LOADS

The voltage across the impedance of the delta load is the line-to-line voltage, and the line current is greater than the phase current by a factor of $\sqrt{3}$. Hence,

$$P = 3P_{ph} = 3V_{ph}I_{ph}\cos\theta = 3V_{ll}\frac{I_l}{\sqrt{3}}\cos\theta \tag{8.32}$$

Equation 8.32 can be rewritten as

$$P = \sqrt{3}\,V_{ll}I_l\cos\theta \tag{8.33}$$

Similarly, the three-phase reactive power is

$$Q = \sqrt{3}\,V_{ll}I_l\sin\theta \tag{8.34}$$

Note that Equations 8.30 and 8.33 are identical. Also, Equations 8.31 and 8.34 are identical. Therefore, the power of a three-phase circuit can be computed by the same equation regardless of the load connection. Keep in mind that θ is the angle between the phase voltage and the phase current of the load in both cases. In other words, θ is the angle of the load impedance.

EXAMPLE 8.8

A three-phase source of $\overline{V}_{an} = 120\angle 0°$ V is powering a delta-connected load of $\overline{Z} = 15\angle 20°$ Ω. Compute the real and reactive powers consumed by the load.

Solution

Compute the magnitude of the line-to-line voltage V_{ll}.

$$V_{ll} = \sqrt{3}\,V_{an} = 207.84 \text{ V}$$

Now compute the line current, but first calculate the magnitude of the phase current I_{ab}.

$$I_{ab} = \frac{V_{ab}}{Z} = \frac{V_{ll}}{Z} = \frac{207.84}{15} = 13.85 \text{ A}$$

$$I_l = \sqrt{3}\,I_{ab} = \sqrt{3}\,13.85 = 24 \text{ A}$$

The phase angle of the load is given as 20°. Hence, the real and reactive powers are

$$P = \sqrt{3}\,V_{ll}I_l\cos\theta = \sqrt{3}\times 207.84 \times 24 \times \cos 20° = 8.12 \text{ kW}$$

$$Q = \sqrt{3}\,V_{ll}I_l\sin\theta = \sqrt{3}\times 207.84 \times 24 \times \sin 20° = 2.95 \text{ kVAr}$$

EXAMPLE 8.9

A three-phase source of $\overline{V}_{an} = 120\angle 0°$ V delivers 10 kW and 5 kVAr to a wye-connected load. Compute the load impedance.

Solution

Compute the power factor angle θ.

$$\theta = \tan^{-1}\left(\frac{Q}{P}\right) = \tan^{-1}\left(\frac{5}{10}\right) = 26.56°$$

Compute the line-to-line voltage

$$V_{ll} = \sqrt{3}\, V_{an} = 207.84 \text{ V}$$

Use the power equation to compute the line current.

$$P = \sqrt{3}\, V_{ll} I_1 \cos\theta$$
$$10,000 = \sqrt{3} \times 207.84 \times I_1 \times \cos 26.56°$$
$$I_1 = 31.05 \text{ A}$$

The line current is the same as the phase current in wye loads. Now compute the magnitude of the load impedance.

$$Z = \frac{V_{ph}}{I_{ph}} = \frac{120}{31.05} = 3.86 \ \Omega$$

The complex value of the load impedance is

$$\bar{Z} = Z(\cos\theta + j\sin\theta) = 3.86(\cos 26.56 + j\sin 26.56) = 3.45 + j1.73 \ \Omega$$

EXERCISES

1. A wye-connected balanced three-phase source is feeding a balanced three-phase load. The phase voltage and the phase current of the source are

$$v(t) = 340 \sin(377t + 0.5236) \text{ V}$$

$$i(t) = 100 \sin(377t + 0.87266) \text{ A}$$

Calculate the following:
(a) rms Phase voltage
(b) rms Line-to-line voltage
(c) rms Phase current
(d) rms Line current
(e) Frequency of the supply
(f) Power factor at the source side, state leading or lagging
(g) Three-phase real power delivered to the load
(h) Three-phase reactive power delivered to the load
(i) Load impedance, if the load is connected in delta configuration

2. The current and voltage of a wye-connected load are $\bar{V}_{ca} = 480\angle -60°$ V, $\bar{I}_b = 20\angle 120°$ A. Compute the following:
(a) $\bar{V}_{an}$
(b) $\bar{I}_a$
(c) Power factor angle
(d) Real power of the load

3. A three-phase, 480 V system is connected to a balanced three-phase load. The line current I_a is 10 A and is in phase with the line-to-line voltage V_{bc}. Calculate the impedance of the load for the following cases:
   (a) If the load is wye-connected
   (b) If the load is delta-connected
4. The waveforms of the line-to-line voltage and line current of a three-phase delta-connected load are

$$v_{ll} = V_{max} \sin \omega t$$
$$i_l = I_{max} \sin(\omega t - 50°)$$

   Calculate the power factor angle.
5. A balanced wye-connected load with a per-phase impedance of $4 + j3 \ \Omega$ is connected across a three-phase source of 173 V (line to line).
   (a) Find the line current, power factor, complex power, real power, and the reactive power consumed by the load.
   (b) With V_{ab} as the reference phasor, sketch the phasor diagram that shows all voltages and currents.
6. Two three-phase wye-connected loads are in parallel across a three-phase supply. The first load draws a phase current of 20 A at 0.9 power factor leading, and the second load draws a phase current of 30 A at 0.8 power factor lagging. Calculate the following:
   (a) Line current from the source side and its power factor
   (b) Real power supplied by the source if the supply voltage is 400 V
7. The line-to-line voltage of a three-phase system is $\overline{V}_{bc} = 340 \angle 20° $ V.
   (a) Calculate the phase voltage $\overline{V}_{an}$.
   (b) If the load impedance is connected in wye and the impedance per phase is $Z = 10 \angle 60° \ \Omega$, calculate the current $\overline{I}_b$.
   (c) Calculate the current in the neutral line $\overline{I}_n$.
8. The three-phase motor is rated at 5.0 hp. What is the power of the motor per phase in kilowatt?
9. The waveform of an ac voltage can be expressed by $V = 180 \sin(300t + 3)$.
   Calculate the following:
   (a) rms Value of the voltage
   (b) Frequency of the supply
   (c) Phase shift in degrees
10. Following are the voltage and current measured for a wye-connected load.

$$\overline{V}_{ab} = 200 \angle 50°$$
$$\overline{I}_c = 10 \angle 140°$$

   (a) Calculate the power factor angle.
   (b) Calculate the real power consumed by the load.
11. A delta-connected source energizes two parallel loads. One of the loads is connected in delta and the other in wye. The line-to-line voltage of the source is 208 V. The delta load has a phase impedance of $\overline{Z}_\Delta = 10 \angle -25° \ \Omega$. The wye load has a phase impedance of $\overline{Z}_Y = 5 \angle 40° \ \Omega$. Compute the line current of phase a.
12. A three-phase wye-connected source energizes a delta-connected load. The phase voltage of the source is $\overline{V}_{bn} = 120 \angle 0° $ V and the phase impedance of the load is $\overline{Z} = 9 \angle 30° \ \Omega$. Compute $\overline{I}_a$ and the power consumed by the delta load.

9 Electric Safety

Electricity is one of the best forms of energy known to man; it is clean, readily available, quiet, and highly reliable. Equipment powered by electricity is pollution free and is more compact than those powered by other energy forms such as gas or oil. Because of the overwhelming advantages of electricity, it became widely available by just flipping a switch.

Electricity is completely safe if used properly. However, when the electric equipment is partially damaged or improperly used, hazardous conditions can develop. For example, loose-fitting plugs can overheat and lead to fire, water intrusion inside a plugged-in appliance can lead to electrical shock, and cracks or damage in electrical insulations can lead to fires or electrical shocks.

Since the early days of the electrical revolution, electricity has been recognized as hazardous to humans and animals. Today, even with safe products in the market, more than 1000 people are killed each year in the United States alone due to electric shocks and several thousands more are injured. The popular myth that only high voltages are dangerous to humans makes some of us unfortunately careless when we use regular household equipment. Indeed, most of the electric shocks occur at the low household voltage level.

The potential hazards of electricity make the manufacturers of electric equipment very sensitive to the safety of their products. Several regulations and standards are developed to address every electric safety issue known to man so far. In the United States, Occupational Safety and Health Administration (OSHA) enforces the basic safety standards. On a more global level, the Institute of Electrical and Electronics Engineers (IEEE) sets several standards for various electric safety issues that are normally adopted by OSHA and other similar agencies all over the world.

9.1 ELECTRIC SHOCK

Electric shocks occur when people become part of electrical circuits and accordingly electrical currents flow through their bodies. The electric current can cause a wide range of harmful effects on humans and animals ranging from minor sensations to death. The current flowing inside a body can overheat the cells leading to internal and external burns. The most sensitive organs to this effect are the lungs, brain, and heart. The degree of injuries depends on several factors such as the magnitude of the current, the duration of the shock, the pathway of the current, and its frequency. These factors are discussed in more detail in Section 9.1.2.

Generally, electric shocks are divided into two categories: *secondary shocks* and *primary shocks*. The secondary shock is due to low currents that may cause pain without direct physical harm. The primary shock, however, is due to higher currents that produce direct physical harm or death.

Most of us probably experience the secondary shock when we use some types of appliance or equipment. Sometimes, when we lightly touch a charged object, we may experience tingling effects due to the small current passing through our fingers. If we grip that object, the current is spread out over a wide contact area and we may not feel anything.

The incredible question is how to find out the safe limit of electrical current; it is hard to find volunteers for primary shock experiments. Early researchers, however, carried out their secondary shock studies on human volunteers and the primary shock experiments on animals with similar weight and general biological characteristics as humans. It was also reported that some bizarre research on primary shocks was done on inmates condemned to death.

FIGURE 9.1 Volunteer during secondary electric shock test. (From Dalziel, C.F., *IEEE Spectr.*, 9, 41–50, 1972. Courtesy of IEEE.)

Among the highly respected researchers in this area was Charles Dalziel from the University of California, Berkeley. Dalziel conducted extensive lab tests during the mid-twentieth century. In one of his papers published in 1972, Dalziel summarizes the results of his various lab tests, which until today are used as the de facto standard. In his study, he had several men and a few women volunteers. Figure 9.1 shows a photo of one of Dalziel's tests published in his paper. The caption of the figure states that "Muscular reaction at the let-go current value is increasingly severe and painful, as shown by the subject during laboratory tests." Today, many people find this type of test incredible.

9.1.1 CURRENT LIMITS OF ELECTRIC SHOCKS

The results of Dalziel's work as well as those of other researchers are summarized in Tables 9.1 and 9.2. Table 9.1 shows the levels of direct current (dc) and alternating current (ac) that produce various types of secondary shocks on men and women. Although secondary shock can be painful, it is not life threatening. Table 9.2 shows the effect of primary shocks and their corresponding ac limits. As seen in Table 9.1, the following conclusions regarding secondary shocks can be made:

- Women are more sensitive to electric shocks than men.
- It takes less of ac current than dc to produce the same effect on people.
- It takes as little as 1.2 mA for women and 1.8 mA for men to produce uncomfortable effects.
- For currents less than 6 mA in women and 9 mA in men, the person, while in pain, can still control his/her muscles.

TABLE 9.1

Effects of ac and dc Secondary Shock Current

| | Current (mA) | | | |
| | dc | | ac | |
Reaction	Men	Women	Men	Women
No sensation on hand	1.0	0.6	0.4	0.3
Tingling (threshold of perception)	5.2	3.5	1.1	0.7
Shock: uncomfortable, muscular control not lost	9.0	6.0	1.8	1.2
Painful shock, muscular control is not lost	62.0	41.0	9.0	6.0

Source: IEEE Standard 524a-1993, 1994.

The more critical statistics are the ones related to primary shock shown in Table 9.2. In this table, the population is divided into sensitive and average sets. The sensitive population, about 0.5% of the total population, is harmed at lower currents than the rest of the population. The average population forms 50% of the total population. The table shows three key effects for the primary shocks: let go, respiratory tetanus, and ventricular fibrillation.

For 50% of the male population, when a person grips an energized conductor and the current passing through his muscles is about 16 mA, the person may not be able to control his muscles and thus may not release the gripped conductor. In this case, the current passes through the person's body for as long as the circuit is energized. This is known as the let-go current.

A current above the let-go threshold passing through the chest can cause involuntary contraction of the muscles, which will arrest breathing as long as the current continues to flow. This is known as respiratory tetanus. If the current passing through the chest disturbs the heart's own electrical stimulation, the heart experiences an uncontrolled vibration and may even cease to beat. This is known as ventricular fibrillation.

9.1.2 FACTORS DETERMINING THE SEVERITY OF ELECTRIC SHOCKS

The harmful effects of electric shocks depend mainly on six factors:

1. Voltage level of the gripped or touched equipment
2. Amount of current passing through the person's body
3. Resistance of the person's body
4. Pathway of the current inside the body
5. Duration of the shock
6. Frequency of the source

TABLE 9.2

Threshold Limit of ac Primary Shock Current

| | Current (mA) | | | |
| | 0.5% of Population | | 50% of Population | |
Threshold	Men	Women	Men	Women
Let go: worker cannot release wire	9	6	16	10.5
Respiratory tetanus: breathing is arrested			23	15
Ventricular fibrillation: arrhythmia	100	67		

Source: IEEE Standard 1048-1990, 1991.

9.1.2.1 Effect of Voltage

Most electric shocks occur at 100–400 V because it is readily available to everyone, it is high enough to produce significant current in the body, and can cause muscles to contract tightly to the energized object.

A higher voltage, in kilovolts range, is even more lethal because it causes high currents to pass through the person, but the access to this voltage is normally limited to professionals such as linemen. The general public rarely comes in contact with high-voltage wires, unless a wire is downed, a person climbs a high object and touches the wire, etc. Therefore, the number of deaths from high voltage is much less than that from low voltage.

When a person touches a high-voltage wire, fierce involuntary muscle contractions may throw the person away from the hazard. However, it may cause the person to fall from the high elevation of the power line to the ground.

9.1.2.2 Effect of Current

Contrary to popular belief, it is the current, not voltage, that causes death. However, the current is a function of the voltage and the impedance of the body; the current is equal to the voltage across the body divided by the impedance of the body.

Electric currents can interfere with the normal operation of the heart and lungs causing the heart to beat out of step and the lungs to function irregularly. Moreover, electric currents passing through tissues produce thermal heat, which is a form of energy that is proportional to the square of the current. This thermal energy E can permanently damage tissues and organs in the body.

$$E = I^2 R t \tag{9.1}$$

where
I is the current in the body
R is the body resistance
t is the time duration of the current

9.1.2.3 Effect of Body Resistance

The higher the body resistance, the lower is the current flow through the person. The body resistance is highly nonlinear and is a function of several factors such as the hydration condition of the body, skin condition, and fat concentration. Palm resistance, for example, can range from 100 Ω to 1 MΩ depending on the skin condition of the person. Dry skin tends to have higher resistance, and sweat tends to lower the resistance. Nerves, arteries, and muscles are low in resistance, while bones, fat, and tendons are relatively high in resistance.

The IEEE established the ranges for body resistances shown in Table 9.3. These numbers can be used to roughly estimate the current through the body. However, the variability in human body

TABLE 9.3
Body Resistance in Ohms

Resistance	Hand to Hand		Hand to Feet
	Dry Condition	Wet Condition	Wet Condition
Maximum	13,500	1,260	1,950
Minimum	1,500	610	820
Average	4,838	865	1,221

Source: IEEE Standard 1048-1990, 1991.

resistance could make the results inaccurate for people with body resistances outside the specified range. Therefore, when designing electric safety equipment, the lower values of body resistance should always be used.

9.1.2.4   Effect of Current Pathway

A current passing through the skin is not as harmful as a current passing through vital organs. Fatal currents often pass through the heart, lungs, and brain. A mere 10 μA passing directly through the heart can cause cardiac arrest. At lower currents, the heart muscles could beat out of step resulting in insufficient blood being pumped throughout the body. A current in the spinal cord may also alter the respiratory control mechanism.

9.1.2.5   Effect of Shock Duration

The longer the duration of the shock, the higher is the likelihood of death. This is because the thermal heat inside the tissues is a form of energy and therefore proportional with time, as shown in Equation 9.1. Also, if the current interferes with the operation of the heart or lungs, the longer the duration, the longer is the chance for death from respiratory or cardiac arrest. When the current is above the let-go threshold, the person is incapable of releasing his/her grip on the wire and the shock duration is therefore long.

Charles Dalziel carried out research on the time–current relationship for primary shocks. He and other researchers obtained their results by experimenting on animals of weights and organ sizes similar to humans. Although inconclusive, the data of these experiments are the best available information to date. Based on these studies, Dalziel developed the following empirical current duration formula for ventricular fibrillation.

$$I = \frac{K}{\sqrt{t}} \tag{9.2}$$

where
  I is the ventricular fibrillation current (mA)
  t is the time duration of the current (s)
  K is a constant that depends on the weight of the test subject: for people weighing less than 70 kg (154 lb), $K = 116$; and for people weighing more than 70 kg, $K = 157$

Figure 9.2 is a graph of Dalziel's formula. As seen in the figure, it takes a very short time to induce ventricular fibrillation. For instance, if the current is about 80 mA, it takes about 4 s to kill a

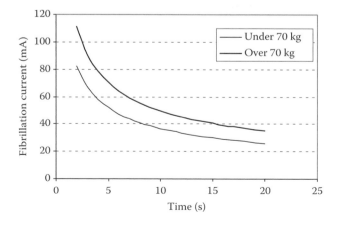

FIGURE 9.2   Ventricular fibrillation current as a function of shock duration.

person over 70 kg, while a person weighting less than 70 kg may survive for just 2 s. In either case, the survival duration is very short for such a small current.

9.1.2.6 Effect of Frequency

As seen in Table 9.1, ac current is more dangerous than dc. The current at power frequencies (50–60 Hz) has a much greater ability to cause ventricular fibrillation than dc. Also, at power frequencies, involuntary muscle contractions could be so severe that the individual cannot let go of the energized object. Higher frequencies in the gigahertz range (microwaves, X-rays, and gamma rays) are extremely dangerous as they are ionizing frequencies, which means they have enough energy to break molecules apart.

EXAMPLE 9.1

A child climbs a tree to retrieve his kite that is tangled on a utility line. The voltage of the line is 240 V. The resistance of the wire from the source to the location of the kite is 0.2 Ω. The tree has a resistance of 500 Ω. The soil resistance is 300 Ω. Assume that the child's resistance is 2000 Ω. If the child touches the power line with his hand, estimate the current through his body. Also, estimate the time it takes to induce ventricular fibrillation.

Solution

If you trace the path of the current starting from the source, you can obtain the electrical circuit shown in Figure 9.3. The current passing through the child starts at the source, passes through the line to the child, then to the tree, and finally to the ground. The current in this case is

$$I = \frac{V}{R_1 + R_b + R_t + R_g}$$
$$= \frac{240}{0.2 + 2000 + 500 + 300} = 85.7 \text{ mA}$$

According to Dalziel's formula, the child will survive only for

$$t = \left(\frac{K}{I}\right)^2 = \left(\frac{116}{85.7}\right)^2 = 1.83 \text{ s}$$

Note that we used $K = 116$, as the child is assumed to weigh less than 70 kg. Also notice that the wire resistance has very little effect on the current since it is much smaller than the other resistances. One important resistance is the ground resistance. It varies widely and can have a major effect on the value of the current.

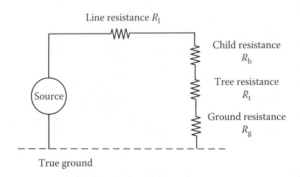

FIGURE 9.3   Model representation of the accident.

9.2 GROUND RESISTANCE

The center of the earth is the only absolute zero potential spot; any other location inside the earth has a nonzero potential. Hence, objects on the surface of the earth such as water pipes, building foundations, and steel structures have nonzero potentials. Figure 9.4 shows a simple schematic of the earth with an object at the earth's surface. The object, which is in contact with the soil, has a potential higher than zero. Hence, the potential difference V between the object and the center of the earth can be represented by Ohm's law $V = IR$, where I is the current flowing from the object to the center of the earth, and R is the resistance between the object and the center of the earth. This resistance R is called the ground resistance of the object.

The ground resistance of an object is nonlinear and is frequency dependent. It is a function of several factors such as the shape of the object, the type of ground soil, and the dampness of the soil. An exact computation of the ground resistance of an object is a very tedious task that requires the knowledge of highly varying parameters that are hard to measure under all possible conditions. Nevertheless, a simplified computation of the ground resistance of an object can be made with a reasonable degree of accuracy as shown below.

Assume the simple case in Figure 9.5 of a hemisphere buried in soil. Assume that the soil is homogeneous with a resistivity ρ. If the hemisphere is connected to a conductor that carries a current I, the current enters the hemisphere and is then dispersed uniformly through the earth. The current density at the surface of the hemisphere J is the current leaving the hemisphere divided by the surface area of the hemisphere.

$$J = \frac{I}{2\pi r^2} \tag{9.3}$$

Assuming the soil to be homogeneous, the current density at any point x from the center of the hemisphere and outside the hemisphere can be computed as

$$J(x) = \frac{I}{2\pi x^2}; \quad x \geq r \tag{9.4}$$

Ohm's law states that the current in a medium creates an electric field (EF) intensity E, which is equal to the current density multiplied by the resistivity of the medium. Hence, the EF intensity $E(x)$ at any distance x outside the hemisphere is

$$E(x) = \rho J(x); \quad x \geq r \tag{9.5}$$

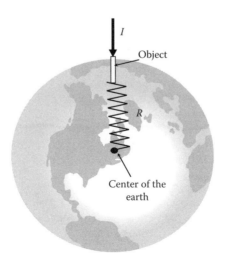

FIGURE 9.4 Ground resistance.

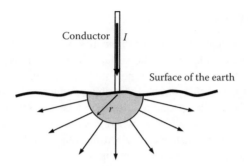

FIGURE 9.5 Current dispersion of a hemisphere.

For homogeneous soil, the potentials of all points located at the same distance from the center of the hemisphere are equal. The surface of these points is known as the *equipotential surface*, shown in Figure 9.6. The potential difference between any two points (such as a and b) inside the earth can be computed by integrating the EF intensity between these two points. Hence, V_{ab} in the figure is the potential difference between two equipotential surfaces where a and b are located. This voltage can be computed as

$$V_{ab} = \int_{x=a}^{x=b} E(x)\,dx = \int_{x=a}^{x=b} \rho J(x)\,dx = \frac{\rho I}{2\pi}\left[\frac{1}{r_a} - \frac{1}{r_b}\right] \tag{9.6}$$

where r_a and r_b are the distances of points a and b from the center of the hemisphere, respectively. Hence, the resistance between a and b is

$$R_{ab} = \frac{V_{ab}}{I} = \frac{\rho}{2\pi}\left[\frac{1}{r_a} - \frac{1}{r_b}\right] \tag{9.7}$$

The ground resistance of the hemisphere R_g is the resistance between the surface of the hemisphere and the center of the earth. This can be computed using Equation 9.7 by setting r_a equal to the radius of the hemisphere and r_b to ∞.

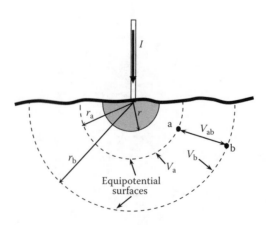

FIGURE 9.6 Equipotential surface.

TABLE 9.4

Ground Resistance of Common Objects

Object	Ground Resistance	Parameters
Rod	$\dfrac{\rho}{2\pi l}\ln\left(\dfrac{2l+r}{r}\right)$	l is the length of the rod r is the radius of the rod
Circular plate (disk) at the surface	$\dfrac{\rho}{4r}$	r is the radius of the disk
Buried wire	$\dfrac{\rho}{2\pi l}\left[\ln\left(\dfrac{l}{r}\right)+\ln\left(\dfrac{l}{2d}\right)\right]$	l is the length of the wire r is the radius of the wire d is the depth at which the wire is buried

TABLE 9.5

Soil Resistivity

	Soil Composition			
	Wet Organic	**Moist**	**Dry**	**Bedrock**
Resistivity ρ (Ωm)	10	100	1,000	10,000

$$R_g = \frac{\rho}{2\pi r} \tag{9.8}$$

The ground resistance can also be computed for other shapes; however, the process is more involved. Table 9.4 shows the ground resistance of some common objects. Table 9.5 shows the resistivity of various soil compositions. Note that the value of the resistivity is substantially reduced if the soil is wet or organic.

EXAMPLE 9.2

A hemisphere 2 m in diameter is buried in wet organic soil. Compute the ground resistance of the hemisphere. Also compute the ground resistance at 2, 10, and 100 m away from the hemisphere.

Solution

$$R_g = \frac{\rho}{2\pi r} = \frac{10}{2\pi \times 1} = 1.6\ \Omega$$

The resistance between two points can be computed using Equation 9.7.

$$R_{ab} = \frac{\rho}{2\pi}\left(\frac{1}{r_a} - \frac{1}{r_b}\right)$$

At 2 m:

$$R_{ab2} = \frac{10}{2\pi}\left(\frac{1}{1} - \frac{1}{2}\right) = 0.8\ \Omega$$

At 10 m:

$$R_{ab10} = \frac{10}{2\pi}\left(\frac{1}{1} - \frac{1}{10}\right) = 1.43\ \Omega$$

At 100 m:

$$R_{ab100} = \frac{10}{2\pi}\left(\frac{1}{1} - \frac{1}{100}\right) = 1.57\ \Omega$$

Figure 9.7 shows the ground resistance between the center of the hemisphere and points at various distances. Notice that the change in ground resistance is insignificant when the distance from the center of the hemisphere increases beyond 10 m. Therefore practically, the ground resistance of an object is a function of the immediate distance, rather than the distance to the center of the earth.

9.2.1 Measuring Ground Resistance

The ground resistance can be measured using a technique known as the fall-of-potential method. The method, which is also known as the three-point test, is explained in Figure 9.8. The setup consists of the object whose ground resistance is to be determined, a current electrode, and a potential probe. The electrode is a small copper rod that is driven into the soil. A voltage source is connected between the object and the current electrode. A voltmeter is connected between the object and the potential probe. The potential probe moves along the line between the object and the current electrode. At each position, the current (of the current electrode) and the voltage (of the potential probe) are recorded and plotted, as shown at the bottom of Figure 9.8. When the potential probe touches the object under test, the measured voltage is zero, and when it touches the current electrode, the voltage is equal to the source voltage. The magnitude of the measured voltage is a nonlinear function with respect to the distance x. However, the voltage is often unchanged for a wide range of x as shown in the flat region in the figure. At the middle of this region, V_f is recorded at the distance x_f. The ratio V_f/I is the ground resistance of the object.

9.2.2 Ground Resistance of People

A person standing on the ground has a ground resistance under his/her feet as shown in Figure 9.9. Each foot has a ground resistance R_f between the bottom of the foot and the center of the earth. The trick is how to model the bottom of a shoe. An approximation is often made by using a circular plate that has the same area as the footprint of an average person. Using the ground resistance equation of the circular plate given in Table 9.4, the ground resistance of a single foot R_f is

$$R_f = \frac{\rho}{4r} \tag{9.9}$$

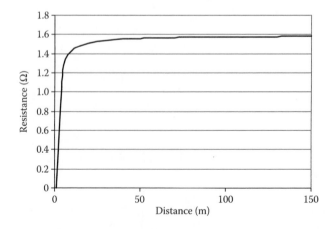

FIGURE 9.7   Ground resistance between the center of the hemisphere and points at various distances.

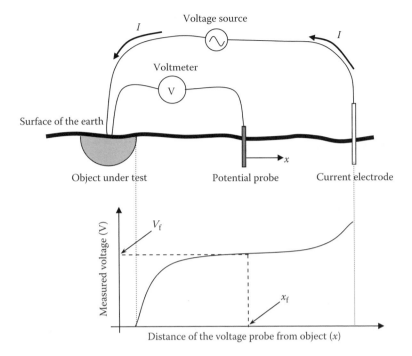

FIGURE 9.8   Fall-of-potential method to measure ground resistance.

The area of the circular plate is

$$A = \pi r^2 \tag{9.10}$$

Assuming an approximate area for a footprint of about 0.02 m², we can compute the ground resistance in Equation 9.9 as

$$R_f = \frac{\rho}{4\sqrt{\frac{A}{\pi}}} = \frac{\rho}{4\sqrt{\frac{0.02}{\pi}}} \approx 3\rho \tag{9.11}$$

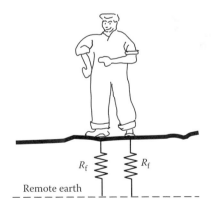

FIGURE 9.9   Feet resistance.

The total ground resistance of a walking person and standing person is different. If you assume that the standing person has his feet close enough to each other, then the total ground resistance R_g is the parallel combination of two R_f.

$$R_g = \frac{R_f \times R_f}{R_f + R_f} = 0.5\,R_f \approx 1.5\rho \tag{9.12}$$

9.3   TOUCH AND STEP POTENTIALS

The direct hazard of electricity is due to an energized object touching a person. However, there are indirect hazards that are less obvious but equally dangerous, such as that due to excessive touch and step potentials. The touch potential could be present if a person touches a metallic structure that attains a charge. If the person is standing on the ground, he may have a potential difference between his hand and feet. This voltage could be hazardous.

When a person is walking adjacent to an object that discharges current into the ground, the person could have a potential difference between his two feet. This voltage could also be hazardous.

9.3.1   TOUCH POTENTIAL

To explain the hazards of touch potentials, let us study a simple but well-known structure of the tower of the power line shown in Figure 9.10. The tower is built out of steel trusses with several cross arms. On the cross arms, insulators are mounted whereas on their other sides, high-voltage wires are attached. In areas known for their lightning activities, ground wires (also known as static wires or overhead ground wire OHGW) are installed on the top of the tower to protect the energized lines from being hit by lightning bolts, thus saving the system from severe damages and blackouts. The ground wires are often bonded to the tower structure, which is grounded through a local ground under the tower footage.

Although made of steel and grounded locally, the tower structure is not always at zero potential because the structure could carry current due to a number of reasons such as the following:

Ground wire discharge: Because of the presence of the energized lines in the vicinity of a ground wire, a voltage is induced on the ground wire. The ground wire discharges its acquired energy

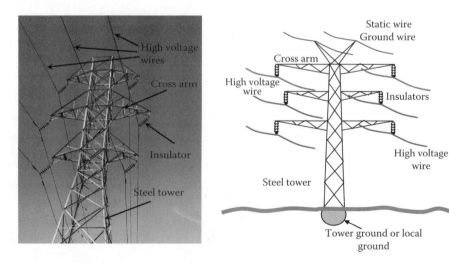

FIGURE 9.10   Main components of a steel power line.

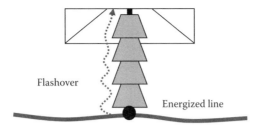

FIGURE 9.11   Insulation leakage or flashover due to conductive depositions on the insulator.

through all adjacent grounds including the tower ground. This may result in currents passing through the structure hardware to the local ground underneath the tower.

Insulator leakage: In humid environments, especially close to seas, the surface of an insulator could become contaminated with salty moisture. This mix can provide a path for the current from the energized line to the tower structure as shown in Figure 9.11. This is known as insulator leakage or flashover.

Accident: While linemen are working on power lines, accidents could happen when an energized conductor comes in contact with the structure of the tower.

Now assume that a person is touching the steel tower while standing on the ground as shown in Figure 9.12. The person's resistance and part of the tower resistance are in parallel. If the structure carries a current I, part of the current passes through the structure and goes to the ground of the tower I_{tg}, the other part goes through the man touching the tower I_{man}. Keep in mind that the ground resistance of the man is $0.5\ R_f$ as given in Equation 9.12.

Figure 9.13 shows a circuit diagram representing the case in Figure 9.12. Let us assume that we need to compute the current going through the man. This can be done using the current divider equation

$$I_{man} = I\left(\frac{R_g}{R_g + R_{man} + 0.5\ R_f}\right) \tag{9.13}$$

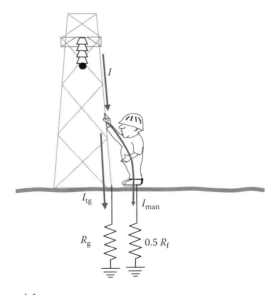

FIGURE 9.12   Touch potential.

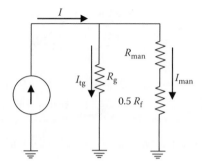

FIGURE 9.13 Circuit diagram of Figure 9.12.

The potential of the structure under the above condition is known as the ground potential rise (GPR). It is the voltage that the grounded object attains with respect to the remote earth.

$$\text{GPR} = I_{tg} R_g \qquad (9.14)$$

EXAMPLE 9.3

A power line insulator has partially failed and 10 A passes through the tower structure to the ground. Assume that the tower ground is a hemisphere with a radius of 0.5 m, and the soil surrounding the hemisphere is moist.

(a) Compute the voltage of the tower.
(b) Assume that a man with a body resistance of 3 kΩ touches the tower while standing on the ground. Compute the current passing through the man.
(c) Use Dalziel's formula and compute the man's survival time.

Solution

(a) The ground resistance of the hemisphere R_g is

$$R_g = \frac{\rho}{2\pi r} = \frac{100}{2\pi \times 0.5} = 32 \ \Omega$$

The voltage of the tower is the GPR of the tower.

$$V = IR_g = 10 \times 32 = 320 \text{ V}$$

(b) To compute the current through the man, first compute R_f.

$$R_f = 3\rho = 300 \ \Omega$$

The current through the man is given in Equation 9.13

$$I_{man} = I\left(\frac{R_g}{R_g + R_{man} + 0.5 \ R_f}\right) = 10\left(\frac{32}{32 + 3000 + 150}\right) = 100 \text{ mA}$$

(c) According to Dalziel's formula, the man can survive for

$$t = \left(\frac{K}{I_{man}}\right)^2 = \left(\frac{157}{100}\right)^2 = 2.5 \text{ s}$$

EXAMPLE 9.4

Repeat Example 9.3 assuming that a grounding rod is used instead of the hemisphere. Assume that the rod is 4 cm in diameter and is driven 1 m into ground.

Solution

(a) The ground resistance of the rod R_g is

$$R_g = \frac{\rho}{2\pi l} \ln\left(\frac{2l+r}{r}\right) = \frac{100}{2\pi \times 1} \ln\left(\frac{2+0.02}{0.02}\right) = 73.45 \ \Omega$$

The voltage of the tower is the GPR of the tower

$$V = IR_g = 10 \times 73.45 = 734.5 \ \text{V}$$

Notice that the GPR of a ground rod is much higher than the GPR of a hemisphere.

(b) $\qquad I_{\text{man}} = I\left(\frac{R_g}{R_g + R_{\text{man}} + 0.5R_f}\right) = 10\left(\frac{73.45}{73.45 + 3000 + 150}\right) = 227.9 \ \text{mA}$

(c) According to Dalziel formula, the man can survive for

$$t = \left(\frac{K}{I_{\text{man}}}\right)^2 = \left(\frac{157}{227.9}\right)^2 = 475 \ \text{ms}$$

If the ground rod is driven 1 m into the soil, it provides less protection than the hemisphere. Repeat the problem by assuming that the ground rod is driven 2 m into the soil. Can you develop a conclusion?

9.3.2 STEP POTENTIAL

A person walking adjacent to a structure that passes large current into ground could be vulnerable to electric shocks. This is particularly hazardous in substations, near power line towers that carry high currents, or during lightning storms.

Figure 9.14 shows a person walking near a structure that discharges current into ground. The ground current may cause a high enough potential difference between the person's feet resulting in current through his legs and abdomen. This current can be computed using Thevenin's theorem, where Thevenin's voltage V_{th} is the open circuit voltage between point a and b. This is the same

FIGURE 9.14   Current through a person due to step potential.

FIGURE 9.15 Thevenins's impedance.

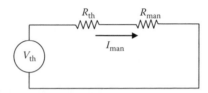

FIGURE 9.16 Equivalent circuit for the step potential.

voltage V_{ab} in Equation 9.6 for a hemisphere grounding object. r_a is the distance from the center of the hemisphere to the person's rear foot assuming the grounding hemisphere is behind the walking person. r_b is the distance from the center of the hemisphere to his front foot. Thevenin's resistance R_{th} is the open circuit resistance between point a and b (excluding the person) as shown in Figure 9.15.

$$R_{th} = 2R_f \tag{9.15}$$

Figure 9.16 shows the equivalent circuit for the step potential calculations. The body resistance of the person is the resistance of his legs and abdomen. The current through the man can be computed using Thevenin's theorem as given in Equation 9.16.

$$I_{man} = \frac{V_{th}}{R_{th} + R_{man}} = \frac{V_{th}}{2R_f + R_{man}} \tag{9.16}$$

EXAMPLE 9.5

A short circuit current of 1000 A passes through a hemisphere grounding object. A person is walking 5 m away from the center of the hemisphere. Assume that the leg-to-leg body resistance of the person is 2 kΩ, and the soil surrounding the hemisphere is moist. Compute the current through the person and his step potential.

Solution

For moist soil, $\rho = 100$ Ωm. Assume that the step of the person is about 0.6 m. Thevenin's voltage can be computed using Equation 9.6.

$$V_{th} = \frac{I\rho}{2\pi}\left(\frac{1}{r_a} - \frac{1}{r_b}\right) = \frac{1000 \times 100}{2\pi}\left(\frac{1}{5} - \frac{1}{5.6}\right) = 341 \text{ V}$$

$$R_f = 3\rho = 300 \ \Omega$$

The current through the man can be computed by Equation 9.16.

$$I_{man} = \frac{V_{th}}{2R_f + R_{man}} = \frac{341}{600 + 2000} = 131 \text{ mA}$$

The step voltage is the voltage between the person's feet

$$V_{step} = I_{man} \times R_{man} = 131 \times 2000 = 262 \text{ V}$$

EXAMPLE 9.6

Repeat Example 9.5 assuming that a grounding rod is used instead of the hemisphere. Assume the rod is inserted 1 m into ground.

Solution

Using the rod equation in Table 9.4, we can obtain Thevenin's voltage

$$V_{th} = \frac{I\rho}{2\pi l}\left[\ln\left(\frac{2l + r_a}{r_a}\right) - \ln\left(\frac{2l + r_b}{r_b}\right)\right] = \frac{1000 \times 100}{2\pi}\left[\ln\left(\frac{2+5}{5}\right) - \ln\left(\frac{2+5.6}{5.6}\right)\right] = 495 \text{ V}$$

Notice that the voltage on the ground is higher for the ground rod than for the hemisphere.
The current through the man is given in Equation 9.16.

$$I_{man} = \frac{V_{th}}{2R_f + R_{man}} = \frac{495}{600 + 2000} = 190 \text{ mA}$$

Notice that the ground rod provides less protection than the ground hemisphere.
The step potential of the man is

$$V_{step} = I_{man} \times R_{man} = 190 \times 2000 = 380 \text{ V}$$

As seen in Examples 9.5 and 9.6, the current through the person is very high. However, it may not be fatal since it does not pass through the heart, lungs, or brain. However, it could be painful enough to cause the person to fall on the ground, and then a lethal current can flow through his vital organs.

EXAMPLE 9.7

During a weather storm, an atmospheric discharge hits a lightning pole. The pole is grounded through a hemisphere, and the maximum lightning current through the pole is 20 kA.

(a) A person is playing golf 30 m away from the center of the hemisphere. The distance between his feet is 0.3 m, and his leg-to-leg resistance is 2 kΩ. Assume that the soil surrounding the hemisphere is moist. Compute the current through the person and his step potential.

(b) Another person is 3 m away from the center of the hemisphere. The distance between his feet is also 0.3 m, and his leg-to-leg resistance is 2 kΩ as well. Compute the current through the person and his step potential.

Solution

For moist soil, $\rho = 100~\Omega\text{m}$.

(a) Thevenin's voltage can be computed using Equation 9.6.

$$V_{\text{th}} = \frac{I\rho}{2\pi}\left[\frac{1}{r_{\text{a}}} - \frac{1}{r_{\text{b}}}\right] = \frac{20{,}000 \times 100}{2\pi}\left[\frac{1}{30} - \frac{1}{30.3}\right] = 105~\text{V}$$

$$R_{\text{f}} = 3\rho = 300~\Omega$$

The current through the man can be computed by Equation 9.16

$$I_{\text{man}} = \frac{V_{\text{th}}}{2R_{\text{f}} + R_{\text{man}}} = \frac{105}{600 + 2000} = 40.4~\text{mA}$$

The step voltage is the voltage between the person's feet

$$V_{\text{step}} = I_{\text{man}} \times R_{\text{man}} = 40.4 \times 2000 = 80.8~\text{V}$$

(b) For the person 3 m away

$$V_{\text{th}} = \frac{I\rho}{2\pi}\left(\frac{1}{r_{\text{a}}} - \frac{1}{r_{\text{b}}}\right) = \frac{20{,}000 \times 100}{2\pi}\left(\frac{1}{3} - \frac{1}{3.3}\right) = 9.646~\text{kV}$$

The current through the man can be computed by Equation 9.16

$$I_{\text{man}} = \frac{V_{\text{th}}}{2R_{\text{f}} + R_{\text{man}}} = \frac{9646}{600 + 2000} = 3.71~\text{A}$$

$$V_{\text{step}} = I_{\text{man}} \times R_{\text{man}} = 3.71 \times 2000 = 7.42~\text{kV}$$

The step voltage for the second person is extremely high, and it is unlikely the person can maintain his balance. If he falls on the ground, a lethal current could pass through his vital organs.

9.4 ELECTRIC SAFETY AT HOME

Before we discuss electric safety at home, let us examine the three most common power outlets shown in Figure 9.17. The outlets are also known as receptacles or sockets. The world has 13 different types of receptacles, but most of them are exclusively used by a few countries. Each of the outlets in the figure has three terminals: hot, neutral, and ground. The hot terminal in these sockets is connected to the hot side of the service transformer (100–240 V depending on the country). The neutral terminal is grounded at the service transformer. The ground terminal is grounded locally at the customer's area through ground rods or water pipes.

In North America, the hot and neutral terminals accept plugs with flat prongs, and the ground terminal accepts a rounded prong. Also, the opening of the neutral terminal is often wider than that for the hot terminal. This is known as a polarized socket. The wires of the terminals are color coded; hot is black, neutral is white, and ground is either green or uninsulated.

For type 2, the socket has two ground terminals; one at the top of the socket and the other at the bottom. The hot and neutral terminals are round shaped. In the British system, type 3, all terminals are flat.

Household plugs come in various shapes such as the ones in Figure 9.18 for type-1 socket and in Figure 9.19 for type-2 receptacles. In Figure 9.18a, the plug is unpolarized with two equal-width prongs; therefore either prong can be connected to the neutral terminal of the outlet. In Figure 9.18b,

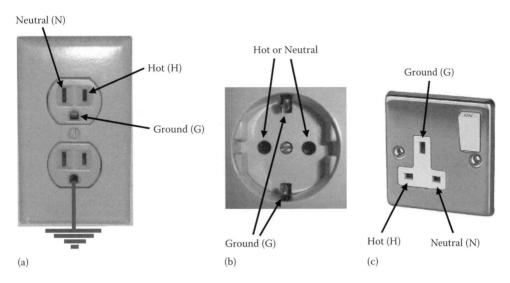

FIGURE 9.17 Household power outlets in two regions. (a) Type 1: most of America, (b) Type 2: most of Europe, and (c) Type 3: England, parts of middle east, and Asia.

the plug has two prongs as well, but it is polarized where the neutral prong is wider than the hot prong. This way, the neutral side of the plug can only fit the neutral side of the outlet. Figure 9.18c shows the three prongs plugs where two of them are flat and the third is rounded. The ground prong is longer than the other two. In this configuration, when the plug is inserted into the outlet, the ground terminal is connected first. Also, in this configuration, the neutral and hot terminals cannot be interchanged.

The international plug versions are shown in Figure 9.19. In Figure 9.19a, the prongs are rounded and unpolarized. In Figure 9.19b, the ground terminals are embedded on both sides of the

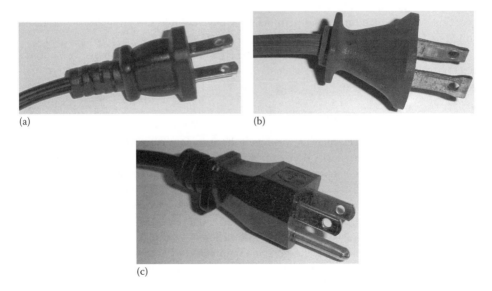

FIGURE 9.18 Common household plugs in North America. (a) Type 1: unpolarized, (b) Type 1: polarized, and (c) Type 1: three prongs.

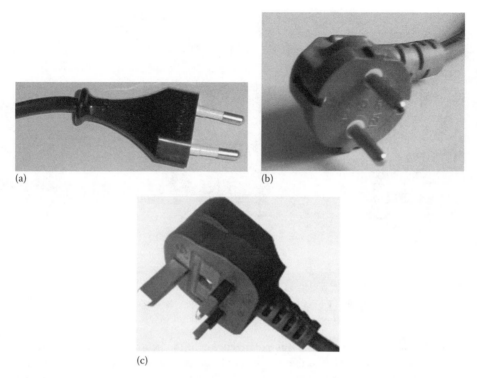

FIGURE 9.19 Common international household plugs. (a) Type 2: unpolarized, (b) Type 2: two prongs plus ground, and (c) Type 3: three prongs.

plug. This way, when the plug is inserted into the socket either straight or upside down, the ground connection is established. Notice that type-2 socket is deep. This way when the plug is inserted into the socket, the ground connection is established before any of the neutral or hot prongs are inserted into the socket. In Figure 9.19c, the ground prong is longer than the other two to ensure that the ground terminal of the plug is connected before its other two terminals. In some type-3 receptacles, the neutral and hot terminals are blocked until the ground terminal of the plug is inserted first.

The reasons for these arrangements may not be obvious at first. On this basis, two questions are raised: (1) why do we have polarized terminals? and (2) why do we have two ground terminals? The answers become apparent next.

9.4.1 Neutral versus Ground

One of the most confusing issues in electric safety is the difference between the neutral and the ground terminals. Both are grounded, so why do we sometimes use both of them? To answer this question, let us examine the generic representation of the electric equipment shown in Figure 9.20. The equipment consists of an internal electric circuit housed inside a chassis. The internal circuit is electrically isolated from the chassis. If the chassis is metallic (conductive), three problems could occur:

1. Any two adjacent conductive elements with different potentials have parasitic (leakage) capacitance between them. In the generic equipment in Figure 9.20, the potential of the internal electric circuit is different from the potential of the chassis. Hence, the capacitive coupling between the internal circuit and the chassis could elevate the potential of the chassis.

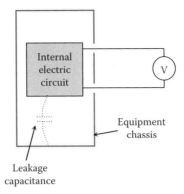

FIGURE 9.20   Generic representation of an appliance.

2. Current of the internal circuit produces magnetic fields (MFs) that link the chassis, thus inducing voltage on the chassis.
3. Faults inside the electric circuit could result in the chassis touching circuit components at elevated voltages.

If a person standing on a grounded object touches the chassis as shown in Figure 9.21, the person closes an electrical circuit and a current flows through his body. The current path is from the source to the circuit, then part of it passes through the leakage capacitance to the person, and finally to the ground. This current could be small and the person may feel nothing or just skin sensations. However, if the circuit is faulty, the current could reach unsafe levels. To protect people from this hazard, the neutral terminal of the receptacle in some equipment is connected to the chassis as shown in Figure 9.22. This way, the chassis is at the potential of the neutral wire, and no current flows through the person. This, however, requires the polarized plug shown in Figure 9.18b, where the neutral is always connected to the chassis, and not the other way.

The system in Figure 9.22 is not always safe. Consider, for example, the case in Figure 9.23 where the service transformer is far from the equipment. In this case, the cables connecting the transformer to the equipment are long and therefore have some resistances. In the figure, the hot wire resistance is R_h and the neutral wire resistance is R_n. The current of the hot wire I starts from the source to the circuit of the electric equipment, then returns to ground through two parallel paths: one through the neutral wire I_n and the other through the man I_{man}. The current through the man is determined by the current divider equation

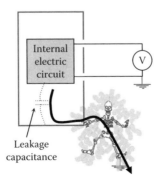

FIGURE 9.21   Person touching a floating chassis.

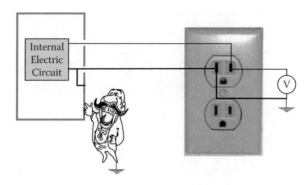

FIGURE 9.22   Person touching a grounded chassis.

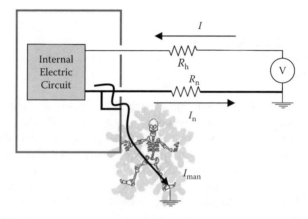

FIGURE 9.23   Person touching a chassis where the neutral wire is long.

$$I_{man} = I\left(\frac{R_n}{R_n + R_{man}}\right) = I\left[\frac{1}{1 + (R_{man}/R_n)}\right] \tag{9.17}$$

R_{man} is the body resistance of the man plus his ground resistance. As seen in the equation, the current through the man depends on the resistance of the neutral wire. When the resistance is small, the current through the man is also small. But when it is high, the current could be high enough to cause a secondary shock in most cases and a primary shock under faulty conditions as demonstrated in the following examples.

EXAMPLE 9.8

For the case in Figure 9.23, assume that the equipment is powered by 240 V and the resistance of the neutral wire is 0.5 Ω. Assume that the resistance of the man plus his ground resistance is 1500 Ω. If the equipment draws 20 A, compute the current through the man.

Solution

The current through the man is given in Equation 9.17.

$$I_{man} = I\left(\frac{R_n}{R_n + R_{man}}\right) = 20\left(\frac{0.5}{0.5 + 1500}\right) = 6.7 \text{ mA}$$

This current produces secondary shocks as given in Table 9.1.

Example 9.9

For the case in Example 9.8, assume that the equipment has an internal fault and draws 50 A from the line without tripping the circuit breaker. Compute the current through the man.

Solution

The current through the man is given in Equation 9.17

$$I_{man} = I\left(\frac{R_n}{R_n + R_{man}}\right) = 50\left(\frac{0.5}{0.5 + 1500}\right) = 16.75 \text{ mA}$$

This current is within the primary shock level and may induce respiratory tetanus as given in Table 9.2.

The last two examples show that the neutral wire may not provide protection from electric shocks. To correct this problem, the chassis must be connected directly to the local ground. This can be done by using the three-prong plug in Figures 9.18c and 9.19b and c. The local ground can be established by a ground rod outside the house, or by a wire connected to a well-grounded water pipe. In this case, the current can have two paths as shown in Figure 9.24: one through the neutral wire I_n and the other through the local ground I_g. The resistance of the man is much higher than the resistance of the local ground, and almost no current flows through his body.

The system in Figure 9.24, although safer than the previous ones, has current I_g passing through its chassis because the neutral and ground terminals are bonded by the chassis. This current could elevate the voltage of the chassis, which is known as stray voltage (stray voltage is discussed in detail in Section 9.4.2). A better system is the one shown in Figure 9.25 where the neutral wire is isolated from the chassis. Hence, the ground current is just the leakage current I_{leak} of the leakage capacitance. This is the safest system.

Needless to say that if the chassis is nonconductive, the problems described above vanish. In some cases, however, even if the chassis is made of isolated material (such as plastic), the knobs, levers, screws, or displays could be made of metal that is connected together by conductive material inside the chassis. In this case, touching any of these components could have the same effects as touching a conductive chassis.

9.4.2 Stray Voltage

When a grounded object carries current, the voltage of the object is nonzero. This voltage is referred as GPR (see Section 9.3.1). Stray voltage is also a GPR, but the term is often used when the voltage

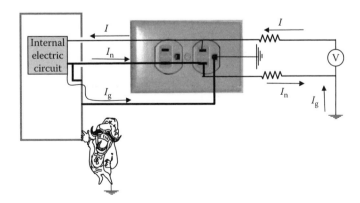

FIGURE 9.24 Person touching a grounded chassis with long neutral wire—chassis carrying high current.

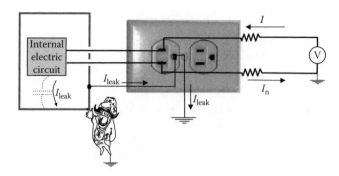

FIGURE 9.25 Person touching a grounded chassis with long neutral wire—chassis carrying low current.

is due to the bonding of the ground and neutral wires. The current that produces the stray voltage is called stray current. Although stray voltage is often very small, it can cause behavioral problems in farm animals and may cause sensitive hospital equipment to malfunction. In severe cases, it could reach lethal levels if the neutral wire is broken.

Let us assume that the ground wire is bonded to the neutral wire at the service box of a building as shown in Figure 9.26. In this circuit, we have the source connected to the load by two wires, the hot wire and the neutral wire whose resistance is R_n. The neutral wire at the source side is grounded and its ground resistance is R_{g1}. At the load side, the chassis is connected to the neutral wire and is also grounded through the grounding resistance R_{g2}. Assume that the load current of the hot wire is I. When this current leaves the load, it branches out into two paths: one through the neutral wire I_n and the other I_g strays through the chassis as well as the local ground path. These two currents must add up to I at the source side.

$$I = I_n + I_g \tag{9.18}$$

The magnitude of the stray current I_g depends on the resistance of the neutral wire R_n and the two ground resistances R_{g1} and R_{g2}.

$$I_g = I\left(\frac{R_n}{R_n + R_{g1} + R_{g2}}\right) = \frac{I}{1 + \left(\frac{R_{g1}+R_{g2}}{R_n}\right)} \tag{9.19}$$

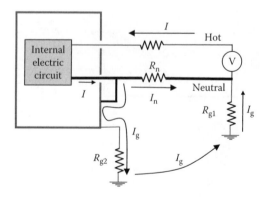

FIGURE 9.26 Stray current.

Hence, the stray voltage V_{stray} on the chassis is

$$V_{stray} = I_g R_{g2} = I\left(\frac{R_n R_{g2}}{R_n + R_{g1} + R_{g2}}\right) = I\frac{R_{g2}}{1+\left(\frac{R_{g1}+R_{g2}}{R_n}\right)} = I\frac{R_n}{\left(\frac{R_n+R_{g1}}{R_{g2}}\right)+1} \qquad (9.20)$$

The magnitude of the stray voltage depends on the following factors:

1. Magnitude of the load current I—the higher the current, the higher the stray voltage.
2. Value of the neutral wire resistance R_n—the higher the resistance, the higher the stray voltage.
3. Resistance of the local ground electrodes R_{g2}—the higher the resistance, the higher the stray voltage.

EXAMPLE 9.10

For the system in Figure 9.26, the neutral wire resistance is 1 Ω, the ground resistance at the source side is 20 Ω, and the ground resistance at the load side is 30 Ω. Assume that the load current is 10 A. Compute the stray voltage at the load side.

Solution

Use Equation 9.19 to compute the stray current I_g

$$I_g = I\left(\frac{R_n}{R_n + R_{g1} + R_{g2}}\right) = 10\left(\frac{1}{1 + 20 + 30}\right) \approx 0.2 \text{ A}$$

The stray voltage can be computed by Equation 9.20

$$V_{stray} = I_g R_{g2} = 0.2 \times 30 = 6 \text{ V}$$

The stray voltage is not high enough to cause any harm to humans, but as we shall see later, it can cause other problems.

Although stray voltage is often very small, it impacts negatively on various industries such as medical and farm businesses. In the medical field, the stray voltage can affect sensitive equipments, such as cardiograms, that rely on small voltages to measure health conditions. In addition, because of the stray current, the electromagnetic field (EMF) created by the unequal currents in the hot and neutral wires can induce unwanted voltage in the sensitive electronic circuits of various medical equipment.

In the farm industry, livestock avoid drinking or eating from metallic containers with stray voltage and are stressed when forced to enter areas with stray voltage. These behaviors make livestock management difficult as well as cause animal production to diminish; weight is reduced and less milk is produced. Several studies carried out by the U.S. Department of Agriculture and other organizations have found that as little as 5 mA may cause livestock to refuse to eat, drink, or produce the normal amount of milk. The summary of these studies is given in Table 9.6.

EXAMPLE 9.11

For the system in Example 9.10, assume that a cow is touching the chassis while standing on conductive surface whose ground resistance is 20 Ω. Assume that the cow body resistance is 500 Ω. Compute the stray current through the animal.

Solution

The stray current through the farm animal can be computed by using the circuit in Figure 9.27. There are two parallel paths for the stray current: one through ground resistance R_{g2} and the other through the cow.

Hence,

$$I_g + I_{cow} = I\left(\frac{R_n}{R_n + R_{g1} + R_{g2}//(R_{cow} + R_{g3})}\right) = 10\left(\frac{1}{1 + 20 + 30//(500 + 20)}\right) = 0.203\,\text{A}$$

The stray current through the cow is

$$I_{cow} = 0.203\left(\frac{R_{g2}}{R_{g2} + R_{g3} + R_{cow}}\right) = 0.203\left(\frac{30}{30 + 20 + 500}\right) = 11\,\text{mA}$$

According to Table 9.6, this level of stray current will have long-term changes in the animal's feed and water consumption as well as milk production.

Besides the stray voltage, the bonding of the ground and neutral wires can make the system unsafe in case the neutral current is broken. This can be explained by Example 9.12.

TABLE 9.6

Effect of Stray Current on Livestock

Stray Current (mA)	Effect on Livestock
1 to 3	Signs of awareness by livestock, but no milk production is lost
3 to 4	Animal may become more difficult to manage
5 to 6	Short-term changes in feed/water consumption or milk production
>6	Long-term changes in feed/water consumption or milk production

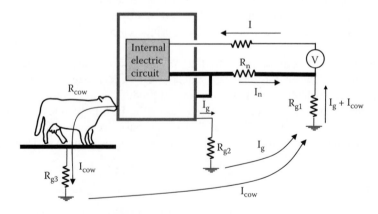

FIGURE 9.27   Stray current in a farm animal.

EXAMPLE 9.12

Repeat Example 9.10, but assume that the neutral wire is broken.

Solution

If the neutral wire is broken, the ground current (stray current) is equal to the load current. This can be seen from the circuit in Figure 9.28 or by examining Equation 9.19.

$$I_g = I\left(\frac{R_n}{R_n + R_{g1} + R_{g2}}\right) = I\left[\frac{1}{1 + \left(\frac{R_{g1}+R_{g2}}{R_n}\right)}\right] = 10\,\text{A}$$

The stray voltage in this case is

$$V_{stray} = I_g \times R_{g2} = 10 \times 30 = 300\,\text{V}$$

The level of stray voltage is lethal.

The stray voltage can affect local areas as well as remote areas connected to the same power circuit (Figure 9.29). The figure shows a system with two loads powered by the same hot and neutral wires. Assume that the source voltage is at the right side of the feeders. In each load, the neutral wire is connected to the chassis of the equipment and also connected to local grounds (R_{g1} and R_{g2}). The resistance of the neutral wire between the source and the first load is R_{n1}, and between the first and second loads is R_{n2}. The current flow is shown in the figure. As explained earlier, the magnitude of the stray currents I_{g1} and I_{g2} is dependent on the load currents I_1 and I_2, all ground resistances, and the resistances of the neutral wire sections. Now let us assume that the neutral wire is broken between the two loads, and let us even assume that the first load is disconnected ($I_1 = 0$). In this case, the current of the second load I_2 returns back to the source via the ground path as shown in Figure 9.30. The stray current here is the entire I_2, which is divided into two paths: one through the ground resistance of the first load I_{s1} and the other through the ground resistance of the source I_{s2}. Because of the high currents through the ground resistances of both loads, the stray voltage is likely to reach a lethal level. This scenario is analyzed in Example 9.13.

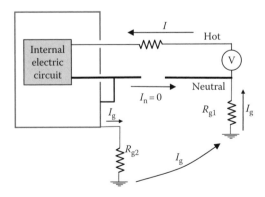

FIGURE 9.28   Stray current when the neutral wire is broken.

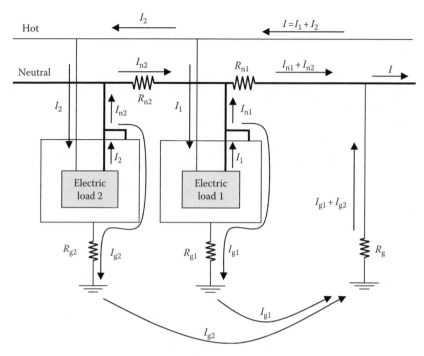

FIGURE 9.29 Stray current in a system with bonded neutral and ground wires.

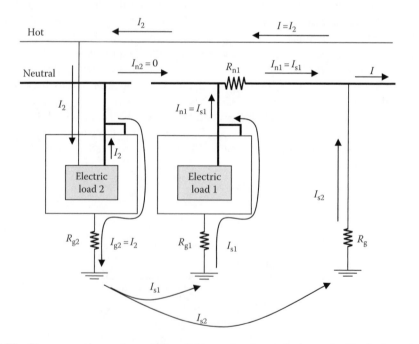

FIGURE 9.30 Stray current in a system with bonded ground and neutral wires and with a broken neutral wire.

Example 9.13

For the system in Figure 9.30, the load current $I_2 = 5$ A, $R_g = 20$ Ω, $R_{g1} = R_{g2} = 30$ Ω, and $R_{n1} = 1$ Ω. Assume that the neutral current between the two loads is broken and load 1 is not energized. Compute the stray voltage at both loads.

Solution

The current of the second load can be written as

$$I_2 = I_{s1} + I_{s2}$$

where

$$I_{s1} = I_2 \left(\frac{R_g}{R_{n1} + R_g + R_{g1}} \right) = 5 \left(\frac{20}{1 + 20 + 30} \right) \approx 2\,\text{A}$$
$$I_{s2} = I_2 - I_{s1} = 3\,\text{A}$$

The stray voltage at the first load is

$$V_{\text{stray1}} = I_{s1} R_{g1} = 2 \times 30 = 60\,\text{V}$$

And the stray voltage at the second load is

$$V_{\text{stray2}} = I_2 R_{g2} = 5 \times 30 = 150\,\text{V}$$

Notice that although the first load is de-energized, its stray voltage is 60 V due to the broken neutral wire. This example shows that electrical safety depends on the entire network connection instead of only the configuration of the local circuit. Repeat the example for the case when the neutral wire between the source and the first load is broken.

Because of the potential hazards due to bonding the neutral and ground wires, several countries have mandated that they be separated as shown in Figure 9.25. Let us examine the same scenario in the previous example, but assume that the neutral and ground wires are separated as shown in Figure 9.31. Notice that the chassis is not connected to the neutral wire anymore. Therefore, the load

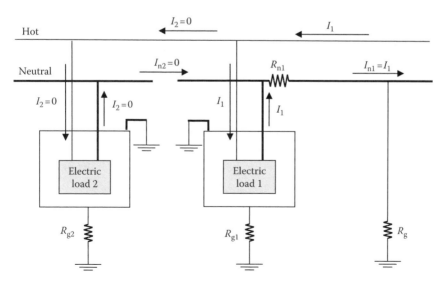

FIGURE 9.31   Stray current in a system with separated ground and neutral wires and with a broken neutral wire.

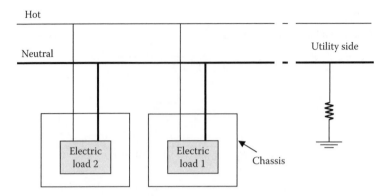

FIGURE 9.32   System without ground wire.

current cannot pass through the chassis. When a neutral wire is broken at any load, the load is simply disconnected and there is no effect on any other load in the system. Just examine the current flow in the circuit.

9.4.3   World's Residential Grounding Practices

When it comes to the wiring practice worldwide, there is a great deal of confusion about the roles of the neutral and ground wires. It is usual to find no ground wires in some countries while others bond the neutral and ground wires to reduce the cost; three-wire cables are more expensive than two-wire cables. As we have seen in Section 9.4.2, the neutral and ground wires must be separated to achieve the maximum safety for people touching grounded surfaces. The system with the worst safety configuration is the one shown in Figure 9.32, as no ground wire is available. This system is discussed earlier in Figure 9.21.

For the system in Figure 9.33, the chassis is bonded to the neutral wire. In this case, the utility provides only two wires: hot and neutral, where the neutral also acts as a ground wire. As we have seen in Figures 9.23, 9.29, and 9.30, this system is not safe under all conditions.

The system in Figure 9.34 has three separate wires coming from the utility: hot, neutral, and ground. The neutral and ground wires are separated. Furthermore, the neutral is only grounded at the utility's service transformer. This system is the safest as discussed in Figures 9.25 and 9.31.

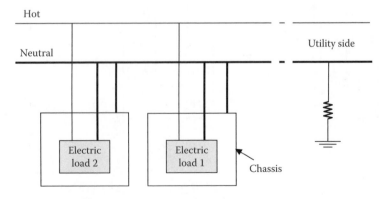

FIGURE 9.33   System with the neutral is also used as a ground wire.

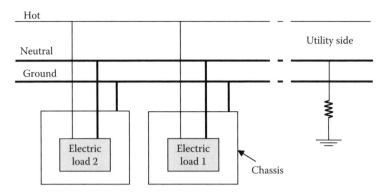

FIGURE 9.34 System with separated neutral and ground wires.

Sometimes, utilities elect to use water pipes as local grounds as seen in Figure 9.35. This is quite common in areas where water pipes are made of copper or iron alloy and are buried in soils with low resistivity.

9.4.4 GROUND FAULT CIRCUIT INTERRUPTER

One of the most common electric shock scenarios is when electric equipment is mixed with water. Tap water is conductive, and its intrusion allows the circuit components to be electrically connected to the chassis. This is as bad a situation as having the circuit components arcing to the chassis.

Assume that a hair drier is plugged into an electric outlet. If the hair dryer falls into a bathroom tub filled with water, the high-voltage wires and the circuit components may come in contact with the drier's chassis. If a person touches the wet drier while part of his/her body is in contact with a grounded surface, such as faucets, the person could be electrocuted as depicted in Figure 9.36. The fundamental question is why does not the housebreaker interrupt the circuit?

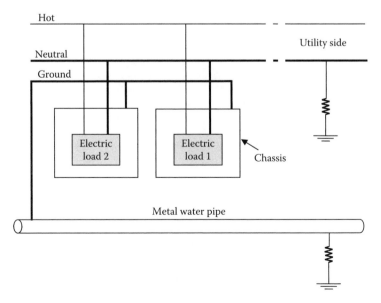

FIGURE 9.35 System with separated neutral and ground wires; ground wire is bonded to a metal pipe.

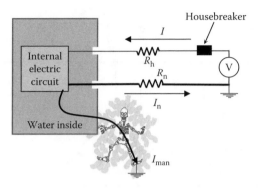

FIGURE 9.36 Water inside a device can create a hazardous condition.

After all, the breakers are installed to disconnect faulty circuits. In fact, the housebreaker may not operate at all if the fault current is below its interruption rating. The current through the man in Figure 9.36, although lethal, could be well below the interruption level of the breaker. For this reason, a ground fault circuit interrupter (GFCI) is used in wet areas such as kitchens, bathrooms, gardens, etc.

The GFCI is similar in shape to the regular outlet as shown in Figure 9.37, but has two buttons and a built-in circuit interrupter. The circuit interrupter of the GFCI detects the currents of the hot and neutral terminals. When a leakage current flows through the chassis to the ground, the currents of the neutral and hot wires are not equal as shown in Figure 9.36. In this case, the GFCI interrupts the circuit by disconnecting the terminal of the hot wire. Of the two buttons of the GFCI, one is used to test the circuit for functionality and the other is used to reset the circuit after interrupting a fault.

The basic component in the circuit interrupter of the GFCI is the current differential detector shown in Figure 9.38. The detector consists of a core with windings wrapped around it. The hot wire and the neutral wire pass inside the core. If the hot wire current I_{hot} and the neutral current $I_{neutral}$ are equal in magnitude, no voltage is induced across the winding. If there is a difference between the two currents, a voltage is induced across the winding triggering a relay that switches off the power.

Notice that the GFCI can also interrupt the circuit when the neutral wire is broken. This unsafe scenario is addressed in Example 9.13.

FIGURE 9.37 GFCI outlet.

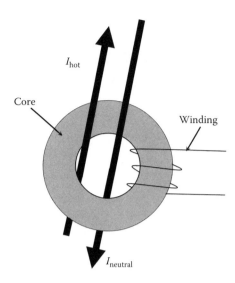

FIGURE 9.38   Current differential detector of the GFCI circuit.

9.5   LOW-FREQUENCY MAGNETIC FIELD AND ITS HEALTH EFFECTS

The health effect of low-frequency MF is a controversial issue that creates strong emotion on all fronts. The interest in the subject started in the Soviet Union in the 1960s after some health problems were observed in personnel working in electric switchyards. These health problems were thought to be associated with MF exposure. The study created widespread scientific interest in the subject among epidemiologist scientists. One of the famous studies was made in Denver, Colorado, in 1979 where epidemiologists suggested that children who had died from cancer were two to three times more likely to have lived in homes near high-current power lines than children in the general population. A large number of additional epidemiological studies followed, but the results were highly inconsistent and contradictory. The outcomes of these studies frustrate the public and spread anxiety and speculation than resolve their concern.

In the opinion of a large segment of the scientific community, most of these epidemiological studies were flat out flawed due to the small samples they used and the single statistical association technique they assumed. In some studies the samples did not exceed 100 people. In most of the studies, no other factor that may affect public health besides the MF was considered. For example, power lines are often located near heavily traveled highways where the pollution from cars and trucks contributes to the health effects. Also, work environments often present many other exposures that could negatively affect the health, such as toxic chemicals.

9.5.1   Low-Frequency Magnetic Fields

Each electrically powered device is surrounded by EMF. An EMF is composed of EF due to the voltage and MFs due to the current. The interest in the potential health effects focused mainly on the MFs.

Besides the MF created by electric currents, MF occurs in nature where the earth's MFs density *B* ranges between 300 and 600 mG (milligauss). The lower number is for South America and South Africa, and the higher number is for areas near the earth's magnetic poles such as northern Canada. Solar flare and cosmic radiation also produce MFs.

In most homes, the MF density *B* in the center of rooms averages to less than 1 mG. Homes close to power lines tend to have higher MF density than average. Home appliances emit EMF that is dependent on their operating voltage and power. Figure 9.39 shows typical values of *B* for

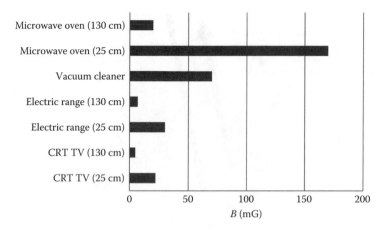

FIGURE 9.39   MF density of home appliances.

some popular home appliances. The figure shows the values of B measured at various distances from the appliances. Notice that the MF is substantially reduced when we move away from the appliance.

For power lines, the MF density varies widely depending on the amount of current passing through their conductors. However, a typical value for 500 kV fully loaded line is about 100 mG under the line and about 5 mG when we move just 70 m away from the energized conductor.

9.5.2   Biological Effects of Magnetic Field

Scientists have tried to pinpoint whether the low-frequency MF affects people, animals, or vegetation. Several cell level laboratory studies have indicated that the MF exposure may cause some temporary biological effects such as changes in hormone levels. The studies have shown that the currents induced in humans due to low-frequency MF are weaker than the natural electrical signals that control the brain and heart. These low-frequency currents pass mostly between the cells and do not penetrate the cell membranes.

The World Health Organization has coordinated several extensive studies on the EMF and its effects on public health and the environment. In their information sheet published in February 2005, several areas were addressed: biological effects on humans and animals, vegetation, and aquatic life. The following are direct quotes from the report.

Animals

Most studies of EMF effects in animals have been conducted to investigate possible adverse health effects in humans. These are usually performed on standard laboratory animals used in toxicological studies, e.g., rats and mice, but some studies have also included other species such as short-living flies for the investigation of genotoxic effects. The subject of this information sheet, however, is whether EMF can have harmful impacts on species of wild and domestic animals. Under consideration are

- Species, in particular certain fish, reptiles, mammals, and migratory birds, which rely on the natural (geomagnetic) static MF as one of a number of parameters believed to be used for orientation and navigational cues.
- Farm animals (e.g., swine, sheep, or cattle) grazing under power lines (50/60 Hz) or in the vicinity of broadcasting antennas.
- Flying fauna, such as birds and insects, which may pass through the main beam of high power radio-frequency antennas and radar beams or through high intensity ELF fields near power lines.

Studies performed to date have found little evidence of EMF effects on fauna at levels below ICNIRP guideline levels. In particular, there were no adverse effects found on cattle grazing below power lines. However, it is known that flight performance of insects can be impaired in EFs above 1 kV/m, but significant effects have only been shown for bees when electrically conductive hives are placed directly under power lines. Un-insulated un-earthed conductors placed in an EF can become charged and cause injury or disrupt the activity of animals, birds, and insects.

Vegetation

Field studies of 50–60 Hz exposure to plants and crops have shown no effects at the levels normally found in the environment, nor even at field levels directly under power lines up to 765 kV. However, the variability of parameters associated with environmental conditions that affect plant growth (e.g., soil, weather) would likely preclude observation of any possible low-level effects of EF exposure. Damage to trees is well known to occur at EF strengths far above ICNIRP's levels due to corona discharge at the tips of the leaves. Such field levels are found only close to the conductors of very high voltage power lines.

Aquatic Life

Although all organisms are exposed to the geomagnetic field, marine animals are also exposed to natural EFs caused by sea currents moving through the geomagnetic field. Electrosensitive fish, such as sharks and rays in oceans and catfish in fresh water, can orient themselves in response to very low EFs by means of electroreceptive organs. Some investigators have suggested that human-made EMF from undersea power cables could interfere with the prey sensing or navigational abilities of these animals in the immediate vicinity of the sea cables. However, none of the studies performed to date to assess the impact of undersea cables on migratory fish (e.g., salmon and eel) and all the relatively immobile fauna inhabiting the sea floor (e.g., molluscs) have found any substantial behavioural or biological impact.

Conclusion

The limited number of published studies addressing the risk of EMF to terrestrial and aquatic ecosystems show little or no evidence of a significant environmental impact, except for some effects near very strong sources. From current information the exposure limits in the ICNIRP guidelines for protection of human health are also protective of the environment.

9.5.3 Standards for Magnetic Field

It is rare to find an enforceable standard for MF exposure at low frequency (50 or 60 Hz). However, IEEE has established the maximum permissible exposure (MPE) levels for controlled (workers) and uncontrolled (unwilling individuals) environments. Table 9.7 shows the highlights of this IEEE standard. The table shows that MPE is substantially reduced when the frequency of the MF increases. This is mainly because high-frequency currents penetrate the body's cells causing internal heat. This is why we use microwaves to cook our food.

TABLE 9.7
Maximum *B* Exposure

Frequency	Maximum Exposure for General Public
dc	1.0 kG
30–100	2.0 G
15,000–300,000	10.0 mG

Source: IEEE Standard C95.1, 1991.

There is no U.S. federal standard limiting occupational or residential exposure to 60 Hz MF. However, some states have set guidelines. At the edge of the right-of-way, Florida limits B to 250 mG and New York limits it to 200 mG.

EXERCISES

1. A person is standing on a wet organic soil; compute his/her ground resistance.
2. Compute the resistance of a 10 m wide stretch of soil that is 2 m away from the edge of a grounded hemisphere. The radius of the hemisphere is 0.5 m. Assume that the soil is dry.
3. A person is working on a steel structure while standing on the ground. An accident occurs when 5 A passes through the structure to the ground. The structure is grounded by a metal rod 6 cm in diameter. The rod is dug 2 m into the ground. The surrounding soil is dry. Assume that the resistance of the man's body is 2000 Ω. Compute the current through the man.
4. Compute the survival time of the man in the previous problem using Dalziel's formula. Assume that the weight of the man is 80 kg.
5. Repeat problems 3 and 4 for wet organic soil.
6. During a weather storm, an atmospheric discharge hits a lightning pole. The pole is grounded through a hemisphere. The maximum value of the lightning current through the pole is 10 kA. The soil of the area is moist. A man who is walking 20 m away from the center of the hemisphere experiences an excessive step potential. The man's body resistance is 1500 Ω. Compute the current through his legs, and his step potential.
7. During a weather storm, an atmospheric discharge hits a lightning pole that is grounded through a hemisphere. The maximum value of the lightning current through the rod is 20 kA. The soil of the area is moist. A man is playing golf 50 m away from the center of the hemisphere. At the moment of the lightning strike, the distance between his two feet is 0.4 m. Compute the current through the person assuming that the resistance between his legs is 1500 Ω.
8. Repeat the previous problem and assume that the person is 5 m away from the center of the hemisphere. What is the effect of the proximity of the person to the grounding hemisphere?
9. A power line insulator partially fails and 10 A passes through the structure to the tower's ground. The tower's ground is a hemisphere with a radius of 0.5 m. The soil resistivity is 100 Ωm. Assume that a man touches the tower while standing on the ground. Compute the current going through the man, assuming his body resistance is 3000 Ω.
10. An electric circuit is powered by an unpolarized 120 V, 60 Hz outlet. The circuit is inside an ungrounded chassis. A 100 nF capacitance exists between the circuit and the chassis. If a man touches the chassis, compute the current through his body. Assume that the body resistance of the man plus his ground resistance is 3000 Ω.
11. An electric circuit is powered by a two-prong polarized outlet through a feeder. The resistance of the neutral wire is 0.2 Ω. The chassis of the circuit is metallic and is connected to the neutral terminal of the outlet. A current of 200 A is drawn from the outlet through the hot wire. If a person with 2 kΩ resistance (including ground resistance) touches the chassis, compute the current through his body. Also state the type of hazard the person is exposed to.
12. State six factors that determine the severity of the electric shock when a person comes in contact with an energized conductor.
13. Compute the survival time of a heavy man receiving an electric shock of 100 mA.
14. Consider a system where the neutral and ground wires are bonded at the customer's side. Assume that the neutral wire from the utility has a resistance of 2 Ω, and the neutral wire is grounded at the source side where the ground resistance is 15 Ω. The chassis of an equipment at the load side is grounded through a ground resistance of 30 Ω. Assume the equipment draws a current of 15 A. Compute the stray voltage on the chassis.

15. Assume that a cow is touching the equipment in the previous example. Assume that the body resistance of the cow plus the ground resistance of the surface it stands on is 300 Ω. Compute the stray current through the cow and its effect.

16. For the system in the following figure, assume that the neutral wire is broken at the utility side. Compute the stray voltage at both loads assuming the load currents are $I_1 = 2$ A and $I_2 = 4$ A. The system has $R_g = 20$ Ω, $R_{g1} = R_{g2} = 30$ Ω, and $R_{n1} = R_{n2} = 1$ Ω. Also compute the GPR at the utility side due to the two load currents.

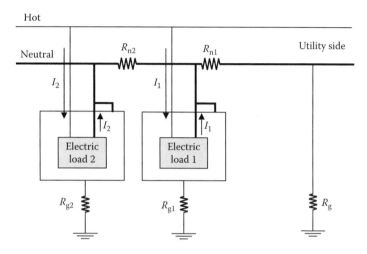

17. Write a technical report on the grounding practices of residential loads in any of the following countries: The United States, England, Japan, India, Australia, or South Africa.

18. Why it is important to separate the ground wire from the neutral wire in residential wiring?

19. What are the advantages and disadvantages of using water pipes as ground path?

20. Can we use natural gas pipes as ground path for current? Why?

REFERENCES

Dalziel, Charles F. 1972. Electric Shock Hazard, *IEEE Spectrum*, 9, 41–50.

IEEE Standard 524a-1993. IEEE guide to grounding during the installation of overhead transmission line conductors, 1994.

IEEE Standard 1048-1990. IEEE guide for protective grounding of power lines, 1991.

IEEE Standard C95.1-1991. IEEE/ANSI standard for safety levels with respect to human exposure to radio frequency electromagnetic field, 1991.

10 Power Electronics

The electric power grid must operate at fairly constant voltage and fixed frequency to ensure that all generators are synchronized and the entire system is stable. If any generator produces voltage or frequency different from the rest of the system, blackouts can occur as discussed in Chapter 14. Because of this limitation, all electrical devices and equipment were designed to operate at the voltage and frequency of the utility's feeders. Accordingly, they were limited in performance, heavy in weight, and inefficient in operation.

With the development of power electronic devices and circuits, the newer apparatuses and equipment have electronic converters that allow users to change the voltage and frequency applied to the equipment to achieve better performance and higher efficiencies. The new washing machine, for example, allows the user to change the speed and torque of the washing cycles by adjusting the voltage and frequency of its electric motor, thus optimizing the washing process. Another example is the air-conditioning systems; the older ones were inefficient because of their switch-in and switch-out operation, whereas the newer ones operate continuously at variable speeds to maximize their efficiencies and to maintain the environment to the users' settings with minimum deviations.

Devices that could not have existed before due to the restrictions on the voltage and frequency are now widely available. This includes microwave ovens whose power electronic circuits operate their magnetrons at very high frequencies (usually above 2 GHz). Other examples include household items such as light dimmers, chargers of electronic devices and computers, variable speed power tools, dishwashers, dryers, heat pumps, and electric stoves.

Besides household usage, power electronics provide the industry with effective methods to save energy and improve performance; the following are some examples:

- Electric load that demands different waveforms from that provided by the utility can still be energized from the utility supply through a proper converter. For example, the motors used in all printers, scanners, and hard disks cannot operate at the utility voltage or frequency, but can still be energized from the utility through power electronic converters.
- Power electronic circuits can increase the efficiency of the system for a wide range of operating conditions.
- Equipment operating by power electronic devices is more reliable and lasts longer.
- Equipment designed to operate by a power electronic converter is lighter in weight and smaller in volume than the ones designed to operate at the utility's fixed voltage and frequency.
- Performance of the electric load can be greatly enhanced, and the load can perform functions that are difficult to implement without power electronics. For example, the smooth speed control and braking of the elevators cannot be achieved without the use of power electronic circuits.
- Performance of electrical equipment can be easily and effectively controlled by power electronic converters.

The history of power electronics started with the vacuum tube invented in 1904 by the British scientist John Fleming. His vacuum tube is quite revolutionary because it allows the electric current to flow in one direction only, acting as a diode. Later developments created tubes that amplify signals and changed the shape of electric waveforms. After the end of World War II, brilliant scientists Bill Shockley, John Bardeen, and Walter Brattain invented the first transistor in 1947. Their invention opened the door wide to the use of solid-state devices in almost all electrical equipment manufactured today. The research team named their device transistor because it is essentially a trans-resistance, that is, a variable resistance. In 1956, these scientists received the Nobel Prize for their marvelous transistor.

Since the invention of the transistor, scientists have worked diligently to invent other semiconductor devices with broader characteristics and higher power ratings. The early transistors were rated at a small fraction of an ampere and at very low voltages. Nowadays, a single solid-state device can be built to withstand about 5 kV and can carry thousands of amperes.

10.1 POWER ELECTRONIC DEVICES

Power electronic devices can be divided into two categories: single component and hybrid component. The single component is composed of one solid-state device, and the hybrid component is a combination of several single components manufactured as a single block. The single devices for power applications are loosely divided into three types: (1) two-layer devices such as the diodes, (2) three-layer devices such as the transistors, and (3) four-layer devices such as the thyristors. The hybrid components include a wide range of devices such as the insulated gate bipolar transistor (IGBT), static induction transistor, and Darlington transistor (DT).

Solid-state devices are the main building components of any converter. Their function is mainly to mimic the mechanical switches by connecting and disconnecting electric loads, but at very high speeds. The characteristic of the ideal mechanical switch is shown in Figure 10.1. When the mechanical switch is open (off), the current through the switch is zero and the voltage across its terminals is equal to the source voltage. When the switch is closed (on), its terminal voltage is zero and its current is determined by the load impedance. A major part of the ongoing research and development in power electronic devices is devoted to improving the characteristics of the solid-state devices to make them as close to the mechanical switch as possible.

10.1.1 SOLID-STATE DIODES

The diode, which is the simplest power electronic device, allows the current to flow only in one direction. It is built of two semiconductor material p- and n-type placed in contact with each other as shown in Figure 10.2a. The symbol for the diode is shown in Figure 10.2b. The p-junction is the anode of the device and the n-junction is the cathode. The current of the diode flows only from the anode to the cathode when the anode-to-cathode voltage V_{AK} is positive (forward biased) as shown in Figure 10.2c. When the voltage V_{AK} is reversed (reverse biased), almost no current flows

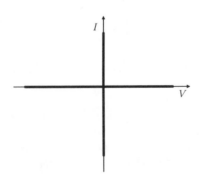

FIGURE 10.1 Current–voltage characteristics of a mechanical switch.

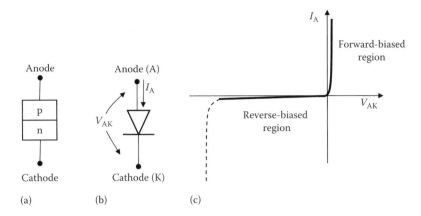

FIGURE 10.2   Solid-state diode.

though the diode. However, if this reverse-biased voltage is increased to an excessive value beyond the device rating, the diode is destroyed by the current avalanche shown by the dashed section of the characteristics. When the diode is forward biased, the voltage drop across the diode V_{AK} is very small (about 0.7 V), and the diode in this case resembles a closed mechanical switch. When the diode is reverse biased (the thick portion of the line), the diode's current is very small and the diode resembles an open mechanical switch.

Figure 10.3 shows a picture of several diodes ranging from 6 V, 10 mA to 5 kV, 5 A. Higher rating diodes can be found for thousands of kiloampere. For higher voltage diodes, several single component diodes are often connected in series to achieve a voltage level as high as 1 MV.

10.1.2   Transistors

In power electronic applications, transistors are mainly used as high-frequency electronic controlled switches. Two main types of transistors are commonly used: the bipolar junction transistor (BJT) and the field effect transistor (FET).

FIGURE 10.3   Several diodes.

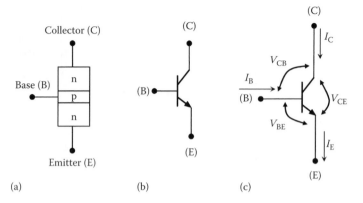

FIGURE 10.4   Bipolar junction transistor.

10.1.2.1   Bipolar Junction Transistor

The BJT consists of three layers of semiconductor material in n-p-n or p-n-p arrangements; the n-p-n shown in Figure 10.4a is more common in power electronic applications. The middle layer of the transistor is thin compared to the other two layers. The three layers of the transistor are called collector (C), base (B), and emitter (E). The bipolar transistor can be viewed as a pair of solid-state diodes joined back to back (Figure 10.4a). But because the middle layer is fairly thin, the transistor behaves differently from just two back-to-back diodes. If we apply a positive voltage between the collector and emitter V_{CE}, the top back-to-back diodes is reverse biased. If we apply a small positive voltage between the base and emitter V_{BE}, the base–emitter junction is forward biased, and a current flows between the base and emitter as if the junction is a forward-biased diode. In this case, electrons move from the emitter to the base. Once the electrons reach the base layer, they are quickly attracted to the collector layer because its voltage is positive. Hence, a current I_C flows through the collector terminal. This collector current is a function of the voltage we apply between the base and the emitter V_{BE}.

Figure 10.5 shows bipolar power transistors of various sizes. The rating of the largest one in the figure is 100 A.

FIGURE 10.5   Various size transistors.

The basic equations of the BJT are

$$I_C = \beta\, I_B \tag{10.1}$$

$$I_E = I_B + I_C \tag{10.2}$$

$$V_{CE} = V_{CB} + V_{BE} \tag{10.3}$$

where
  β is the current gain (the ratio of collector to base currents)
  I_B is the base current
  I_E is the emitter current
  V_{CE} is the collector–emitter voltage
  V_{CB} is the collector–base voltage
  V_{BE} is the base–emitter voltage

The characteristics of the transistor are shown in Figure 10.6. The base characteristic (I_B versus V_{BE}) is very similar to the characteristic of the diode in the forward-bias mode. In the forward direction the base–emitter voltage V_{BE} is below 0.7 V. The collector characteristic is I_C versus V_{CE} for various values of base currents I_B. This characteristic can be divided into three regions: linear, cutoff, and saturation. In the linear region, the transistor operates as an amplifier, where β is almost constant and in the order of a few hundreds. Hence, any base current is amplified a few hundred times in the collector circuit. This is the region where most audio and video amplifiers operate.

The cutoff region is the area of the characteristic where the base current is zero. In this case, the collector current is negligibly small regardless of the value of the collector–emitter voltage. In the saturation region, the collector–emitter voltage is very small compared to the transistor ratings, and the collector current is determined by the load. β in the saturation region is often less than 30.

When a transistor is used to emulate a mechanical switch, it operates in the cutoff and saturation regions. Power transistors cannot operate in the linear region for long periods because V_{CE} in the linear region is high and the current of power electronic circuits is often high. Therefore, the power loss of the power transistor ($V_{CE} \times I_C$) is excessive and eventually will damage the transistor due to the excessive thermal heat.

The circuit in Figure 10.7 explains how the transistor operates as a switch. The transistor is connected to an external circuit consisting of a direct current (dc) source V_{CC} (which is the source of energy) and a load resistance R_L. The base circuit of the transistor is connected to a separate current source to produce the base current I_B. The loop equation of the collector circuit is

$$V_{CC} = V_{CE} + R_L\, I_C \tag{10.4}$$

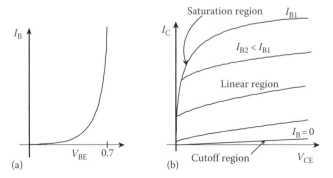

FIGURE 10.6 Characteristics of BJT: (a) base characteristics and (b) collector characteristics.

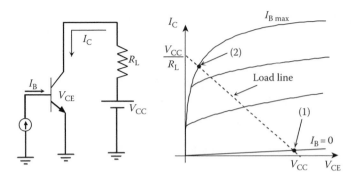

FIGURE 10.7 Switching of a transistor.

The collector equation, which is also known as the *load line* equation, shows a linear relationship between I_C and V_{CE}. The equation has a negative slope and intersects the *x*-axis at V_{CC}, and the *y*-axis at V_{CC}/R_L. If the base current is zero, the operating point of the circuit is in the cutoff region at point 1. The collector current in this case is very small and the collector–emitter voltage of the transistor is almost equal to the source voltage V_{CC}. Hence, the operation of the transistor in the cutoff region resembles an open mechanical switch. If the base current is set to the maximum value, the transistor is in the saturation region at point 2. At this operating point, the voltage drop across the collector–emitter is very small and the collector (or load) current is almost equal to V_{CC}/R_L. The transistor in this case is equivalent to a closed mechanical switch.

The bipolar transistor emulates a normally open mechanical switch; it is always open unless a control current (base current) is injected into the base. The transistor will remain closed as long as the base current exists. To open the transistor, the base current must be removed from the base circuit.

The BJT has several advantages; the best among them are its high switching speeds and high reliability. However, the BJT has two main drawbacks:

1. When the transistor is closed in the saturation region, V_{CE} is about 0.1 V, which is very low for high-voltage circuits. However, if the transistor carries high currents while closed, the losses could be excessive leading to high internal heat. This heat must be dissipated as quickly as it is generated to prevent the transistor from being damaged.
2. BJT is a current-controlled device; the transistor is closed and maintained in the closed position only if the maximum base current is present. Since β is small in the saturation region, the base current is a large percentage of the collector current. This high base current causes high losses in the base circuit. Furthermore, the driving circuit that generates this base current must be capable of producing high currents for as long as the transistor is closed. Such a circuit is large in size and complex to build.

EXAMPLE 10.1

A BJT has a current gain of 10 in the saturation region. When the transistor is closed, the collector current is 100 A. Compute the following:

(a) Minimum base current to keep the transistor closed
(b) Losses of the transistor

Solution

(a)
$$I_B = \frac{I_C}{\beta} = \frac{100}{10} = 10 \text{ A}$$

Notice that the base current is very high and may require an elaborate driving circuit.

(b) Losses inside the transistor are mainly in the base–emitter loops and the collector–emitter loop. The base–emitter loss is

$$P_{BE} = I_B \times V_{BE}$$

Since the base–emitter junction is just a diode in the forward bias, we can assume that V_{BE} is about 0.7 V.

$$P_{BE} = I_B \times V_{BE} = 10 \times 0.7 = 7\,\text{W}$$

The collector–emitter loss is

$$P_{CE} = I_C \times V_{CE}$$

Assume V_{CE} is about 0.1 V when the transistor is fully closed

$$P_{CE} = I_C \times V_{CE} = 100 \times 0.1 = 10\,\text{W}$$

Total losses $= 17$ W.

EXAMPLE 10.2

In the saturation region, the BJT circuit shown in Figure 10.7 has a current gain of 20 and $V_{CE} = 0.5$ V. The source voltage is 100 V and the load resistance is 5 Ω.

(a) Calculate the minimum base current that maintains the transistor closed.
(b) Calculate the losses in the collector–emitter loop.
(c) Assume that the base current is reduced to half the value computed in part 1, and the transistor is pulled out of the saturation region and moved into the linear region where the current gain is 30; compute the losses in the collector–emitter circuit.

Solution

(a) First compute the collector current by using Equation 10.4

$$I_C = \frac{V_{CC} - V_{CE}}{R_L} = \frac{100 - 0.5}{5} = 19.9\,\text{A}$$

Hence, the minimum base current is

$$I_B = \frac{I_C}{\beta} = \frac{19.9}{20} = 0.995\,\text{A}$$

(b) $P_{CE} = I_C \times V_{CE} = 19.9 \times 0.5 = 9.95$ W
(c) When the base current is reduced, V_{CE} increases as shown by the load line in Figure 10.7. But first, let us compute the new collector current in the linear region.

$$I_C = \beta I_B = 30 \times \frac{0.995}{2} = 14.925\,\text{A}$$

$$V_{CE} = V_{CC} - R_L\,I_C = 100 - 5 \times 14.925 = 25.375\,\text{V}$$

The new loss is

$$P_{CE} = I_C \times V_{CE} = 14.925 \times 25.375 = 378.72\,\text{W}$$

Notice that the loss in the collector–emitter loop is much higher when the transistor is outside the saturation region. This is why the power transistor must always be in the saturation region when closed.

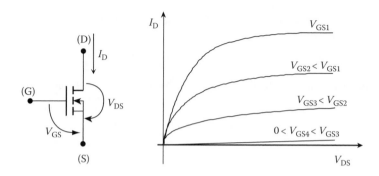

FIGURE 10.8 Enhanced mode MOSFET.

10.1.2.2 Metal Oxide Semiconductor Field Effect Transistor

A metal oxide semiconductor field effect transistor (MOSFET), unlike the BJT, is a voltage-controlled device. This is a great advantage over the BJT as the MOSFET is a more efficient device that requires a simpler driving circuit. The MOSFET is widely used in digital circuits as well as high-power electronic circuits.

The MOSFET is based on the main principle that operates all FET. The field effect theory states that the current near the surface of a semiconductor material can be altered by imposing an electric field on the surface. Since the electric field is produced by voltage, the current of the FET is therefore controlled by a voltage signal. This is different from the BJT whose collector current is controlled by the current in the base circuit.

The MOSFET is structured as a three-layer semiconductor material, and the middle layer is subjected to an external electric field that controls the flow of current between the top and bottom layers. The strength of the electric filed depends on the applied voltage (not current). The three terminals of the MOSFET in Figure 10.8 are the drain (D), source (S), and gate (G). The control signal V_{GS} is applied between the gate and the source, and the load current is the drain current I_D. There are two main types of MOSFETs, the enhanced mode and the enhanced–depletion mode. The characteristic of the enhanced mode is shown in Figure 10.8 where V_{GS} is always positive. The enhanced–depletion mode MOSFET has its V_{GS} either positive or negative as shown in Figure 10.9.

The power MOSFET has several advantages; among them are

- MOSFET behaves as a voltage-controlled resistance; the value of the gate voltage determines the on-resistance of the MOSFET (V_{DS}/I_D). When fully closed, the on-resistance is just a few milliohms.
- Its input resistance is high since the gate current is almost zero.

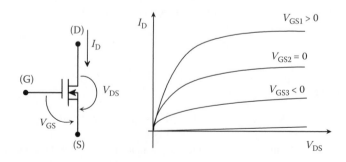

FIGURE 10.9 Enhanced and depletion mode MOSFET.

- Because of its low on-resistance and the lack of current in the gate loop, the losses of the MOSFET are lower than that of the BJT.
- MOSFET can be connected in parallel for heavier loads. This is not possible with the BJT.

The main disadvantage of the MOSFET is its high gate–source capacitance. This is due to the difference in charges between the gate and the source, and the high input resistance of the gate. Therefore, the switching frequency of the MOSFET is limited by this input capacitance. High switching frequency may result in excessive capacitive current in the gate circuit.

10.1.3 Thyristors

The thyristor's family includes several devices such as the silicon-controlled rectifier (SCR), bidirectional switch (Triac), and gate turn-off SCR (GTO). These devices are widely used in high-power applications as they can handle thousands of amperes and can be cascaded to withstand hundreds of kilovolts. Figure 10.10 shows various high-current SCRs with the large one rated at 300 A. The thyristor's family also has several devices in the low-power range such as the silicon diode for alternating current (SIDAC), silicon unilateral switch, and the bilateral diode (Diac).

10.1.3.1 Silicon-Controlled Rectifier

SCR is built of four layers of semiconductor material as shown in Figure 10.11. The device has three terminals: anode (A), cathode (K), and gate (G). When the anode-to-cathode voltage V_{AK} is negative, the SCR is reverse biased and a small leakage of current flows from the cathode to the anode; the device in this case is open. If the reverse-biased voltage reaches the reverse breakdown limit V_{RB} of the device, the SCR is destroyed. When the anode-to-cathode voltage is positive, the SCR is forward biased. If no current pulse is applied at the gate, no current flows in the anode loop as long as V_{AK} is less than the breakover voltage V_{BO} of the SCR. In this case, the SCR is open.

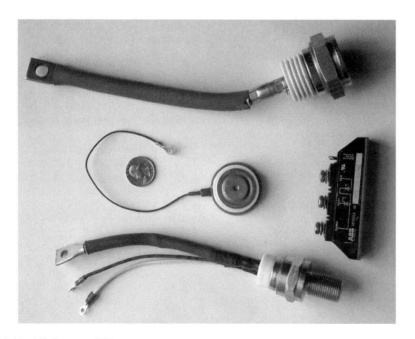

FIGURE 10.10   High-power SCRs.

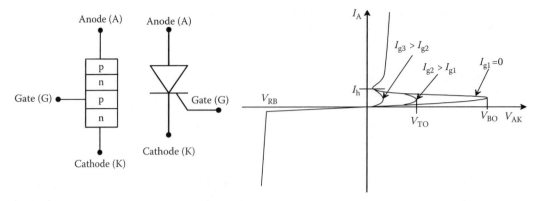

FIGURE 10.11 SCR structure, symbol, and characteristics.

However, if a gate current pulse I_G is applied when $V_{AK} < V_{BO}$, the SCR is closed. The voltage at which the SCR is closed is called turn-on voltage V_{TO}. The higher the gate current, the lower is the turn-on voltage. If we desire to close the SCR when the turn-on voltage is near zero, the gate pulse must be the maximum allowed by the SCR's specification. If the gate signal is removed after the SCR is closed, the SCR remains latched closed unless the anode current falls below the holding value I_h. At or below the holding current, the SCR is opened.

The SCR is highly popular in heavy power applications due to several reasons; among them are the following:

- SCRs are much cheaper than BJT and MOSFETs.
- SCR is triggered by a single pulse, instead of the continuous current needed by the BJT. Thus, the input losses are much less than that for the BJT.
- In ac circuits, the SCR can be self-commutated (opened) without the need for an external commutation circuit. This is because the ac naturally goes through zero values twice in one cycle.
- SCR can be built with much larger current and voltage ratings than the BJT or MOSFET.

10.1.3.2 Silicon Diode for Alternating Current

The SIDAC is a silicon bilateral voltage triggered switch used mainly in low-power control circuits. As seen in the characteristics of the SIDAC in Figure 10.12, when the anode-to-cathode voltage exceeds the breakover voltage of the SIDAC, the device switches on (closes) and the anode current flows in the circuit. The conduction is continuous until the current drops below the minimum

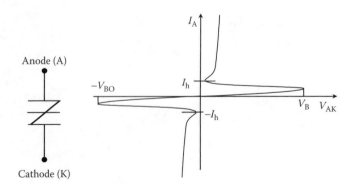

FIGURE 10.12 Characteristics of the SIDAC.

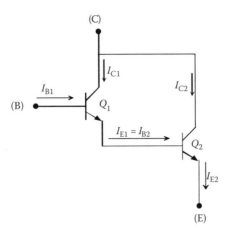

FIGURE 10.13   Darlington transistor.

holding current of the device. The turn-on voltage of the SIDAC is a fixed value and cannot be controlled.

10.1.4 HYBRID POWER ELECTRONIC DEVICES

There are various hybrid devices composed of several components such as BJT and MOSFET, SCR and FET, and Cascaded BJTs. Hybrid structures offer desirable features such as higher switching speeds, higher currents, higher current gains, or higher efficiencies. The hybrid device research is fast growing and newer designs are always developed. Here, the DT and the IGBT are discussed as they are very common in power applications.

10.1.4.1 Darlington Transistor

One of the main problems with the BJT is its low current gain in the saturation region, which demands a high base current. This problem is solved by cascading two bipolar transistors in the configuration shown in Figure 10.13. The collector current of Q_1 is $\beta_1 I_{B1}$, and its emitter current is

$$I_{E1} = I_{B1} + I_{C1} = (1 + \beta_1) I_{B1} \tag{10.5}$$

This emitter current is also equal to the base current of transistor Q_2, and the emitter current of Q_2 is

$$I_{E2} = (1 + \beta_2) I_{B2} = (1 + \beta_2)(1 + \beta_1) I_{B1} \tag{10.6}$$

Hence, the ratio of the emitter current of Q_2 (which is the load current) and the base current of Q_1 (which is the control current) is

$$\frac{I_{E2}}{I_{B1}} = (1 + \beta_1)(1 + \beta_2) \tag{10.7}$$

EXAMPLE 10.3

A BJT has $\beta = 9$ in the saturation region. When the transistor is closed, the emitter current is 100 A. Compute the following:

(a) Base current to keep the transistor closed.
(b) If two of these transistors are connected in Darlington configuration, compute the new base current that produces the same value of emitter current.

Solution

(a)
$$I_B = \frac{I_E}{1+\beta} = \frac{100}{10} = 10\,\text{A}$$

(b) With Darlington connection

$$I_B = \frac{I_{E2}}{(1+\beta_1)(1+\beta_2)} = \frac{100}{10 \times 10} = 1\,\text{A}$$

Notice that the base current of this DT is 10% of the base current of the single BJT.

10.1.4.2 Insulated Gate Bipolar Transistor

BJTs are very reliable devices that can operate at high switching frequencies. However, they are low current gains switching devices that demand high base currents to keep the devices latched in the closed position. This requires triggering circuits that are bulky, expensive, and less efficient. The MOSFETs, on the other hand, are voltage-controlled devices and their triggering circuits are much simpler and less expensive. For these reasons, power electronic engineers have merged these two devices into the configuration shown in Figure 10.14 to produce a high current–voltage triggered device. This device is called IGBT. The symbol and the characteristics of the IGBT are shown in Figure 10.15.

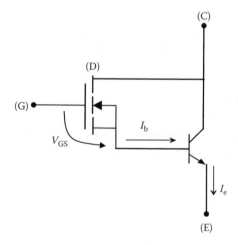

FIGURE 10.14 MOSFET and bipolar circuit.

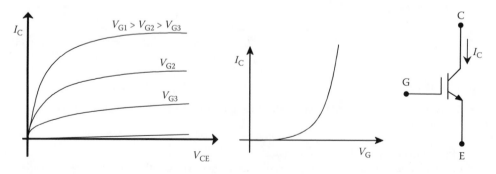

FIGURE 10.15 Characteristics of IGBT.

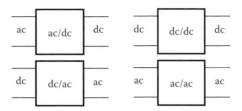

FIGURE 10.16 Four types of converters.

10.2 SOLID-STATE SWITCHING CIRCUITS

Solid-state switching circuits have various configurations and are often called converters. There are four types of converters and are shown in Figure 10.16. The ac/dc converter converts any ac waveform into a dc waveform with adjustable voltage. The dc/dc converter converts a dc waveform into an adjustable voltage dc waveform. The dc/ac converter converts a dc waveform into ac with adjustable voltage and frequency. The ac/ac converter converts the fixed voltage, fixed frequency waveform into waveforms with adjustable voltage and frequency.

10.2.1 AC/DC CONVERTERS

These are very common converters used to drive dc equipment from ac sources such as power supplies and chargers. There are three main types of ac/dc converters: fixed voltage, variable voltage, and fixed current. Fixed voltage circuits produce constant voltage across the load and the circuit is known as a rectifier circuit. For variable voltage across the load, a switching device such as the BJT, MOSFET, or SCR is used to regulate the flow of current in the external circuit, thus controlling the voltage across the circuit. The fixed current circuit is used in applications such as battery chargers and constant torque of electromechanical devices.

10.2.1.1 Rectifier Circuits

The simplest form of an ac/dc converter is the half-wave diode rectifier circuit shown in Figure 10.17. The circuit consists of an ac source of potential v_s, a load resistance R, and a diode connected between the source and the load. Because of the diode, the current flows only in one direction when the voltage source is in the positive half of its cycle; the diode is then forward biased. Hence, the load voltage v_t is only present when the current flows in the circuit as shown in Figure 10.18.

A full-wave rectifier circuit is shown in Figure 10.19. Because the diodes are connected in the bridge configuration as shown in the figure, the current of the load flows in the same direction whether the source voltage is in the positive or negative half of its ac cycle. In the positive half cycle, point A has higher potential than point B. Hence, diodes D_1 and D_2 are forward biased and the current flows as shown by the solid arrows. During the negative half, point B has higher

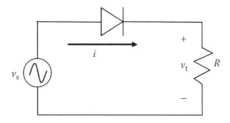

FIGURE 10.17 Half-wave rectifier circuit.

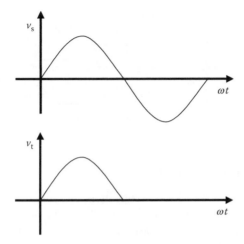

FIGURE 10.18 Waveforms of a half-wave rectifier circuit.

potential than point A. Hence, D_3 and D_4 are forward biased and the current flows as shown by the dashed arrows. In either half of the ac cycle, the current in the load is unidirectional, hence, it is dc. The waveforms of the circuit are shown in Figure 10.20.

Let us assume that the source voltage is a sinusoidal expressed by the equation

$$v_s = V_{max} \sin(\omega t) \qquad (10.8)$$

The dc voltage across the load can be expressed by the average value of v_t. For the half-wave rectifier circuit in Figure 10.17, the average voltage across the load is

$$V_{ave\text{-}hw} = \frac{1}{2\pi} \int_0^{2\pi} v_t \, d\omega t = \frac{1}{2\pi} \int_0^{\pi} v_s \, d\omega t = \frac{1}{2\pi} \int_0^{\pi} V_{max} \sin(\omega t) \, d\omega t = \frac{V_{max}}{\pi} \qquad (10.9)$$

For the full-wave rectifier circuit in Figure 10.19, the average voltage is

$$V_{ave\text{-}fw} = \frac{1}{2\pi} \int_0^{2\pi} v_t \, d\omega t = \frac{1}{\pi} \int_0^{\pi} v_s \, d\omega t = \frac{1}{\pi} \int_0^{\pi} V_{max} \sin(\omega t) \, d\omega t = \frac{2V_{max}}{\pi} \qquad (10.10)$$

Notice that the average voltage across the load for a full-wave circuit is double that for a half-wave circuit.

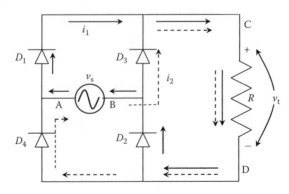

FIGURE 10.19 Full-wave rectifier circuit.

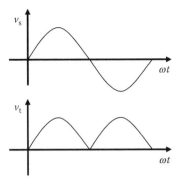

FIGURE 10.20 Waveforms of a full-wave rectifier circuit.

The current and voltage are often expressed by their root mean square (rms) values. As given in Chapter 7, the rms voltage is defined as

$$V_{rms} = \sqrt{\frac{1}{2\pi} \int_0^{2\pi} [v(t)]^2 \, d\omega t} \tag{10.11}$$

Hence, the rms voltage across the load for the half-wave rectifier circuit is

$$V_{rms\text{-}hw} = \sqrt{\frac{1}{2\pi} \int_0^{2\pi} v_t^2 \, d\omega t} = \sqrt{\frac{1}{2\pi} \int_0^{\pi} (V_{max} \sin \omega t)^2 \, d\omega t}$$

$$V_{rms\text{-}hw} = \sqrt{\frac{V_{max}^2}{4\pi} \int_0^{\pi} (1 - \cos 2\omega t) \, d\omega t} = \frac{V_{max}}{2} \tag{10.12}$$

For a full-wave rectifier circuit, the rms voltage across the load is

$$V_{rms\text{-}fw} = \sqrt{\frac{1}{2\pi} \int_0^{2\pi} v_t^2 \, d\omega t} = \sqrt{\frac{1}{\pi} \int_0^{\pi} (V_{max} \sin \omega t)^2 \, d\omega t}$$

$$V_{rms\text{-}fw} = \sqrt{\frac{V_{max}^2}{2\pi} \int_0^{\pi} (1 - \cos 2\omega t) \, d\omega t} = \frac{V_{max}}{\sqrt{2}} \tag{10.13}$$

Notice that the rms voltage across the load for the full-wave rectifier circuit is the same as the rms for the source voltage. The rms current of the load in either circuit can be computed as

$$I_{rms} = \frac{V_{rms}}{R} \tag{10.14}$$

The electric power consumed by the resistive load is

$$P = \frac{V_{rms}^2}{R} = I_{rms}^2 R \tag{10.15}$$

Hence, for half- and full-wave rectifier circuits, the load powers are

$$P_{hw} = \frac{V^2_{rms-hw}}{R} = \frac{V^2_{max}}{4R}$$
$$P_{fw} = \frac{V^2_{rms-fw}}{R} = \frac{V^2_{max}}{2R}$$

(10.16)

EXAMPLE 10.4

A full-wave rectifier circuit converts a 120 V (rms) source into dc. The load of the circuit is a 10 Ω resistance. Compute the following:

(a) Average voltage across the load
(b) Average voltage of the source
(c) rms voltage of the load
(d) rms current of the load
(e) Power consumed by the load

Solution

(a)
$$V_{ave-fw} = \frac{2V_{max}}{\pi} = \frac{2(\sqrt{2}\,120)}{\pi} = 108 \text{ V}$$

(b) Average voltage of the source is zero because the source waveform is symmetrical across the time axis.

(c)
$$V_{rms-fw} = \frac{V_{max}}{\sqrt{2}} = 120 \text{ V}$$

(d)
$$I_{rms-fw} = \frac{V_{rms-fw}}{R} = \frac{120}{10} = 12 \text{ A}$$

(e)
$$P_{fw} = \frac{V^2_{rms-fw}}{R} = 1.44 \text{ kW}$$

10.2.1.2 Voltage-Controlled Circuits

The simple half-wave SCR converter in Figure 10.21 allows us to control the voltage across the load. The circuit consists of an ac source of potential v_s, a load resistance R, and an SCR connected between the source and the load. The triggering circuit of the SCR is not shown in the figure.

The waveforms of the circuit are shown in Figure 10.22. When the SCR is open, the current of the circuit is zero and the load voltage is also zero. When the SCR is triggered at $\omega t = \alpha$, the SCR is closed and the voltage across the load resistance is equal to the source voltage. Since the source is a sinusoidal waveform, the current of the circuit is zero at π. Hence, the SCR is opened at π (the SCR

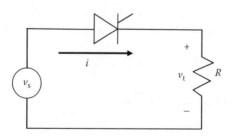

FIGURE 10.21 Half-wave SCR circuit.

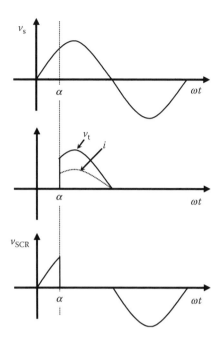

FIGURE 10.22 Waveforms of a half-wave SCR circuit.

opens when the current is below its holding value, which is very small). The period during which the current flows in the circuit is called the conduction period γ. Hence,

$$\gamma = \pi - \alpha \tag{10.17}$$

The full-wave SCR circuit is shown in Figure 10.23. It consists of two SCRs and two diodes in bridge configuration. The waveforms of the circuit are shown in Figure 10.24. The current flow in the full-wave SCR circuit is similar to that explained for the full-wave rectifier circuit except that the current starts flowing after the triggering pulse is applied on S_1 at α during the positive half of the ac cycle, and on S_2 at $\alpha + 180°$ during the negative half of the cycle.

The average voltage across the load for the half-wave SCR circuit is

$$V_{\text{ave-hw}} = \frac{1}{2\pi} \int_0^{2\pi} v_t \, d\omega t = \frac{1}{2\pi} \int_\alpha^\pi v_s \, d\omega t = \frac{1}{2\pi} \int_\alpha^\pi V_{\text{max}} \sin(\omega t) \, d\omega t$$

$$V_{\text{ave-hw}} = \frac{V_{\text{max}}}{2\pi} (1 + \cos \alpha) \tag{10.18}$$

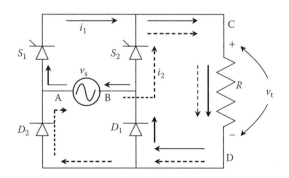

FIGURE 10.23 Full-wave SCR circuit.

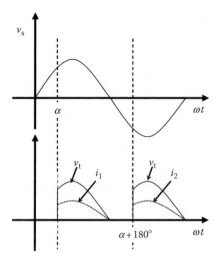

FIGURE 10.24 Waveforms of a half-wave SCR circuit.

For the full-wave SCR circuit, the average voltage across the load resistance is

$$V_{\text{ave-fw}} = \frac{1}{2\pi} \int_0^{2\pi} v_t \, d\omega t = \frac{1}{\pi} \int_\alpha^\pi v_s \, d\omega t = \frac{1}{\pi} \int_\alpha^\pi V_{\max} \sin(\omega t) \, d\omega t$$

(10.19)

$$V_{\text{ave-fw}} = \frac{V_{\max}}{\pi}(1 + \cos\alpha)$$

The rms voltage across the load for the half-wave SCR circuit can be computed as

$$V_{\text{rms-hw}} = \sqrt{\frac{1}{2\pi} \int_0^{2\pi} v_t^2 \, d\omega t} = \sqrt{\frac{1}{2\pi} \int_\alpha^\pi (V_{\max} \sin\omega t)^2 \, d\omega t}$$

(10.20)

$$V_{\text{rms-hw}} = \sqrt{\frac{V_{\max}^2}{4\pi} \int_\alpha^\pi (1 - \cos 2\omega t) \, d\omega t} = \frac{V_{\max}}{2}\sqrt{\left(1 - \frac{\alpha}{\pi} + \frac{\sin 2\alpha}{2\pi}\right)}$$

A similar process can be used to compute the rms voltage across the load for the full-wave SCR circuit

$$V_{\text{rms-fw}} = \sqrt{\frac{1}{2\pi} \int_0^{2\pi} v_t^2 \, d\omega t} = \sqrt{\frac{1}{\pi} \int_\alpha^\pi (V_{\max} \sin\omega t)^2 \, d\omega t}$$

(10.21)

$$V_{\text{rms-fw}} = \frac{V_{\max}}{\sqrt{2}}\sqrt{\left(1 - \frac{\alpha}{\pi} + \frac{\sin 2\alpha}{2\pi}\right)}$$

The rms voltage across the load for the full-wave SCR circuit is shown in Figure 10.25. As you can see, the load voltage is controlled from zero to $V_{\max}/\sqrt{2}$ by adjusting the triggering angle.

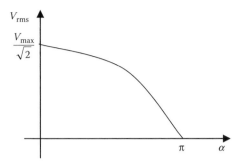

FIGURE 10.25   Root mean square voltage of the load for the full-wave SCR circuit.

The power across the load can be computed as

$$P = \frac{V_{rms}^2}{R} = I_{rms}^2 R$$

$$P_{hw} = \frac{V_{max}^2}{8\pi R}(2(\pi - \alpha) + \sin 2\alpha)$$ (10.22)

$$P_{fw} = \frac{V_{max}^2}{4\pi R}(2(\pi - \alpha) + \sin 2\alpha)$$

EXAMPLE 10.5

A full-wave SCR converter circuit is used to regulate power across a 10 Ω resistance. The voltage source is 120 V (rms). At what triggering angles power across the load 1 kW? Compute the average and rms currents of the load.

Solution

The power expression in Equation 10.22 is nonlinear with respect to the triggering angle α.

$$P_{fw} = \frac{V_{max}^2}{4\pi R}[2(\pi - \alpha) + \sin(2\alpha)] = \frac{(\sqrt{2} \times 120)^2}{4\pi \times 10}[2(\pi - \alpha) + \sin(2\alpha)] = 1000$$

$$2\alpha - \sin(2\alpha) = 1.92$$

The solution of the above nonlinear equation is iterative.

$$\alpha = 71.9°$$

To compute the average current of the load, we need to compute the average voltage across the load as given by Equation 10.19.

$$V_{ave-fw} = \frac{V_{max}}{\pi}(1 + \cos\alpha) = \frac{\sqrt{2} \times 120}{\pi}(1 + \cos 71.9°) = 70.8 \text{ V}$$

$$I_{ave-fw} = \frac{V_{ave-fw}}{R} = 7.08 \text{ A}$$

For the rms current, we can compute the rms voltage first, and then divide the voltage by the resistance of the load. Another simple method is to use the power formula in Equation 10.15.

$$I_{rms} = \sqrt{\frac{P_{fw}}{R}} = 10 \text{ A}$$

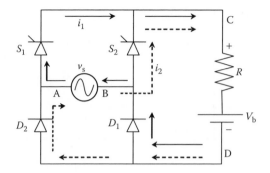

FIGURE 10.26 Full-wave charger circuit.

10.2.1.3 Constant Current Circuits

Constant current circuits are used to charge batteries as well as to drive electric motors in constant torque applications. The current must be controlled during the charging process of the battery to prevent the battery from being damaged by excessive charging currents. The simplest charging circuit is the full-wave shown in Figure 10.26. The load of this circuit is the battery and the resistance R represents the internal resistance of the battery.

The range of the triggering angle of the charger circuit is less than the range for the full-wave SCR circuit in Figure 10.23. This is because the minimum triggering angle and the conduction period depend on the voltage of the battery. Examine the circuit in Figure 10.26 and its waveforms in Figure 10.27. During the first half of the ac cycle, S_1 can only close when the source voltage is higher than the battery voltage (when $\beta > \alpha \geq \alpha_{min}$), where α_{min} is the angle at which the source voltage is equal to the battery voltage. If the triggering angle of the SCRs $\alpha < \alpha_{min}$, the voltage across S_1 or D_1 is negative and the SCRs cannot close. At α_{min},

$$V_b = V_{max} \sin \alpha_{min} \tag{10.23}$$

The current to the battery will flow as long as $V_b < v_s$. Hence, the current is commutated at β, where

$$\beta = 180° - \alpha_{min} \tag{10.24}$$

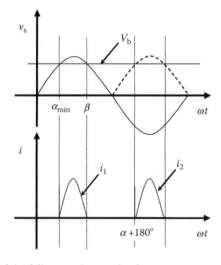

FIGURE 10.27 Waveforms of the full-wave charger circuit.

During the charging process, the voltage of the battery changes, so the triggering angle must change accordingly to maintain the current constant. When the voltage of the battery is low, the triggering angle increases so that the current is reduced to the level safe for charging the battery. When the voltage of the battery increases, the current is reduced so that the triggering angle is reduced to compensate for the current reduction.

EXAMPLE 10.6

A 120 V full-wave battery charger is designed to provide 0.1 A of charging current. A battery set with a 1 Ω internal resistance is connected across the charger. At the beginning of the charging process, the voltage of the battery is 45 V. After 3 h of charging, the voltage of the battery is 50 V. Compute the triggering angle of the converter at these two times. Also compute the conduction periods.

Solution

The instantaneous charging current can be expressed as

$$i = \frac{v_s - V_b}{R}$$

Hence, the average charging current is

$$I_{ave} = \frac{1}{2\pi} \int_0^{2\pi} i \, d\omega t = \frac{1}{\pi} \int_0^{\pi} i \, d\omega t$$

$$I_{ave} = \frac{1}{R\pi} \int_\alpha^\beta (V_{max} \sin \omega t - V_b) \, d\omega t = \frac{1}{R\pi} [V_{max}(\cos \alpha - \cos \beta) - V_b(\beta - \alpha)]$$

At the Beginning of the Charging Process
   We can compute β by using Equations 10.23 and 10.24.

$$\beta = 180° - \alpha_{min} = 180° - \sin^{-1}\left(\frac{V_b}{V_{max}}\right) = 180° - \sin^{-1}\left(\frac{45}{\sqrt{2} \times 120}\right) = 164.62°$$

Hence,

$$I_{ave} = \frac{1}{R\pi}[V_{max}(\cos \alpha - \cos \beta) - V_b(\beta - \alpha)]$$

$$0.1 = \frac{1}{\pi}\left[\sqrt{2} \times 120(\cos \alpha - \cos 164.62) - 45\left(164.62\frac{\pi}{180} - \alpha\right)\right]$$

$$\alpha \approx 161°$$

The conduction period γ is

$$\gamma = \beta - \alpha = 164.62 - 161 = 3.62°$$

After 3 h:

$$\beta = 180° - \alpha_{min} = 180° - \sin^{-1}\left(\frac{V_b}{V_{max}}\right) = 180° - \sin^{-1}\left(\frac{50}{\sqrt{2} \times 120}\right) = 162.86°$$

Hence,

$$I_{ave} = \frac{1}{R\pi}[V_{max}(\cos \alpha - \cos \beta) - V_b(\beta - \alpha)]$$

$$0.1 = \frac{1}{\pi}\left[\sqrt{2} \times 120(\cos \alpha - \cos 162.86) - 50\left(162.86\frac{\pi}{180} - \alpha\right)\right]$$

$$\alpha \approx 159.2°$$

The conduction period γ is

$$\gamma = \beta - \alpha = 162.86 - 159.2 = 3.66°$$

10.2.2 DC/DC CONVERTERS

The dc/dc converters are normally designed to provide adjustable dc voltage waveforms at the output. There are three basic types of dc/dc converters:

- Buck converter: This is a step-down converter where the output voltage is less than the input voltage.
- Boost converter: This is a step-up converter where the output voltage is higher than the input voltage.
- Buck–boost converter: This is a step-down/step-up converter where the output voltage can be made either lower or higher than the input voltage.

10.2.2.1 Buck Converter

Figure 10.28 shows a simple buck converter (also known as chopper). The circuit has a transistor as a switching device and the load is connected between the source and the collector of the transistor. The top waveform in the figure is for the base current of the transistor where the transistor is closed throughout the time segment t_{on} of the period τ of one cycle. Only when the transistor is closed, the voltage across the load V_t is equal to the source voltage V_s; otherwise $V_t = 0$. The current in the circuit flows only when the transistor is closed.

The average voltage across the load V_{ave} is

$$V_{ave} = \frac{1}{\tau} \int_0^{t_{on}} V_s \, dt = \left(\frac{t_{on}}{\tau}\right) V_s = KV_s \tag{10.25}$$

where $K = t_{on}/\tau$ is called the duty ratio. The maximum value of K is 1 when $t_{on} = \tau$. In this case, the transistor is always closed and the average load voltage is equal to the source voltage. For any other

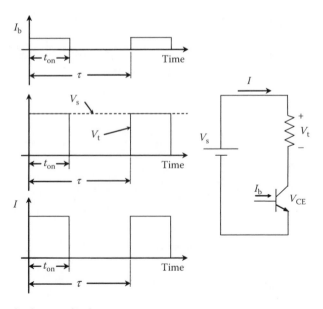

FIGURE 10.28 Simple chopper circuit.

value of K (when $t_{on} < \tau$), the average load voltage is less than the source voltage. The output voltage of the converter can then be controlled by fixing the period τ and adjusting the on-time t_{on}.

EXAMPLE 10.7

The switching frequency of a chopper is 10 kHz, and the source voltage is 40 V. For an average load voltage of 20 V and a load resistance of 10 Ω, compute the following:

(a) Duty ratio
(b) On-time and the switching period
(c) Average voltage across the transistor
(d) Average current of the load
(e) rms voltage across the load
(f) Load power

Solution

(a) As given in Equation 10.25, the duty ratio is

$$K = \frac{V_{ave}}{V_s} = \frac{20}{40} = 0.5$$

(b) Before we compute the on-time, we need to compute the period.

$$\tau = \frac{1}{f} = \frac{1}{10} = 0.1 \text{ ms}$$

$$t_{on} = K\tau = 0.5 \times 0.1 = 0.05 \text{ ms}$$

(c) Average voltage across the transistor $V_{ave\text{-}tr}$ is

$$V_{ave\text{-}tr} = V_s - V_{ave} = 40 - 20 = 20 \text{ V}$$

(d) Average load current is

$$I_{ave} = \frac{V_{ave}}{R} = \frac{20}{10} = 2 \text{ A}$$

(e) rms voltage of the load can be computed using the formula for the rms quantity.

$$V = \sqrt{\frac{1}{\tau} \int_0^{t_{on}} V_s^2 \, dt} = \sqrt{\frac{V_s^2}{\tau} t_{on}} = V_s \sqrt{\frac{t_{on}}{\tau}} = 28.28 \text{ V}$$

(f)
$$P = \frac{V^2}{R} = \frac{28.28^2}{10} = 80 \text{ W}$$

10.2.2.2 Boost Converter

A boost converter can increase the voltage across the load to values higher than the source voltage. A simple circuit for the boost converter is shown in Figure 10.29a. The load of this circuit is the resistance R. The transistor is closed for a time interval t_{on} and is open during t_{off}. When the transistor is closed, the current i_{on} flows and the energy is stored in the inductor as depicted in Figure 10.29b. In this process, the voltage across the inductor v_L is

$$v_L = L\frac{di_{on}}{dt} \approx L\frac{\Delta i_{on}}{t_{on}} \tag{10.26}$$

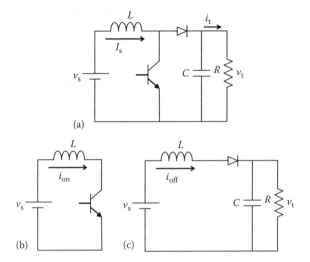

FIGURE 10.29 Simple boost converter.

where Δi_{on} is the change in the current during the period t_{on}. Since the voltage across the inductor is equal to the source voltage when the transistor is closed, Equation 10.26 can be rewritten as

$$V_s = L\frac{\Delta i_{on}}{t_{on}} \tag{10.27}$$

When the transistor is opened, the stored energy in the inductor is transferred to the load via the diode. The inductor current in this case is i_{off} as depicted in the circuit in Figure 10.29c. If we ignore the current through the capacitor, the voltage loop can be written as

$$v_t = V_s - v_L = V_s + L\frac{\Delta i_{off}}{t_{off}} \tag{10.28}$$

Because the inductor is producing energy during t_{off}, and because the current through the inductor does not change its direction, the voltage polarity across the inductance is reversed. This is the reason for the positive sign (+) in front of L in Equation 10.28. At steady state when $\Delta i_{on} = \Delta i_{off}$ as shown in the current waveforms in Figure 10.30, Equation 10.28 can be rewritten as

$$v_t = V_s + L\frac{\Delta i_{off}}{t_{off}} = V_s\left(1 + \frac{t_{on}}{t_{off}}\right) \tag{10.29}$$

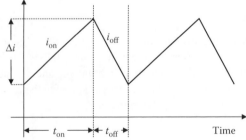

FIGURE 10.30 Waveform of a boost converter.

Equation 10.29 shows that the voltage across the load can be adjusted by adjusting t_{on} and t_{off}. Furthermore, the load voltage is always higher than the source voltage when $t_{on}/t_{off} > 0$.

The capacitor in the circuit is used as a filter to reduce the voltage ripples across the load, and the diode is used to block the capacitor from discharging through the transistor when it is closed. If the capacitance is large enough, v_t can be assumed ripple free. Hence, we can rewrite Equation 10.29 as

$$V_t = V_s + L\frac{\Delta i_{off}}{t_{off}} = V_s\left(1 + \frac{t_{on}}{t_{off}}\right) \tag{10.30}$$

where V_t is the average voltage across the load. If we assume that the average currents and the average voltages are much larger than their ripples, the rms and the average values can be considered equal. Hence, we can compute the input and output powers by using the average values.

$$\begin{aligned} P_{in} &= V_s I_s \\ P_{out} &= V_t I_t \end{aligned} \tag{10.31}$$

where
P_{in} is the input power to the converter
P_{out} is the output power consumed by the load
I_s and I_t are the average currents of the source and load, respectively

If we further assume that the components of the circuit are ideal, the input power to the converter is equal to the output power. Hence,

$$V_t I_t = V_s I_s \tag{10.32}$$

EXAMPLE 10.8

A boost converter is used to step up a 20 V source to 50 V across a load resistance. The switching frequency of the transistor is 5 kHz, and the load resistance is 10 Ω. Compute the following:

(a) Value of the inductance that would limit the current ripple at the source side to 100 mA
(b) Average current of the load
(c) Power delivered by the source
(d) Average current of the source

Solution

(a) We can use Equation 10.27 to compute the inductance. But first, we need to compute t_{on} using Equation 10.30.

$$V_t = V_s\left(1 + \frac{t_{on}}{t_{off}}\right)$$

$$50 = 20\left(1 + \frac{t_{on}}{t_{off}}\right)$$

$$t_{on} = 1.5 \times t_{off}$$

Since the switching frequency is 5 kHz

$$t_{on} + t_{off} = \frac{1}{5} = 0.2 \text{ ms}$$

Then,

$$t_{on} = 1.5 \times t_{off} = 1.5 \times (0.2 - t_{on})$$
$$t_{on} = 0.12 \text{ ms}$$

To compute the value of the inductance, we can use Equation 10.27.

$$V_s = L \frac{\Delta i_{on}}{t_{on}}$$
$$20 = L \frac{100}{0.12}$$
$$L = 24 \text{ mH}$$

(b)
$$I_t = \frac{V_t}{R} = \frac{50}{10} = 5 \text{ A}$$

(c) Power delivered by the source is the same power consumed by the load assuming that the system components are all ideal.

$$P = V_t \times I_t = 50 \times 5 = 250 \text{ W}$$

(d) Average current of the source can be computed using the power equation in the previous step.

$$I_s = \frac{P}{V_s} = \frac{250}{20} = 12.5 \text{ A}$$

10.2.2.3 Buck–Boost Converter

The buck–boost converter has the same components as the boost converter, but is structured differently as shown in Figure 10.31a. When the transistor is closed, as shown in Figure 10.31b, the current i_{on} flows through the inductor, and energy is acquired by the inductor. When the transistor is opened, the inductor delivers its stored energy to the load as shown in Figure 10.31c. The inductor

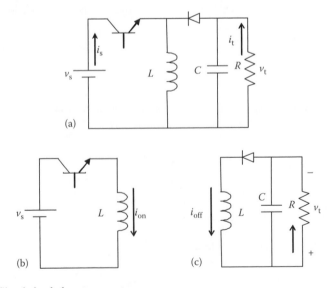

FIGURE 10.31 Simple buck–boost converter.

current in this period is i_{off}. The capacitor is used as a filter to minimize the voltage ripples across the load, and the diode is used to prevent the load from being energized when the transistor is closed. The capacitor current can be assumed very small compared with the load current.

When the transistor is closed, the voltage across the inductor is

$$v_L = L\frac{di_{on}}{dt} \approx L\frac{\Delta i_{on}}{t_{on}} \tag{10.33}$$

Note that when the transistor is closed, the voltage across the inductor is equal to the source voltage. Hence,

$$V_s = L\frac{\Delta i_{on}}{t_{on}} \tag{10.34}$$

When the transistor is open, the voltage across the inductor reverses because the inductor becomes a source of energy and the voltage across the inductor is

$$v_L = -L\frac{\Delta i_{off}}{t_{off}} \tag{10.35}$$

The voltage across the inductor when the transistor is open is also equal to the load voltage. Hence,

$$v_t = -L\frac{\Delta i_{off}}{t_{off}} \tag{10.36}$$

Assume that the capacitor is large enough so that the voltage v_t is equal to the average voltage across the load V_t. At steady state when $\Delta i_{on} = \Delta i_{off}$, Equations 10.34 and 10.36 can be combined as

$$V_t = -V_s\frac{t_{on}}{t_{off}} \tag{10.37}$$

As seen in Equation 10.37, the average value of the load voltage can be controlled by adjusting the ratio t_{on}/t_{off}. If the on-time is zero, the load voltage is zero. If $0 < t_{on} < t_{off}$, the load voltage is lower than the source voltage. If $t_{on} = t_{off}$, the load voltage is equal to the source voltage. If $t_{on} > t_{off}$, the load voltage is higher than the source voltage. Keep in mind that Equation 10.37 is not valid when $t_{off} = 0$; that is, the transistor never opens. The system in this case is unstable as the current i_{on} will reach very high values because the inductor acts as a short circuit when continuous dc flows through it.

If we assume that the circuit's components are ideal, the input power to the converter is equal to the output power. Hence,

$$P_{in} = P_{out} = V_t I_t = V_s I_s \tag{10.38}$$

EXAMPLE 10.9

A buck–boost converter with an input voltage of 20 V is used to regulate the voltage across a 10 Ω load. The switching frequency of the transistor is 5 kHz. Compute the following:

(a) On-time of the transistor to maintain the output voltage at 10 V
(b) Output power
(c) On-time of the transistor to maintain the output voltage at 40 V
(d) Output power

Solution

(a) Period can be computed from the switching frequency f

$$\tau = \frac{1}{f} = \frac{1}{5} = 0.2 \text{ ms}$$

Use Equation 10.37 to compute the time ratio. We need to use only the magnitudes of the voltages.

$$\frac{t_{on}}{t_{off}} = \frac{V_t}{V_s} = \frac{10}{20} = 0.5$$

Hence,

$$t_{on} = 0.0667 \text{ ms}$$

(b) Output power

$$P = \frac{V_t^2}{R} = 10 \text{ W}$$

(c)

$$\frac{t_{on}}{t_{off}} = \frac{V_t}{V_s} = \frac{40}{20} = 2$$

Hence,

$$t_{on} = 0.1333 \text{ ms}$$

(d) Output power

$$P = \frac{V_t^2}{R} = \frac{40^2}{10} = 160 \text{ W}$$

As you have seen, the inductor plays a major role in dc/dc converters. It acts as a temporary storage device that transfers the energy from the source to the load. This inductor can be built using a toroid as explained in Appendix D.

10.2.3   DC/AC CONVERTERS

The dc/ac converter is also known as an inverter. The inverter is used in applications such as uninterruptible power supplies, variable speed drives, and dc transmission lines. It is also common to use this converter to convert the 12 V dc waveform of automobiles into household alternating voltage to power small equipment such as personal computers and televisions.

There are two types of dc/ac converters: single phase and multiphase. The first is used in low-to-medium power applications and the second is used with heavy industrial loads.

10.2.3.1   Single-Phase dc/ac Converter

Figure 10.32 shows a simple dc/ac converter in H-bridge configuration. It consists of a dc source, four transistors, and a load. The desired frequency of the output waveform determines the switching period of the transistors. During the first half of the period, Q_1 and Q_2 are closed, and during the second half, Q_3 and Q_4 are closed. The voltage waveform of the circuit is also shown in the figure. During the first half of the period, the current i_1 (in solid arrows) flows through the load. Similarly, during the second half, the current i_2 (in dashed arrows) flows through the load in the opposite direction to i_1. Thus, the load current is alternating. Notice that the waveform in Figure 10.32 is not sinusoidal, but still considered an ac waveform.

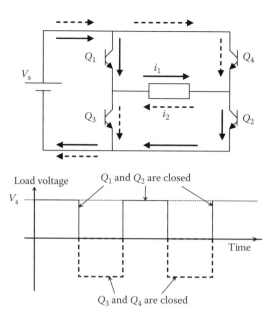

FIGURE 10.32   Single-phase dc/ac converter.

EXAMPLE 10.10

The source voltage of a single-phase dc/ac converter is 100 V. The switching frequency of the transistors is 1 kHz. Compute the following:

(a) rms voltage across the load
(b) Average voltage across the load

Solution

(a) As given in Chapter 7, the rms voltage of an arbitrary waveform v is

$$V = \sqrt{\frac{1}{\tau} \int_0^\tau v^2 \, dt}$$

where $\tau = \dfrac{1}{f} = 1$ ms.

$$V = \sqrt{\frac{1}{\tau} \int_0^\tau v^2 \, dt} = \sqrt{\frac{2}{\tau} \int_0^{\tau/2} V_s^2 \, dt}$$

where V_s is the source voltage.

$$V = \sqrt{\frac{2}{\tau} 10^4 \int_0^{\tau/2} dt} = \sqrt{\frac{2}{\tau} 10^4 \frac{\tau}{2}} = 100 \ \text{V}$$

The rms voltage across the load is equal to the dc voltage of the source. This is because the ac waveform is rectangular and symmetrical.

(b) Average voltage across the load is zero since the waveform is symmetrical around the time axis.

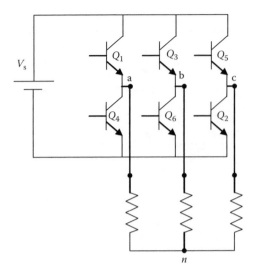

FIGURE 10.33   Three-phase dc/ac converter.

10.2.3.2   Three-Phase dc/ac Converter

Three-phase waveforms can be obtained using the dc/ac converter shown in Figure 10.33. The converter is composed of six transistors, a dc source, and a three-phase load. This circuit is also known as *six-pulse* converter. The transistors are arranged in three legs, each leg has two transistors, and the midpoint of each leg is connected to one of the terminals of the three-phase load. The switching sequence of the transistors is shown at the top part of Figure 10.34. The cycle is divided into six intervals, so that each interval is 60°. Each transistor is closed for three consecutive intervals (180°) and opened for the following three consecutive intervals. The switching of the transistors is based on their ascending numbers. For example, transistor Q_1 is closed first and then Q_2 after one time interval, Q_3 after additional time interval, and so on. During any time interval, one transistor per leg is closed. No two transistors on the same leg are closed at the same time as this creates a short circuit across the supply.

Now let us study the effect of the switching pattern during the first time interval when transistors Q_1, Q_5, and Q_6 are closed. This interval is depicted in Figure 10.35a, which shows only the transistors that are closed. During this interval, the current from the source I is divided into two equal components; one passes through Q_1 and the other through Q_5. At the neutral point n, the two currents are summed up and returned to the source through Q_6. The load currents during this interval are

$$i_a = i_c = \frac{I}{2}$$
$$i_b = -I$$

(10.39)

These load currents during the first interval are shown at the bottom part of Figure 10.34. Now let us study the second interval when Q_1, Q_2, and Q_6 are closed. The flow of the current during this interval is shown in Figure 10.35b. The current from the source passes through Q_1 and at the neutral point branches into two equal components; one goes through Q_2 and the other through Q_6. The load currents during this interval are

$$i_a = I$$
$$i_b = i_c = -\frac{I}{2}$$

(10.40)

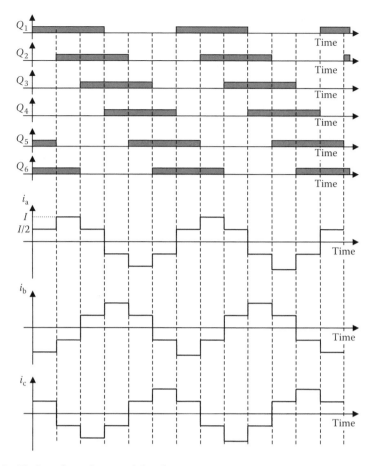

FIGURE 10.34 Timing of transistors and the phase currents.

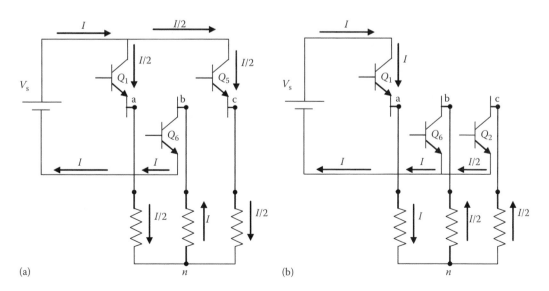

FIGURE 10.35 Active transistors and current flow during the first two intervals.

If you follow the same procedure for the other intervals, you will get the current waveforms in Figure 10.34. Note the following:

- Current waveforms are symmetrical around the time axis.
- All phase currents have the same maximum value.
- i_b lags i_a by two time intervals, that is, 120°.
- i_c lags i_b by two time intervals, that is, 120°.

Hence, the output currents of the converter are balanced three-phase waveforms.

The frequency of the load current is dependent on the interval time t_i. Since the cycle is composed of six segments, the frequency of the ac waveform is

$$f = \frac{1}{6t_i} \tag{10.41}$$

To change the frequency of the ac waveform, the time of the interval must change.

EXAMPLE 10.11

A three-phase dc/ac converter is used to power a three-phase, Y-connected, resistive load of 10 Ω (per phase). The dc voltage is 300 V. Compute the following:

(a) Time interval if the desired frequency of the ac waveform is 200 Hz
(b) Current of the dc source
(c) rms current of the load
(d) rms voltage across each phase of the load

Solution

(a) Use Equation 10.41 to compute the interval t_i

$$t_i = \frac{1}{6f} = \frac{1}{6 \times 200} = 833 \ \mu s$$

(b) During any given interval, the load has two of its phases in parallel, and the combination is in series with the third phase. In the circuit in Figure 10.35a, the load resistances of phases a and c are in parallel, and this combination is in series with the resistance of phase b. Similar combinations are true during all other time intervals. Hence,

$$R_{total} = 10 + \frac{10 \times 10}{10 + 10} = 15 \ \Omega$$

Then the source current is

$$I = \frac{V_s}{R_{total}} = \frac{300}{15} = 20 \ A$$

(c) rms current of phase a is

$$I_a = \sqrt{\frac{1}{6t_i} \int_0^{6t_i} i_a^2 \, dt}$$

If you examine the waveform of i_a for one period, you will discover that the magnitude of the current is equal to I during two time intervals and is equal to $I/2$ during four intervals. Hence, the rms current of phase a can be written as

$$I_a = \sqrt{\frac{1}{6t_i}\left(\int_0^{2t_i} I^2\, dt + \int_0^{4t_i} \left(\frac{I}{2}\right)^2\right)} = \sqrt{\frac{I^2}{6t_i}(3t_i)} = \frac{I}{\sqrt{2}} = 14.14\,\text{A}$$

This rms current is the same as that computed for the purely sinusoidal waveform where I is the maximum current of the instantaneous waveform.

(d) rms phase voltage of the load is

$$V_{an} = I_a R = 14.14 \times 10 = 141.4\,\text{V}$$

10.2.4 AC/AC CONVERTERS

A simple ac/ac converter is shown in Figure 10.36 and the waveforms of the circuit are shown in Figure 10.37. The circuit consists of two SCRs connected in parallel in the back-to-back configuration.

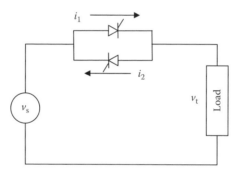

FIGURE 10.36 ac/ac Converter.

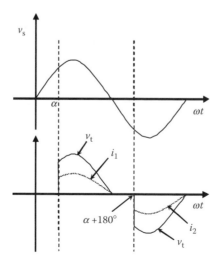

FIGURE 10.37 Waveform of the ac/ac converter in Figure 10.36.

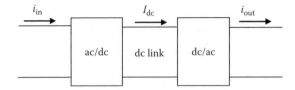

FIGURE 10.38   ac/ac Converter with a dc link.

When the voltage of the source is in the positive half of its ac cycle and the top SCR is triggered, the SCR closes and the current i_1 flows through the load. When the source voltage is at its negative half of the cycle, the bottom SCR can be triggered and the load current is i_2, which is opposite to the direction of i_1.

If the top SCR is triggered at α, and the bottom SCR is triggered at $\alpha + 180°$, the waveform of the load voltage or the load current is symmetrical around the time axis. Therefore, the average voltage or average current of the load is zero.

The rms voltage and current of the load are exactly the same as those discussed for the full-wave ac/dc circuits.

$$
V_{rms} = \sqrt{\frac{1}{2\pi} \int_0^{2\pi} v_t^2 \, d\omega t} = \sqrt{\frac{1}{\pi} \int_\alpha^\pi (V_{max} \sin \omega t)^2 \, d\omega t}
$$

$$
V_{rms} = \frac{V_{max}}{\sqrt{2}} \sqrt{\left(1 - \frac{\alpha}{\pi} + \frac{\sin 2\alpha}{2\pi}\right)}
$$

(10.42)

The power across the load can be computed as

$$
P = \frac{V_{rms}^2}{R} = I_{rms}^2 R
$$

$$
P = \frac{V_{max}^2}{4\pi R} (2(\pi - \alpha) + \sin 2\alpha)
$$

(10.43)

A more elaborate form of ac/ac converter that allows for the adjustment of the output frequency, in addition to the output voltage, is the one shown in Figure 10.38. The system consists of two ac/dc converters connected back-to-back as shown in the figure. In the first converter, the input ac waveform is converted into dc through the first ac/dc converter. The dc is then converted back to ac waveform through the dc/ac converter. The connection between the two converters is called *dc link*.

This type of converter is commonly used as an uninterruptible power supply (UPS). A UPS system is shown in Figure 10.39. The system is similar to that shown in Figure 10.38, but a

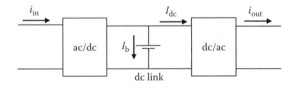

FIGURE 10.39   Normal operation of a UPS system.

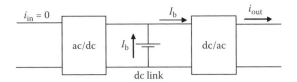

FIGURE 10.40 UPS operation during power outage.

rechargeable battery is added in the dc link. In normal operations, the input ac i_{in} is converted into dc where part of the current I_b charges the battery and the other part I_{dc} is converted into the ac i_{out}. If the input power is lost due to an outage, the energy stored in the battery is used to feed the load through the dc/ac converter as shown in Figure 10.40. Hence, the load power is not interrupted. This system can provide temporary power until the battery is fully discharged.

EXERCISES

1. A bipolar transistor is connected to a resistive load as shown in Figure 10.7. The source voltage $V_{CC} = 40$ V and $R_L = 10$ Ω. In the saturation region, the collector–emitter voltage $V_{CE} = 0.1$ V and $\beta = 5$. While the transistor is in the saturation region, calculate the following:
   (a) Load current
   (b) Load power
   (c) Losses in the collector circuit
   (d) Losses in the base circuit
   (e) Efficiency of the circuit

2. For the transistor in the previous problem, compute the load power and the efficiency of the circuit when the transistor is in the cutoff region. Assume that the collector current is 10 mA in the cutoff region.

3. A BJT operating in the saturation region has a base current of 10 A and a collector current of 50 A. Compute the following:
   (a) Current gain of the transistor in the saturation region
   (b) Losses of the transistor

4. Compute the rms voltage of the following waveform.

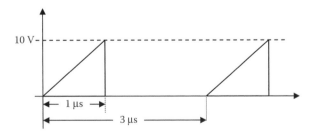

5. A half-wave rectifier circuit converts a 120 V (rms) into dc. The load of the circuit is a 5 Ω resistance. Compute the following:
   (a) Average voltage across the load
   (b) Average voltage of the source
   (c) rms voltage of the load
   (d) rms current of the load
   (e) Power consumed by the load

6. A half-wave SCR converter circuit is used to regulate the power across a 10 Ω resistance. When the triggering angle is 30°, the power consumed by the load is 500 W. Compute the rms voltage of the source and the average and rms currents of the load.

7. A ac/dc half-wave SCR circuit is used to energize a resistive load. At a triggering angle of 30°, the average voltage across the load is 45 V.
   (a) Compute the source voltage in rms.
   (b) If a full-wave circuit is used, while the triggering angle is maintained at 30°, compute the average voltage across the load.

8. A Full-wave SCR converter circuit is used to regulate the power across a 10 Ω resistance. The voltage source is 120 V (rms). At a triggering angle of 72°, compute the power across the load and the rms currents of the load.

9. A 120 V full-wave SCR battery charger is designed to provide 1.0 A of charging current. The battery has a 1 Ω internal resistance. At the beginning of the charging process the voltage of the battery set is 60 V. Compute the triggering angle of the converter.

10. A full-wave, ac/dc converter is connected to a resistive load of 5 Ω. The voltage of the ac source is 120 V (rms). If the triggering angle of the converter is 90°, compute the rms voltage across the load and the power consumed by the load.

11. A boost converter has a source voltage $V_s = 20$ V and a load resistance $R = 10$ Ω. If the duty ratio of the transistor is 50%, compute the voltage across the load and the power consumed by the load.

12. A boost converter is used to step up 25 V into 40 V. The switching frequency of the transistor is 1 kHz, and the load resistance is 100 Ω. Compute the following:
   (a) Current ripple when the inductor is 30 mH
   (b) Average current of the load
   (c) Power delivered by the source

13. A buck–boost converter has an input voltage of 30 V. The switching frequency of its transistor is 5 kHz, and its duty ratio is 25%. Compute the average voltage across the load.

14. A buck–boost converter with an input voltage of 40 V is used to regulate the load voltage from 10 to 80 V. The on-time of the transistor is always fixed at 0.1 ms, and the switching frequency is adjusted to regulate the load voltage. Compute the range of the switching frequency.

15. A three-phase dc/ac converter is used to power a three-phase, Y-connected, resistive load of 50 Ω (per phase). The dc voltage is 150 V. Compute the following:
   (a) Frequency of the ac waveform if the time interval is 100 μs
   (b) Current of the dc source
   (c) rms current of the load
   (d) rms of the line-to-line voltage across the load

16. A full-wave ac/dc SCR converter circuit is used to power a resistive load of 10 Ω. The ac voltage is 120 V (rms), and the triggering angle of the SCR is adjusted to 60°. Calculate the following:
   (a) Conduction period
   (b) Average voltage across the load
   (c) Average voltage across the SCRs
   (d) rms voltage across the load
   (e) Average current
   (f) Load power

17. A dc/dc converter consists of a 100 V dc source in series with a 10 Ω load resistance and a bipolar transistor. For each cycle, the transistor is turned on for 200 μs and turned off for 800 μs. Calculate the following:
   (a) Switching frequency of the converter
   (b) Average voltage across the load
   (c) Average load current

(d) rms voltage across the load

(e) rms current of the load

(f) rms power consumed by the load

18. A dc/dc buck converter has an input voltage of 100 V and a duty ratio of 0.2. Compute the following:

(a) Load voltage

(b) Switching frequency of the converter if the on-time is 0.1 ms

19. The load of the full-wave SCR circuit in Figure 10.23 consumes 130 W. The rms voltage across the load is 80 V, and its average voltage is 50 V.

(a) Compute the average voltage across any SCR.

(b) The triggering circuit of one SCR failed and that SCR does not conduct anymore. If the triggering angle of the other SCR is unchanged, compute the average voltage of the load and the load power.

20. A resistive load of 5 Ω is connected to an ac source of 120 V (rms) through a back-to-back SCR circuit (two parallel SCRs with the anode of each one connected to the cathode of the other). The triggering angle of the forward SCR (the one that conducts at the positive half of the cycle) is 30°. The triggering angle of the other SCR is (180° + 30°). Calculate the following:

(a) Average voltage across the load

(b) Power consumed by the load

11 Transformers

The Italian scientist Antonio Pacinotti who invented the transformer in 1860 granted the power industry with one of its most important devices, without which the power grid would not exist. The transformer is used to step up (increase) or step down (decrease) the voltage. This capability allows us to transmit and distribute massive amounts of power in today's extensive power grid. Because power is the product of current and voltage, a large power at a low voltage leads to a high current. Since the current determines the cross section of the transmission line wires, large currents require unrealistic large cross-section wires. These impractical wires are heavy and expensive to manufacture and require enormous towers to carry them. The transformer provides us with the solution; it allows us to increase the voltage of the transmission lines so that the current is reduced. Thus, the cross section of the wire is reasonably small and the cost of the transmission line is dramatically reduced.

Figure 11.1 shows a schematic of a power system with its main transformers. The system has four types of transformers: transmission, distribution, service, and circuit transformers.

1. *Transmission transformers*: These transformers are connected to both ends of the transmission line (Figure 11.2a). At the generating power plants, the transformer steps up the voltage of the generators to very high levels (220 kV–1 MV) so that the current can be reduced substantially. At the other end of the transmission line, the transformer steps down the voltage to a lower level suitable for distribution to load centers. Since transmission transformers carry the bulk power of the system, they are very large in size. The internal losses of the transformer, which are due to the high currents in the windings and core losses, can cause excessive heat inside the container of the transformer. Therefore, the transformer is immersed in a cooling medium such as oil and the container is equipped with a system of pipes and fans to extract heat from the cooling medium much like radiators in automobiles. The medium also increases the dielectric strength of the windings so that the transformer can withstand higher voltages.
2. *Distribution transformers*: These transformers are installed in distribution substations near the load centers (Figure 11.2b). They are designed to reduce the output voltage of the transmission transformers to a lower level of 15–25 kV as shown in Figure 11.1.
3. *Service transformers*: These transformers are located near customers' loads (Figure 11.2c). They are designed to lower the distribution voltage to the household level (120/240 V in the United States). They are normally mounted on power poles, placed inside vaults, or installed inside large buildings.
4. *Circuit transformers*: These are small transformers extensively used in power supplies and electronic circuits where the household voltage is stepped down to the circuit voltage level of a few volts (Figure 11.2d). Other applications of the circuit transformers include impedance matching, filters, and electrical isolation.

11.1 THEORY OF OPERATION

The basic components of a transformer are shown in Figure 11.3. The transformer consists of at least two windings wrapped around a laminated iron alloy core. These windings are electrically isolated from the core by a varnished insulation material capable of withstanding the voltage of the windings. The core provides a low reluctance path for the magnetic flux. An alternating current (ac) voltage

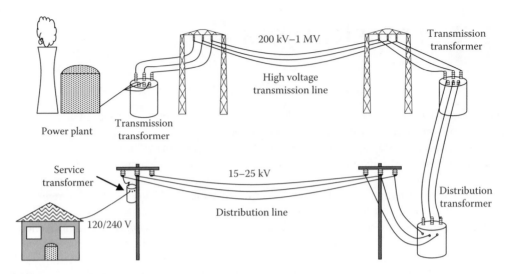

FIGURE 11.1 Various transformers in a power system.

source is connected across one of the windings, and the current of the winding produces flux that passes through the core and links the second winding. Since the current is alternating, the flux in the core changes its direction 120 times/s for the 60 Hz system. This change in flux induces a voltage on the second winding.

The rapid change in the flux induces a voltage in the core itself, thus creating currents that move within the core in circular paths. This current, which is called eddy current, creates heat in the core and results in wasted energy. To reduce eddy current, the electrical resistance of the core must increase. This is done by laminating the core as shown in Figure 11.3. Each laminated plate is insulated by oxidized material or varnish. The cross section of a laminated sheet is much smaller than the total cross section of the core, thus the circular eddy current is substantially reduced.

Assume that an ac voltage source is connected across one of the windings and a load is connected across the other winding as shown in Figure 11.4. The winding connected to the source is called *primary winding*, and the winding connected to the load is called *secondary winding*. The number of turns of the primary winding is N_1 and that for the secondary winding is N_2. The analysis of this transformer is given in the following sections.

11.1.1 VOLTAGE RATIO

According to Faraday's law, the instantaneous voltage e_1 across the primary winding produces a flux ϕ, where

$$e_1 = +N_1 \left(\frac{\mathrm{d}\phi}{\mathrm{d}t} \right)$$

$$\phi = \frac{1}{N_1} \int e_1 \, \mathrm{d}t$$

(11.1)

Assume that the core captures all the flux produced by e_1, and the flux links the secondary winding. According to Faraday's law, the flux induces voltage e_2 across the secondary winding, where

$$e_2 = -N_2 \left(\frac{\mathrm{d}\phi}{\mathrm{d}t} \right)$$

(11.2)

(a)

(b)

(c1)

(c2)

(d)

FIGURE 11.2 (See color insert following page 300.) Various types of transformers. (a) Transmission transformer, (b) distribution transformer, (c) service transformers, and (d) circuit transformer.

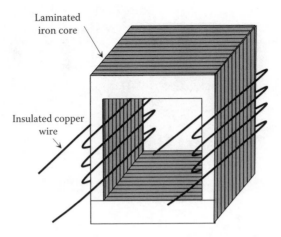

FIGURE 11.3   Main components of a transformer.

The positive sign in Equation 11.1 indicates that e_1 induces the flux, and the negative sign in Equation 11.2 indicates that the flux induces e_2.

The dot at one end of each winding indicates that the potentials at the dotted terminals are in phase; that is, e_1 and e_2 are in phase. Moreover, the current i_1 flowing into the dotted terminal of the primary winding is in phase with the current i_2 leaving the dotted terminal of the secondary windings.

If we are interested in the magnitudes of e_1 and e_2, Equations 11.1 and 11.2 can be rewritten as

$$\frac{e_1}{N_1} = \frac{e_2}{N_2} = \frac{d\phi}{dt} \tag{11.3}$$

Equation 11.3 shows that the voltage per turn is the same for the primary or secondary winding. This is a key equation for the transformer and is only true if no flux escapes the core into air.

In root mean square (rms) quantities, we can rewrite Equation 11.3 as

$$V_T = \frac{E_1}{N_1} = \frac{E_2}{N_2} \tag{11.4}$$

where
  V_T is the voltage per turn
  E is the rms value of e

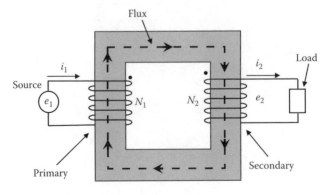

FIGURE 11.4   Flux linkage of the transformer.

Equation 11.4 can also be rewritten as

$$\frac{E_1}{E_2} = \frac{N_1}{N_2} \tag{11.5}$$

The above equation shows that the voltage ratio of the transformer is equal to the turns ratio. If $N_2 < N_1$, the secondary voltage E_2 is less than the primary voltage E_1, and the transformer is called a *step-down transformer*. When $N_2 > N_1$, the secondary voltage E_2 is higher than the primary voltage E_1, and the transformer is called a *step-up transformer*.

EXAMPLE 11.1

A transformer has 4000 turns in the primary winding and 100 turns in the secondary winding. If 120 V is applied to the primary winding, compute the voltage across the secondary winding.

Solution

The voltage ratio of the transformer is

$$\frac{E_1}{E_2} = \frac{N_1}{N_2}$$

Hence,

$$E_2 = E_1 \left(\frac{N_2}{N_1}\right) = 120 \left(\frac{100}{4000}\right) = 3 \text{ V}$$

EXAMPLE 11.2

Compute the voltage per turn for the transformer in Example 11.1.

Solution

The voltage per turn is

$$V_T = \frac{E_1}{N_1} = \frac{E_2}{N_2} = \frac{120}{4000} = 30 \text{ mV/turn}$$

11.1.2 CURRENT RATIO

If we assume that the transformer is ideal without losses, the input power to the transformer is equal to the output power consumed by the load. In complex number form, the apparent power of the primary winding S_1 is equal to the apparent power of the secondary winding S_2.

$$\overline{S}_1 = \overline{S}_2$$
$$\overline{S}_1 = \overline{E}_1 \overline{I}_1^* \tag{11.6}$$
$$\overline{S}_2 = \overline{E}_2 \overline{I}_2^*$$

$\overline{I}^*$ is the conjugate of the current $\overline{I}$. Based on Equations 11.4 and 11.6, we can write the current relationship as

$$\frac{I_1}{I_2} = \frac{E_2}{E_1} = \frac{N_2}{N_1} \tag{11.7}$$

Notice that the current ration is the inverse of the voltage ratio. Hence, the winding with higher voltage carries lower current. Equation 11.7 can be rearranged as

$$\Im = I_1 N_1 = I_2 N_2 \tag{11.8}$$

$\Im$ is known as the *magnetomotive force* of the primary or secondary winding. It is the current of the winding multiplied by its number of turns and its unit is ampere turn (At). Equation 11.8 is another key relationship for the transformer; the ampere turns (or magnetomotive force) of the primary and secondary windings are equal.

If you open a transformer, you will find two sets of windings: one has a larger number of turns made of a small cross-section wire, and the other has a smaller number of turns made of a thicker wire. The number of turns is determined by the voltage/turn ratio of the transformer, and the cross section of the wire is determined by the amount of current passing through the windings. Can you tell which winding has the higher voltage and which one has the higher current?

11.1.3 REFLECTED LOAD IMPEDANCE

Assume that the load impedance in Figure 11.4 is Z_{load}, where

$$Z_{\text{load}} = \frac{E_2}{I_2} \tag{11.9}$$

This load impedance when seen from the primary winding is Z'_{load}, where

$$Z'_{\text{load}} = \frac{E_1}{I_1} \tag{11.10}$$

Hence,

$$\frac{Z'_{\text{load}}}{Z_{\text{load}}} = \left(\frac{E_1}{E_2}\right)\left(\frac{I_2}{I_1}\right) = \left(\frac{N_1}{N_2}\right)^2$$
$$Z'_{\text{load}} = Z_{\text{load}} \left(\frac{N_1}{N_2}\right)^2 \tag{11.11}$$

Z'_{load} is known as the reflected impedance of the load, or the load impedance referred to the primary winding.

EXAMPLE 11.3

If the load of a transformer in Example 11.1 is 3 VA, compute the reflected impedance of the load as seen from the primary winding.

Solution

The magnitude of the load current is

$$I_2 = \frac{S}{E_2} = \frac{3}{3} = 1 \text{ A}$$

The magnitude of the load impedance is

$$Z_{\text{load}} = \frac{E_2}{I_2} = \frac{3}{1} = 3 \ \Omega$$

The magnitude of the reflected impedance of the load is

$$Z'_{\text{load}} = Z_{\text{load}} \left(\frac{N_1}{N_2}\right)^2 = 3\left(\frac{4000}{100}\right)^2 = 4.8 \text{ k}\Omega$$

EXAMPLE 11.4

A transformer with a turns ratio $N_1/N_2 = 2$ is connected to a load impedance of $(3 + j4)$ Ω. Compute the reflected impedance of the load.

Solution

$$\overline{Z}'_{\text{load}} = \overline{Z}_{\text{load}} \left(\frac{N_1}{N_2}\right)^2 = (3 + j4)\,4 = (12 + j16)\ \Omega$$

11.1.4 TRANSFORMER RATINGS

The transformer has electrical and mechanical ratings. The electrical ratings include voltages, currents, and powers. The mechanical ratings include thermal limits, container dimensions, weight, volume, etc. The rated voltages and currents are the values at which the transformer can operate continuously without any damage to its components. Exceeding these values can cause instant or gradual damage to the transformer. Excessive voltage can cause the insulation of the windings to fail very rapidly leading to internal short circuits. Excessive current causes heat buildup inside the transformer, leading to eventual meltdown of the insulation varnish of the winding leading to short circuits.

Two important electrical ratings are of particular importance to the analysis of the transformer and are always included in the nameplate data: the voltage ratio and the apparent power. For example, the nameplate data may read "10 kVA, 8 kV/240 V." With this data, you can extract the following information:

1. Rated apparent power of the transformer is 10 kVA.
2. Voltage ratio is $E_1/E_2 = 8000/240 = 33.33$. Keep in mind that the voltage ratio is the same as the turns ratio N_1/N_2. However, we cannot assume that $N_1 = 8000$ turns or $N_2 = 240$ turns, but we can assume that the ratio $N_1/N_2 = 33.33$.
3. Rated current of the primary winding is $I_1 = S/E_1 = 10,000/8,000 = 1.25$ A.
4. Rated current of the secondary winding is $I_2 = S/E_2 = 10,000/240 = 41.67$ A.
5. At full load, meaning the current and voltage are at their rated values, the magnitude of the load impedance is $Z_{\text{load}} = E_2/I_2 = 240/41.67 = 5.76$ Ω.
6. Magnitude of the reflected impedance of the load is $Z'_{\text{load}} = Z_{\text{load}}(N_1/N_2)^2 = 5.76(33.33)^2 = 6.4$ kΩ.

11.2 MULTI-WINDING TRANSFORMER

The transformer can have multiple secondary windings to provide different voltage levels to different loads. An example of a transformer with two secondary windings is shown in Figure 11.5. The numbers of turns of the two secondary windings are N_2 and N_3, respectively.

Since the voltage per turn V_T for the transformer is constant, hence

$$V_T = \frac{E_1}{N_1} = \frac{E_2}{N_2} = \frac{E_3}{N_3} \tag{11.12}$$

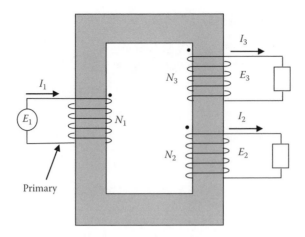

FIGURE 11.5 Transformer with multiple windings.

The computations of the current ratios require more attention. If we use the superposition theorem and assume that the load of winding N_3 is removed, then the current of the primary winding I_{12} due to I_2 alone can be computed by using the magnetomotive force in Equation 11.8.

$$I_{12}N_1 = I_2N_2 \qquad (11.13)$$

Now let us assume that the load of winding N_2 is removed and that of N_3 is connected. The primary current I_{13} due to I_3 is

$$I_{13}N_1 = I_3N_3 \qquad (11.14)$$

Hence, the total primary current I_1 is the phasor sum of I_{12} and I_{13}.

$$\bar{I}_1 = \bar{I}_{12} + \bar{I}_{13} = \bar{I}_2\left(\frac{N_2}{N_1}\right) + \bar{I}_3\left(\frac{N_3}{N_1}\right)$$
$$\bar{I}_1N_1 = \bar{I}_2N_2 + \bar{I}_3N_3 \qquad (11.15)$$

Equation 11.15 shows that the magnetomotive force of the primary winding is equal to the sum of the magnetomotive forces of all secondary windings. Hence, the currents are shared between the windings based on their respective number of turns.

For this multi-winding transformer, the source of power is connected to the primary winding and the loads are connected to the secondary windings. Hence, the sum of the apparent powers of all loads in the secondary windings is equal to the apparent power delivered by the source in the primary winding.

$$\bar{S}_1 = \bar{S}_2 + \bar{S}_3$$
$$\bar{E}_1\bar{I}_1^* = \bar{E}_2\bar{I}_2^* + \bar{E}_3\bar{I}_3^* \qquad (11.16)$$

EXAMPLE 11.5

The transformer shown in Figure 11.6 consists of one primary winding and two secondary windings. The numbers of turns of the windings are $N_1 = 4000$, $N_2 = 1000$, and $N_3 = 500$.

A voltage source of 120 V is applied across the primary winding, and purely resistive loads are connected across the secondary windings. A wattmeter placed in the primary circuit measures 300 W. Another wattmeter placed in the secondary winding N_2 measures 90 W. Compute the following:

(a) Voltages of the secondary windings
(b) Current in N_3
(c) Power of the load connected across N_3

Solution

(a) Compute the voltage per turn as given in Equation 11.4.

$$V_T = \frac{E_1}{N_1} = \frac{120}{4000} = 30 \text{ mV/turn}$$

Now, we can compute the voltages of the secondary windings.

$$E_2 = V_T N_2 = 30 \times 1000 = 30 \text{ V}$$
$$E_3 = V_T N_3 = 30 \times 500 = 15 \text{ V}$$

(b) Before we compute the currents of the secondary winding N_3, we need to compute I_1 and I_2. Since the loads are purely resistive, the power factor everywhere is unity.

$$I_1 = \frac{P_1}{E_1 \cos \theta_1} = \frac{300}{120} = 2.5 \text{ A}$$
$$I_2 = \frac{P_2}{E_2 \cos \theta_2} = \frac{90}{30} = 3 \text{ A}$$

Since all loads are resistive, we do not need to include the phase angles in Equation 11.15.

$$I_1 N_1 = I_2 N_2 + I_3 N_3$$
$$2.5 \times 4000 = (1000 \times 3) + 500 I_3$$
$$I_3 = 14 \text{ A}$$

We can arrive at the same result by using Equation 11.16.

$$\overline{S}_1 = \overline{S}_2 + \overline{S}_3$$

Since the loads are resistive, the angles of all currents are zero.

$$\overline{E}_1 \overline{I}_1^* = \overline{E}_2 \overline{I}_2^* + \overline{E}_3 \overline{I}_3^*$$
$$E_1 I_1 = E_2 I_2 + E_3 I_3$$
$$120 \times 2.5 = (30 \times 3) + 15 I_3$$
$$I_3 = 14 \text{ A}$$

(c) Since the power factor is unity and the transformer is lossless,

$$P_1 = P_2 + P_3$$
$$300 = 90 + P_3$$
$$P_3 = 210 \text{ W}$$

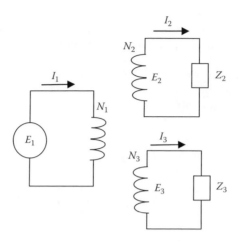

FIGURE 11.6   Three-winding transformer.

EXAMPLE 11.6

For the transformer in Example 11.5, assume that the load impedances connected across the secondary windings are $\bar{Z}_2 = (3 + j4)\ \Omega$ and $\bar{Z}_3 = (5 + j0)\ \Omega$
A voltage source of 120 V is applied across the primary winding. Compute the following:

(a) Currents in all windings
(b) Real and reactive powers in the primary windings

Solution

(a) Assume that the voltage of the primary winding is the reference voltage, thus all voltages have zero phase angles. Hence, the currents in the secondary windings are

$$\bar{I}_2 = \frac{\bar{E}_2}{\bar{Z}_2} = \frac{30\angle 0°}{3 + j4} = 6\angle -53.1°\ \text{A}$$

$$\bar{I}_3 = \frac{\bar{E}_3}{\bar{Z}_3} = \frac{15\angle 0°}{5 + j0} = 3\angle 0°\ \text{A}$$

Using Equation 11.15

$$\bar{I}_1 N_1 = \bar{I}_2 N_2 + \bar{I}_3 N_3$$
$$\bar{I}_1 4000 = (6\angle -53.1°) \times 1000 + (3 \times 500)$$
$$\bar{I}_1 = 1.276 - j1.2 = 1.75\angle -43.24\ \text{A}$$

(b) The complex power of the primary winding is

$$\bar{S}_1 = \bar{E}_1 \bar{I}_1^* = (120\angle 0°)(1.75\angle 43.24) = 153.12 + j144\ \text{VA}$$

Hence,

$$P_1 = 153.12\ \text{W}$$
$$Q_1 = 144\ \text{VAr}$$

11.3   AUTOTRANSFORMER

An autotransformer has its primary and secondary windings connected in series as shown in Figure 11.7. The primary terminal X_2 is connected to the secondary terminal Y_1, and the source voltage V_1 is connected between X_1 and Y_2. The load is connected between Y_1 and Y_2. This way, almost any transformer can be connected as an autotransformer.

The wiring diagram of an autotransformer is shown in Figure 11.8. Let us assume that the rated voltage of the primary winding is E_1 and that for the secondary winding is E_2. Also, assume that the rated current of the primary winding is I_1 and that for the secondary winding is I_2.

As stated in Equation 11.4, the volt per turn ratio V_T is constant for any winding. Hence,

$$V_T = \frac{E_1}{N_1} = \frac{E_2}{N_2} = \frac{V_1}{(N_1 + N_2)} = \frac{V_2}{N_2} \qquad (11.17)$$

The voltage ratio (input voltage versus output voltage) is

$$\frac{V_1}{V_2} = \frac{N_1 + N_2}{N_2} \qquad (11.18)$$

Using the magnetomotive force relationship in Equation 11.8, we can write the current equation as

$$\Im = N_1 I_1 = N_2 I_2 \qquad (11.19)$$

As shown in Figure 11.8, the load and source currents are

$$I_{\text{load}} = I_1 + I_2$$
$$I_s = I_1 \qquad (11.20)$$

Hence, the current ratio of the autotransformer (output current versus input current) is

$$\frac{I_{\text{load}}}{I_s} = \frac{I_1 + I_2}{I_1} = \left(1 + \frac{I_2}{I_1}\right) \qquad (11.21)$$

Substituting Equation 11.19 into 11.21 yields

$$\frac{I_{\text{load}}}{I_s} = 1 + \left(\frac{I_2}{I_1}\right) = 1 + \left(\frac{N_1}{N_2}\right) = \frac{N_1 + N_2}{N_2} \qquad (11.22)$$

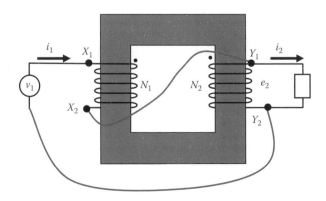

FIGURE 11.7   Connecting a regular transformer as autotransformer.

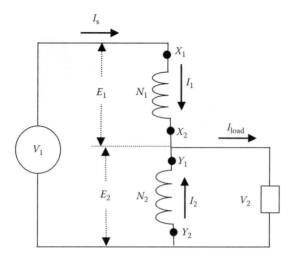

FIGURE 11.8 Wiring connection of the autotransformer in Figure 11.7.

The apparent power can be computed for the input or output circuits of the autotransformer.

$$\overline{S} = \overline{V}_1 \overline{I}_s^* = (\overline{E}_1 + \overline{E}_2) \overline{I}_1^* = \overline{E}_1 \overline{I}_1^* + \overline{E}_2 \overline{I}_1^* \qquad (11.23)$$

Note that $E_1 I_1$ is the apparent power of the regular transformer as given in Equation 11.6. Hence, the autotransformer can handle more power than the regular transformer without exceeding the rated voltage or the rated current of any winding.

EXAMPLE 11.7

A 1 kVA transformer has a voltage ratio of 120/240 V. It is desired to change the voltage ratio of the transformer to 360/240 V by connecting it as an autotransformer. Compute the new power rating of the autotransformer.

Solution

If you assume that the voltage of winding $X_1 - X_2$ is 120 V in Figure 11.7 and $Y_1 - Y_2$ is 240 V, the voltage ratio is

$$\frac{V_1}{V_2} = \frac{E_1 + E_2}{E_2} = \frac{120 + 240}{240} = \frac{360}{240}$$

As a regular transformer, the rated currents of the windings are

$$I_1 = \frac{S}{E_1} = \frac{1000}{120} = 8.33 \text{ A}$$

$$I_2 = \frac{S}{E_2} = \frac{1000}{240} = 4.167 \text{ A}$$

The power of the autotransformer is

$$S = E_1 I_1 + E_2 I_1 = 1000 + (240 \times 8.33) = 3.0 \text{ kVA}$$

This new power rating of the autotransformer is three times the power rating of the regular transformer.

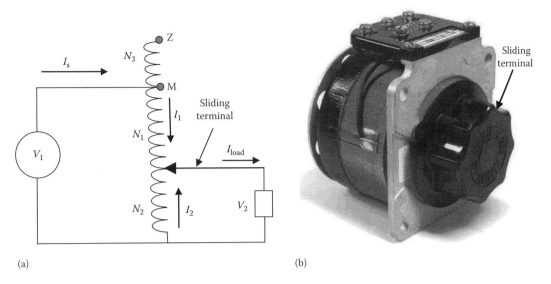

(a)   (b)

FIGURE 11.9 Autotransformer for adjustable output voltage.

The autotransformer can be built to provide adjustable output voltage. This type of autotransformer, shown in Figure 11.9, consists of one winding where the primary voltage is applied across a part of the winding. The number of turns of the secondary winding can be changed by sliding one of the secondary terminals over the winding. This requires a mechanism by which the winding and the sliding terminal are always connected electrically. In the position shown in the figure, the secondary winding is N_2. Hence, the secondary voltage is

$$V_2 = V_1 \left(\frac{N_2}{N_1 + N_2} \right) = V_T N_2 \qquad (11.24)$$

In Equation 11.24, $V_2 < V_1$. When the slider is moved to position M, the voltage across the load is

$$V_2 = V_1 \left(\frac{N_1 + N_2}{N_1 + N_2} \right) = V_T (N_1 + N_2) \qquad (11.25)$$

Hence, $V_2 = V_1$. To make $V_2 > V_1$, we move the slider to point Z, where

$$V_2 = V_1 \left(\frac{N_1 + N_2 + N_3}{N_1 + N_2} \right) = V_T (N_1 + N_2 + N_3) \qquad (11.26)$$

This autotransformer, which is also known as *variac*, can adjust the load voltage from zero to greater than the supply voltage. The variac is very useful in laboratories and test setups where the load voltage is easily adjusted to any desired value. It can also be used in speed control and start-up for electric motors.

11.4 THREE-PHASE TRANSFORMER

Three-phase transformers can be constructed of a three-legged core as shown in Figure 11.10. The primary and secondary windings of each phase are placed on the same leg of the core. The primary windings are labeled a a', b b', and c c'. The secondary windings are labeled A A', B B', and

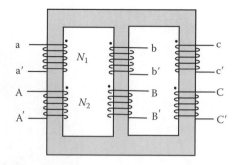

FIGURE 11.10 Configuration of a three-phase transformer.

C C'. N_1 is the number of turns of any primary winding, and N_2 is the number of turns of any secondary winding. A more convenient schematic representation of three-phase windings is shown in Figure 11.11.

11.4.1 THREE-PHASE TRANSFORMER RATINGS

The ratings of three-phase transformers are given in three-phase quantities. The nameplate data can read, for example, "60 kVA, 8 kV(Δ)/416 V(Y)." This nameplate data can provide us with the following information:

1. Rated apparent power of the three phases is 60 kVA.
2. Rated apparent power of each phase is $60/3 = 20$ kVA.
3. Primary windings are connected in delta.
   (a) Rated line-to-line voltage of the primary circuit is 8 kV.
   (b) Voltage across any primary winding is 8 kV.
4. Secondary windings are connected in Y.
   (a) Rated line-to-line voltage of the secondary circuit is 416 V.
   (b) Voltage across any secondary winding is $416/\sqrt{3} = 240$ V.
5. Line-to-line voltage ratio of the transformer is $8000/416 = 19.23$.
6. Voltage per turn is constant in any winding of the transformer. Hence, $V_T = V_{\text{phase of primary}}/N_1 = V_{\text{phase of secondary}}/N_2$. The turns ratio is the ratio of the phase voltages.

11.4.1.1 Wye–Wye Transformer

A three-phase transformer connected in wye–wye (Y–Y) is shown in Figure 11.12. The top figure shows the wiring of the transformer where terminals a', b', and c' of the primary windings are connected to a common point called neutral n. The secondary windings are similarly connected. A more convenient schematic is the one at the bottom of Figure 11.12.

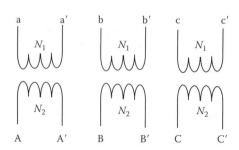

FIGURE 11.11 Schematic of a three-phase transformer.

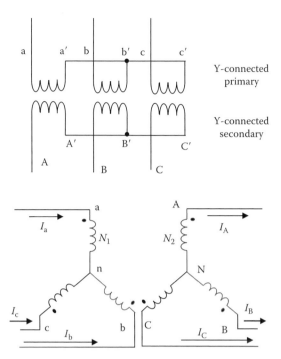

FIGURE 11.12   Y–Y transformer.

The turns ratio of the transformer is N_1/N_2, while the line-to-line voltage ratio of the transformer is V_{ab}/V_{AB}. These two ratios are equal only if the connection of the primary and secondary windings are the same, that is, both windings are connected in wye (Y) or delta. The best way to compute the turns ratio is to use the voltage per turn V_T.

$$V_T = \frac{V_{an}}{N_1} = \frac{V_{AN}}{N_2}$$

$$\frac{N_1}{N_2} = \frac{V_{an}}{V_{AN}} = \frac{\sqrt{3}V_{an}}{\sqrt{3}V_{AN}} = \frac{V_{ab}}{V_{AB}}$$

(11.27)

The current ratio of the phase voltages is the inverse of the turns ratio. We can arrive at this conclusion by using the magnetomotive force equation, where any winding current multiplied by its number of turns is constant for any winding.

$$\Im = I_a N_1 = I_A N_2$$

$$\frac{N_1}{N_2} = \frac{I_A}{I_a}$$

(11.28)

EXAMPLE 11.8

A 12 kVA transformer has a voltage ratio of 13.8 kV(Y)/416 V(Y). Compute the following:

(a) Ratio of the phase voltages
(b) Turns ratio
(c) Rated current of the primary and secondary windings

Solution

(a) The transformer is connected in Y–Y. Consider Figure 11.12; the phase voltage ratio is V_{an}/V_{AN}. Hence,

$$\frac{V_{an}}{V_{AN}} = \frac{\left(\frac{V_{ab}}{\sqrt{3}}\right)}{\left(\frac{V_{AB}}{\sqrt{3}}\right)} = \frac{V_{ab}}{V_{AB}} = \frac{13,800}{416} = 33.17$$

Note that the line-to-line voltage ratio is the same as the phase voltage ratio. This is true only if both the primary and secondary windings have the same connection.

(b) The turns ratio is N_1/N_2, which is the same as the phase voltage ratio V_{an}/V_{AN}.

$$\frac{N_1}{N_2} = \frac{V_{an}}{V_{AN}} = 33.17$$

(c) The current of the primary winding is

$$I_a = \frac{S}{\sqrt{3}V_{ab}} = \frac{12}{\sqrt{3} \times 13.8} = 0.502 \text{ A}$$

The current of the secondary winding is

$$I_A = \frac{S}{\sqrt{3}V_{AB}} = \frac{12,000}{\sqrt{3} \times 416} = 16.65 \text{ A}$$

11.4.1.2 Delta–Delta Transformer

The Δ–Δ connection of a three-phase transformer is shown in Figure 11.13. Each phase in the delta configuration is connected between two lines of the three-phase circuit. Hence, the phase voltage of

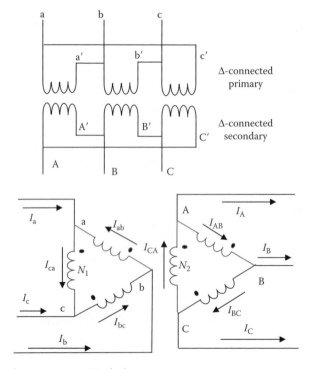

FIGURE 11.13 Transformer connected in Δ–Δ.

any winding is equal to the line-to-line voltage of the circuit. Keep in mind that the currents in the transformer windings are phase currents, and the currents of the circuit feeding the transformers are the line currents.

The current inside the windings follows the same rule discussed in Section 11.1.1. That is, the current inside the primary winding is flowing from the dotted terminal and that for the secondary winding is flowing into the dotted terminal.

The turns ratio of the transformer is N_1/N_2. The voltage ratio of the transformer is V_{ab}/V_{AB}, which are the same as the ratio of the line-to-line voltages in the delta configuration. Keep in mind that the voltage per turn is constant in any winding. Hence,

$$V_T = \frac{V_{ab}}{N_1} = \frac{V_{AB}}{N_2}$$
$$\frac{N_1}{N_2} = \frac{V_{ab}}{V_{AB}}$$

(11.29)

Using the magnetomotive force equation, we can compute the current ratio.

$$\Im = I_{ab}N_1 = I_{AB}N_2$$
$$\frac{N_1}{N_2} = \frac{I_{AB}}{I_{ab}}$$

(11.30)

EXAMPLE 11.9

A 15 kVA transformer has a voltage ratio of 25 kV(Δ)/5 kV(Δ). Compute the following:

(a) Turns ratio
(b) Line current in the primary circuit
(c) Currents of the primary and secondary windings

Solution

(a) The transformer is connected in Δ–Δ. As shown in Figure 11.13, the turns ratio is the ratio of the phase voltages, which are equal to the line-to-line voltages.

$$\frac{N_1}{N_2} = \frac{V_{ab}}{V_{AB}} = \frac{25}{5} = 5$$

(b) The line current of the source is

$$I_a = \frac{S}{\sqrt{3}V_{ab}} = \frac{15}{\sqrt{3} \times 25} = 0.346 \text{ A}$$

(c) The current of the primary winding is

$$I_{ab} = \frac{I_a}{\sqrt{3}} = \frac{0.346}{\sqrt{3}} = 0.2 \text{ A}$$

The current of the secondary winding is

$$\Im = I_{ab}N_1 = I_{AB}N_2$$
$$I_{AB} = \left(\frac{N_1}{N_2}\right)I_{ab} = 5 \times 0.2 = 1 \text{ A}$$

11.4.1.3 Wye–Delta Transformer

The Y–Δ transformer has one set of windings (primary or secondary) connected in wye and the other windings connected in delta as shown in Figure 11.14. The analysis of the wye–delta transformer requires some attention when computing the various ratios. In Figure 11.14, the voltage across N_1 is the phase voltage of the primary circuit V_{an}, and the voltage across N_2 is the line-to-line voltage of the secondary circuit V_{AB}. Since the voltage per turn is constant in any winding, we can write the turns ratio as

$$V_T = \frac{V_{an}}{N_1} = \frac{V_{AB}}{N_2}$$

$$\frac{N_1}{N_2} = \frac{V_{an}}{V_{AB}}$$

(11.31)

Similarly, the current through N_1 is the line current of the primary circuit I_a and the current through N_2 is the phase current of the secondary winding I_{AB}. Hence, we can compute the current ratio using the Ampere-turn equation.

$$\Im = I_a N_1 = I_{AB} N_2$$

$$\frac{N_1}{N_2} = \frac{I_{AB}}{I_a}$$

(11.32)

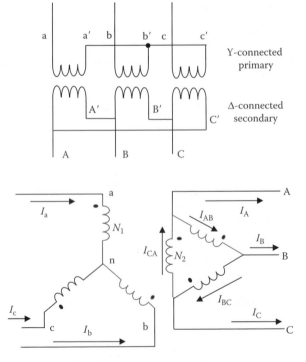

FIGURE 11.14   Y–Δ transformer.

EXAMPLE 11.10

A 25 kVA transformer has a voltage ratio of 20 kV(Y)/10 kV(Δ). Compute the following:

(a) Phase voltages
(b) Turns ratio
(c) Line current of the primary and secondary circuits
(d) Phase currents of the transformer

Solution

(a) The primary windings of the transformer are connected in wye. Hence, the phase voltage is

$$V_{an} = \frac{V_{ab}}{\sqrt{3}} = \frac{20}{\sqrt{3}} = 11.55 \text{ kV}$$

The secondary windings are connected in delta. Hence, the phase voltage is equal to the line-to-line voltage.

$$V_{AB} = 10 \text{ kV}$$

(b) The turns ratio is the ratio of the phase voltages across the windings.

$$\frac{N_1}{N_2} = \frac{11.55}{10} = 1.155$$

(c) The line current of the primary circuit is

$$I_a = \frac{S}{\sqrt{3}V_{ab}} = \frac{25}{\sqrt{3} \times 20} = 0.7217 \text{ A}$$

The line current of the secondary circuit is

$$I_A = \frac{S}{\sqrt{3}V_{AB}} = \frac{25}{\sqrt{3} \times 10} = 1.4434 \text{ A}$$

(d) Since the primary windings are connected in wye, the phase current of the primary winding is the same as the line current of the primary circuit.
The secondary windings are connected in delta, hence the phase current of the secondary winding is

$$I_{AB} = \frac{I_A}{\sqrt{3}} = \frac{1.4434}{\sqrt{3}} = 0.8333 \text{ A}$$

11.4.2 TRANSFORMER BANK

A transformer bank is two or three single-phase transformers connected as a three-phase transformer. These transformer banks are often used in distribution systems. Consider the cases in Figure 11.15. Figure 11.15a shows a power pole with one single-phase transformer. The load served by this transformer is small enough and one transformer is adequate. If more loads are added to the circuit, an additional single-phase transformer is installed, as shown in Figure 11.15b. Further increase in loads may demand a third single-phase transformer, as shown in Figure 11.15c.

(a)

(b)

(c)

FIGURE 11.15   Transformer banks. (a) Single-phase transformer, (b) two-transformer bank, and (c) three-transformer bank.

The three single-phase transformers can be connected in several configurations. One of these is the Δ–Y connection shown in Figure 11.16. The analysis of the transformer bank is the same as that of the three-phase transformers.

Example 11.11

To provide electric power to a residential area, three single-phase transformers are connected in the Δ–Y configuration shown in Figure 11.16. Each single-phase transformer is rated at 10 kVA, 15 kV/240 V. Compute the ratings of the transformer bank and the turns ratio.

Solution

The ratings of the transformer bank are the same as the ratings of the three-phase transformer; that is, the power rating is for the three phases and the voltage ratings are for the line-to-line voltage.

The power rating of the transformer bank is

$$S = 3 \times 10 = 30 \text{ kVA}$$

The primary windings are connected in delta. Therefore, the voltage across N_1 is the same as the line-to-line voltage. Hence,

$$V_{ab} = 15 \text{ kV}$$

The secondary windings are connected in wye. The voltage across N_2 is the phase voltage; hence,

$$V_{AB} = \sqrt{3} \, V_{AN} = \sqrt{3} \times 240 = 415.7 \text{ V}$$

The ratings of the transformer bank are 30 kVA, 15 kV(Δ)/415.7 V(Y)
The turns ratio of the transformer bank is the ratio of its phase voltages.

$$\frac{N_1}{N_2} = \frac{V_{ab}}{V_{AN}} = \frac{15,000}{240} = 62.5$$

11.5 ACTUAL TRANSFORMER

The transformers we discussed so far are ideal and do not have any internal losses. Actual transformers have parasitic parameters such as the resistances and inductances of the windings and core. The resistance and inductance of the windings can be modeled as shown in Figure 11.17. The dashed box shows the ideal transformer discussed here. R_1 and X_1 are the resistance and inductive reactance of the primary winding, respectively. R_2 and X_2 are the resistance and inductive reactance of the secondary winding, respectively. The input voltage of the transformer is V_1 and the load voltage is V_2. V_1 and V_2 are known as *terminal voltages*. One major difference between the ideal transformer and actual transformer is that the turns ratio of the actual transformer is not equal to the ratio of its terminal voltages. Let us explain this by looking at the volt per turn, V_T, of the actual transformer, which is

$$V_T = \frac{E_1}{N_1} = \frac{E_2}{N_2} \tag{11.33}$$

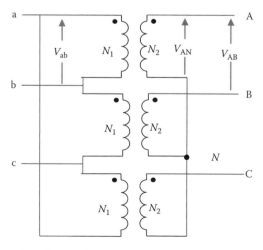

FIGURE 11.16 Δ–Y connection of a transformer bank.

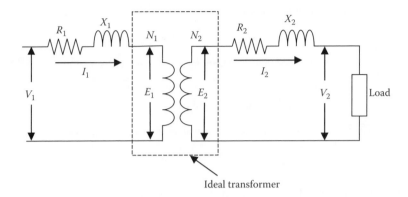

FIGURE 11.17 Winding impedance of the transformer.

Due to the presence of the resistances and inductive reactances, E_1 and V_1 are not equal, and neither are E_2 and V_2. Then,

$$\frac{N_1}{N_2} = \frac{E_1}{E_2} \neq \frac{V_1}{V_2} \tag{11.34}$$

Modeling the core of the transformer is a more involved process because of the nonlinear behavior of the flux in the core. Consider Figure 11.18a where a coil is wrapped around a core. Let us assume that the current of the coil is sinusoidal as shown in the middle figure. This current produces a sinusoidal flux ϕ in the core that has a flux density B, where

$$B = \frac{\phi}{A} \tag{11.35}$$

A is the cross-sectional area of the core. The relationship between the flux density and the magnetic field intensity in the core H is

$$B = \mu_o \mu_r H \tag{11.36}$$

where $\mu_o = 4\pi \times 10^{-7}$ H/m. It is a constant value known as the absolute permeability. μ_r is the relative permeability; its magnitude depends on the material of the core. For air $\mu_r = 1$, and for iron $\mu_r \approx 5000$.

   The relationship between the flux density B and the flux intensity H is nonlinear as shown in Figure 11.18b. The curve is known as the *hysteresis loop*. Let us assume that the core has never been

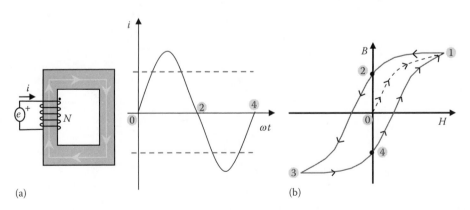

(a)   (b)

FIGURE 11.18   (a) Winding current and (b) *B–H* relationship.

used before. If the current in the winding starts from zero (point 0), the flux starts to flow in the core as shown by the dashed line in the hysteresis loop. Unlike air, the core cannot carry unlimited amounts of flux. When the current exceeds a certain limit (e.g., above or below the dashed lines in the middle figure), no appreciable increase occurs in the flux, and the core is said to be saturated. This is shown in the area around point 1 in the figure. When the current starts to fall, the flux starts to decrease. At point 2, the current is zero, but the flux density is nonzero. This is known as *residual flux*. The core at this point acts as a magnet. When the current is further reduced, the flux is also reduced to zero and then reverses its direction. After reaching point 3, if the current increases, the flux increases along the rightmost path of the hysteresis loop. This hysteresis loop is repeated in every ac cycle.

To model the hysteresis loop, we can use an approximate process with linear elements: resistance and inductance. Using Faraday's law, we can write the expression for the flux as

$$e = \frac{d\phi}{dt} \tag{11.37}$$

Hence, the flux density is

$$B = \frac{\phi}{A} \sim \int e \, dt \tag{11.38}$$

Equation 11.38 shows that the flux density is directly proportional to the integral of the voltage across the winding. Keep in mind that the magnetic field intensity is directly proportional to the current. Hence, the B versus H characteristics in Figure 11.18 can be approximated by $\int e \, dt$ versus i relationship. This relationship can be obtained by the two circuits in Figures 11.19 and 11.20. In Figure 11.19a, the resistive circuit is connected across a sinusoidal voltage source.

$$e = E_{max} \sin \omega t \tag{11.39}$$

The voltage integral and currents of this circuit are

$$\int e \, dt = -\left(\frac{E_{max}}{\omega}\right) \cos \omega t \tag{11.40}$$

$$i_R = \frac{e}{R} = \left(\frac{E_{max}}{R}\right) \sin \omega t \tag{11.41}$$

Note that $\int e \, dt$ and i_R are phase shifted by 90°. The relationship of $\int e \, dt$ versus i_R is shown in Figure 11.19b. Because of the phase shift, the relationship has an elliptical shape with two radii that are functions of the resistance and the angular frequency.

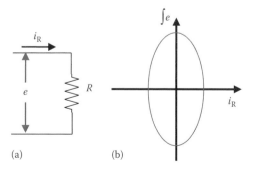

(a)   (b)

FIGURE 11.19 Voltage integral versus current of a resistive element.

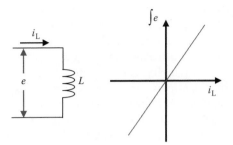

FIGURE 11.20 Voltage integral versus current of an inductive element.

For the inductive element in Figure 11.20, $\int e\,dt$ and i_L are computed as follows:

$$\int e\,dt = -\left(\frac{E_{max}}{\omega}\right)\cos\omega t \tag{11.42}$$

$$e = L\left(\frac{di_L}{dt}\right)$$

$$i_L = \frac{1}{L}\int e\,dt = -\left(\frac{E_{max}}{\omega L}\right)\cos\omega t \tag{11.43}$$

where

L is the inductance

$\int e\,dt$ and i_L are in phase, forming a straight-line relationship as shown in Figure 11.20

Now let us add the two elements in parallel as shown in Figure 11.21. The total current in this case is

$$i = i_R + i_L \tag{11.44}$$

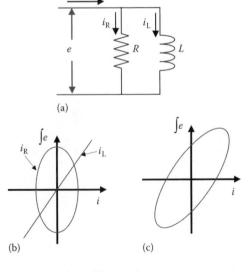

FIGURE 11.21 Approximate representation of hysteresis.

FIGURE 1.2 Pixii's generator. (Image courtesy of the Smithsonian Museum.)

(a)

(b)

FIGURE 2.2 Power plants. (a) Byron nuclear power plant. (Image courtesy of the U.S. Department of Energy.) (b) Itaipu hydroelectric power plant.

(*continued*)

(c)

FIGURE 2.2 (continued) (c) Coal-fired thermal power plant.

FIGURE 2.7 Control center. (Image courtesy of Tennessee Valley Authority.)

FIGURE 4.1 The Grand Coulee Dam and Franklin D. Roosevelt Lake. (Image courtesy of U.S. Army Corps of Engineers.)

FIGURE 4.2 Fox River diversion hydroelectric power plant, Wisconsin. (Image courtesy of U.S. Army Corps of Engineers.)

FIGURE 4.5 Hydroelectric turbine–generator units at the Lower Granite power plant, Walla Walla, Washington. (Image courtesy of U.S. Army Corps of Engineers.)

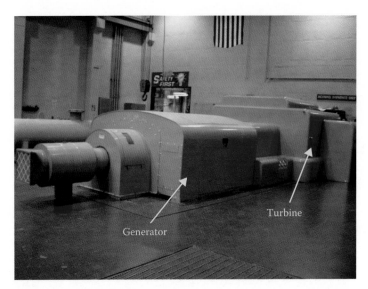

FIGURE 4.10   Inside a small-size thermal power plant.

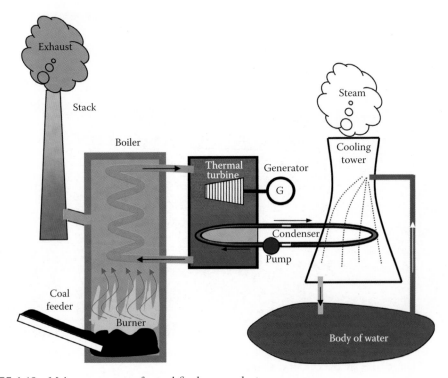

FIGURE 4.12   Main components of a coal-fired power plant.

FIGURE 4.13 Weston coal-fired power plant. (Image courtesy of the U.S. Department of Energy.)

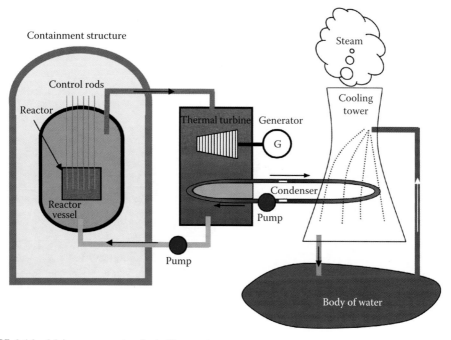

FIGURE 4.16 Main components of a boiling water reactor.

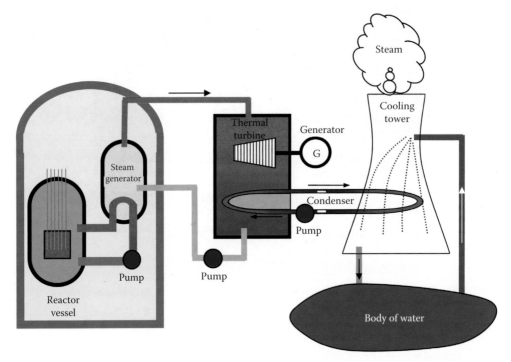

FIGURE 4.17 Pressurized water reactor.

(a)

(b)

FIGURE 4.18 Pressurized water reactor nuclear power plants.

(a)   (b)

FIGURE 5.1   Effects of acid rain on (a) stones and (b) trees.

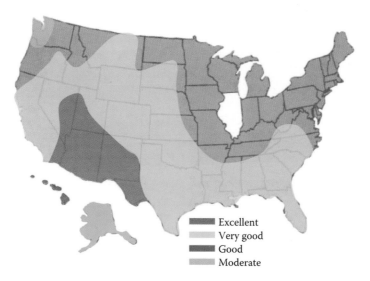

Excellent
Very good
Good
Moderate

FIGURE 6.2   Solar power density in the United States.

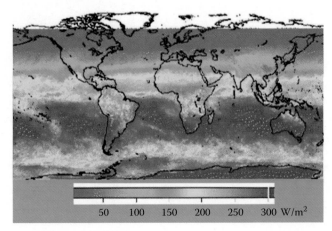

FIGURE 6.3 Sample of solar power density worldwide in December. (Image courtesy of NASA.)

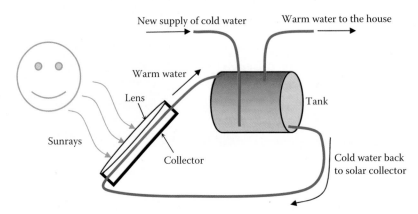

FIGURE 6.6 Thermosiphon hot water system.

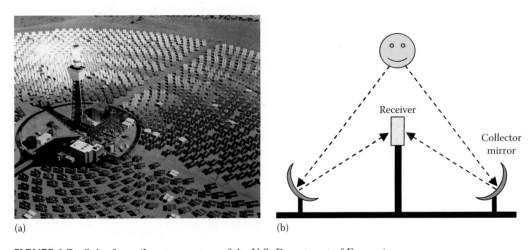

(a)   (b)

FIGURE 6.7 Solar farm. (Images courtesy of the U.S. Department of Energy.)

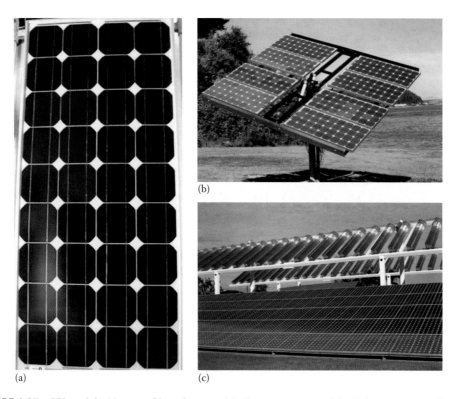

FIGURE 6.25 PV module (a), array (b), and system (c). (Images courtesy of the U.S. Department of Energy.)

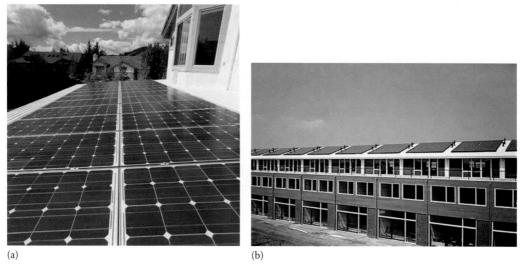

FIGURE 6.33 Various PV systems. (Images courtesy of the U.S. Department of Energy.)

(*continued*)

(c)

(d)

FIGURE 6.33 (continued)

(a)

(b)

FIGURE 6.37 High-power PV systems. (Images courtesy of the U.S. Department of Energy.)

(a)   (b)

FIGURE 6.39 Basic components of a wind-generating system. (a) Horizontal design and (b) mechanical structure.

(a)

(b)

FIGURE 6.40 (a) Housing and (b) blade of a 2 MW wind-generating system.

(a)

(b)

FIGURE 6.44 Wind farm located in California. (Images courtesy of the U.S. Department of Energy.)

FIGURE 6.45 2 MW offshore wind turbine farm in Denmark. (Image courtesy of the LM Glasfiber Group.)

FIGURE 6.51 PEM FC module.

FIGURE 6.65 Steam generated from rain even in a cold environment.

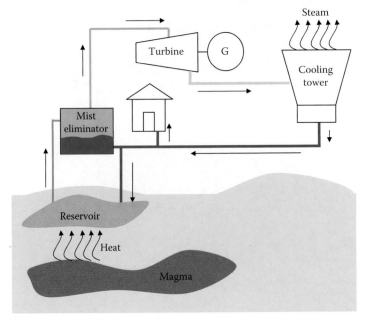

FIGURE 6.67 Geothermal power plant.

FIGURE 6.68 The Geysers in northern California, first geothermal power plant in the United States.

(a)

FIGURE 6.69 Free-flow tidal energy system. (a) Free-flow tidal turbine and

(*continued*)

(b)

FIGURE 6.69 (continued) (b) conceptual design of a farm. (Images courtesy of Marine Current Turbines Limited.)

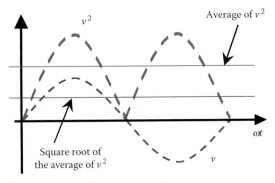

FIGURE 7.2 Concept of rms.

(a)   (b)

FIGURE 7.30 Capacitors for power factor correction, (a) on a distribution pole and (b) in a distribution substation.

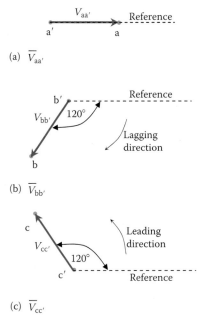

FIGURE 8.3 Phasor diagram of balanced three phases. (a) Phasor diagram of phase $\overline{V}_{aa'}$, (b) phasor diagram of phase $\overline{V}_{bb'}$, and (c) phasor diagram of phase $\overline{V}_{cc'}$.

FIGURE 8.5 Single-circuit distribution line.

(a)

(b)

(c1)

(c2)

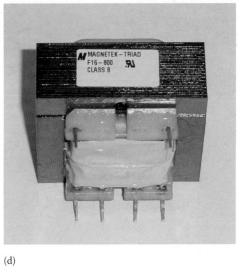

(d)

FIGURE 11.2 Various types of transformers. (a) Transmission transformer, (b) distribution transformer, (c) service transformers, and (d) circuit transformer.

FIGURE 12.6 Induction machines are the prime movers of these vehicles. (a) Mass rover. (Image courtesy of NASA.) (b) Underwater unmanned robot. (c) Electric propulsion ferry. (d) Hybrid electirc vehicle. (e) City transportation. (f) City electric bus. (g) Maglev train. (Image courtesy of Transrapid International.) (h) Electric train. (i) Monorail.

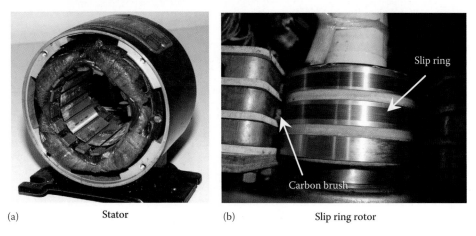

FIGURE 12.7 Main parts of an induction machine. (a) Stator and (b) slip-ring rotor.

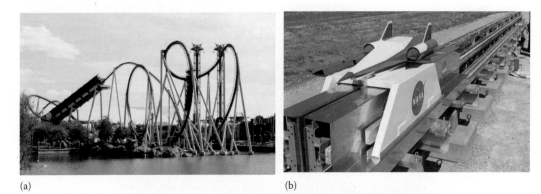

(a)   (b)

(c)

FIGURE 12.22 Examples of LIM applications. (a) Roller coaster. (b) Maglev launcher. (Image courtesy of NASA Marshall Space Flight Center.). (c) Maglev rapid transportation. (Image courtesy of Transrapid International.)

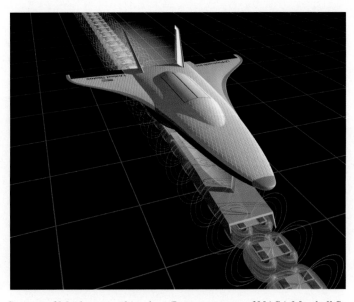

FIGURE 12.29 Concept of Maglev space launcher. (Image courtesy of NASA Marshall Space Flight Center.)

FIGURE 12.33   Slip-ring arrangement.

FIGURE 12.56   Hybrid electric vehicle using a permanent magnet synchronous motor.

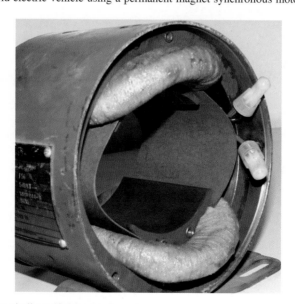

FIGURE 12.64   Stator windings (field) of a dc motor.

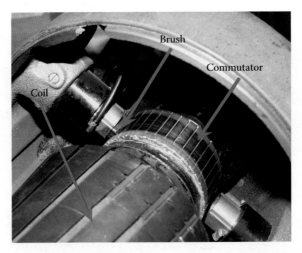

FIGURE 12.65 Rotor of a dc motor.

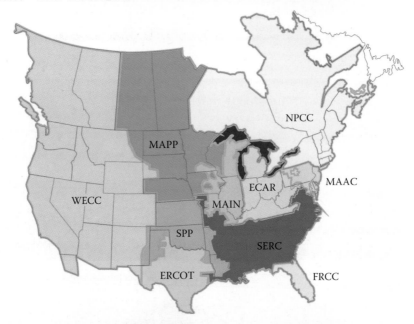

FIGURE 14.12 U.S. regional reliability councils.

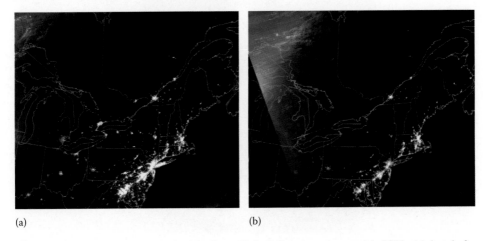

(a)   (b)

FIGURE 14.20 Satellite pictures of the Northeast United States on August 14, 2003, (a) just before the blackout and (b) just after the blackout. (Images courtesy of NASA.)

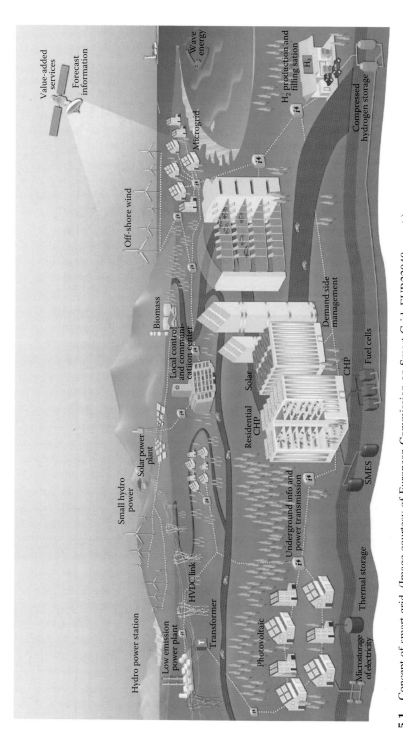

FIGURE 15.1 Concept of smart grid. (Image courtesy of European Commission on Smart Grid, EUR22040 report.)

FIGURE 15.2 International Space Station. (Image courtesy of NASA.)

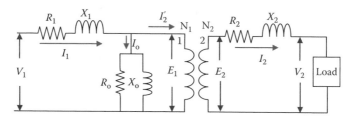

FIGURE 11.22 Equivalent circuit of a transformer.

Combining the two characteristics in Figures 11.19 and 11.20 gives the elliptically bent shape shown in Figure 11.21c. This shape is similar enough to the hysteresis loop in Figure 11.18 and is used as an approximate representation of the core. The parameters of the core model are often computed from the primary side of the transformer. Hence, the core is often included in the primary circuit as shown in Figure 11.22. I_o is known as the *excitation current* representing the magnetic field intensity of the hysteresis loop. R_o and X_o are the equivalent core resistance and core inductive reactance of the hysteresis model, respectively.

11.5.1 ANALYSIS OF ACTUAL TRANSFORMER

The transformer model in Figure 11.22 can be used to compute the relationships between all currents and voltages. For example, the primary current I_1 of the transformer is

$$\bar{I}_1 = \bar{I}_o + \bar{I}_2' \tag{11.45}$$

The current I_2' is equal to the load current as seen from the primary side. This is also known as the reflected load current. The relationship between I_2 and I_2' is the turns ratio of the transformer, as given by the equation of the magnetomotive force.

$$\Im = I_2' N_1 = I_2 N_2$$
$$\frac{I_2'}{I_2} = \frac{N_2}{N_1} \tag{11.46}$$

The voltage equations of the primary and secondary circuits are

$$\bar{V}_1 = \bar{E}_1 + \bar{I}_1 \ (R_1 + jX_1) \tag{11.47}$$

$$\bar{E}_2 = \bar{V}_2 + \bar{I}_2 \ (R_2 + jX_2) \tag{11.48}$$

We can rewrite Equation 11.47 by substituting Equation 11.48 into Equation 11.47.

$$\bar{V}_1 = \bar{E}_2 \left(\frac{N_1}{N_2}\right) + \bar{I}_1 (R_1 + jX_1)$$
$$\bar{V}_1 = \left[\bar{V}_2 + \bar{I}_2 (R_2 + jX_2)\right] \frac{N_1}{N_2} + \bar{I}_1 (R_1 + jX_1)$$
$$\bar{V}_1 = \left[\bar{V}_2 + \bar{I}_2' \left(\frac{N_1}{N_2}\right) (R_2 + jX_2)\right] \frac{N_1}{N_2} + \bar{I}_1 (R_1 + jX_1) \tag{11.49}$$
$$\bar{V}_1 = \left[\bar{V}_2 \left(\frac{N_1}{N_2}\right) + \bar{I}_2' \left(\frac{N_1}{N_2}\right)^2 (R_2 + jX_2)\right] + \bar{I}_1 (R_1 + jX_1)$$

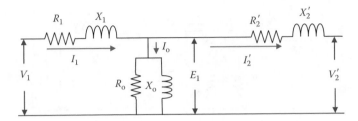

FIGURE 11.23 Modified equivalent circuit of the transformer.

The following variables can be defined as

$$\overline{V}_2' = \overline{V}_2 \left(\frac{N_1}{N_2} \right)$$

$$R_2' = R_2 \left(\frac{N_1}{N_2} \right)^2 \qquad (11.50)$$

$$X_2' = X_2 \left(\frac{N_1}{N_2} \right)^2$$

where

V_2' is known as the reflected voltage of the secondary terminal voltage (or secondary voltage referred to primary)

R_2' is the reflected resistance of the secondary winding (or secondary resistance referred to the primary)

X_2' is the reflected inductive reactance of the secondary winding (or secondary inductive reactance referred to the primary)

Substituting Equation 11.50 into Equation 11.49 yields

$$\overline{V}_1 = \overline{V}_2' + \overline{I}_2' (R_2' + jX_2') + \overline{I}_1 (R_1 + jX_1) \qquad (11.51)$$

The transformer model representing Equation 11.51 is shown in Figure 11.23. This circuit is easier to use than the one shown in Figure 11.22. Further approximation of the transformer model can be made since the excitation current I_o is often less than 5% of the rated primary current. In this case, moving the core branch anywhere in the model, or even eliminating it altogether, as shown in Figure 11.24, should not introduce significant errors.

EXAMPLE 11.12

A single-phase transformer has the following parameters:

$$\frac{N_1}{N_2} = 10; \quad R_{eq} = R_1 + R_2' = 1 \ \Omega; \quad X_{eq} = X_1 + X_2' = 10 \ \Omega; \quad R_o = 1000 \ \Omega; \quad X_0 = 5000 \ \Omega$$

The rated voltage of the primary winding is 1000 V. A $0.5 \angle 30° \ \Omega$ load is connected across the secondary terminals. Compute the load voltage.

Solution

Let us use the equivalent circuit in Figure 11.25. The magnitude of the load impedance referred to the primary side is

$$Z'_L = Z_L \left(\frac{N_1}{N_2}\right)^2 = 50 \ \Omega$$

The current $\bar{I}'_2$ is

$$\bar{I}'_2 = \frac{\bar{V}_1}{(R_{eq} + jX_{eq}) + Z'_L} = \frac{1000\angle 0°}{(1+j10) + 50\angle 30°} = 17.7\angle -38.31° \ A$$

The load voltage referred to the primary side V'_2 is

$$\bar{V}'_2 = \bar{I}'_2 \ Z'_L = (17.7\angle -38.31°)(50\angle 30°) = 885\angle -8.31°$$

Use Equation 11.50 to compute the actual voltage across the load.

$$V_2 = V'_2 \left(\frac{N_2}{N_1}\right) = 885 \left(\frac{1}{10}\right) = 88.5 \ V$$

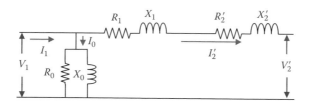

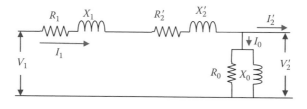

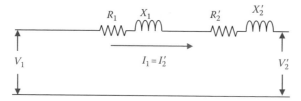

FIGURE 11.24 Acceptable equivalent circuits for the transformer.

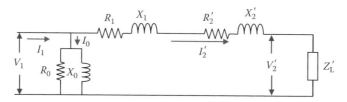

FIGURE 11.25 Equivalent circuit of the transformer with load.

11.5.2 Transformer Efficiency

The efficiency η of any equipment is defined by

$$\eta = \frac{\text{Output power}}{\text{Input power}} = \frac{\text{output energy}}{\text{input energy}} = \frac{\text{output power}}{\text{output power} + \text{losses}} = \frac{\text{input power} - \text{losses}}{\text{input power}} \quad (11.52)$$

The losses of the transformer are due to R_1, R_2, and R_o. The losses in R_1 and R_2 are the winding losses and are named *copper losses* P_{cu} (the windings are made of copper material). The loss in R_o is known as *iron loss* P_{iron} (the core is made of iron). Using Figure 11.22, the losses can be computed as

$$P_{cu} = I_1^2 \, R_1 + I_2^2 \, R_2 \quad (11.53)$$

$$P_{iron} = \frac{E_1^2}{R_o} \quad (11.54)$$

These losses can also be computed using the approximate equivalent circuits in Figure 11.24. A good approximation for the losses is given below:

$$P_{cu} \approx I_1^2(R_1 + R_2') \approx (I_2')^2 \, (R_1 + R_2') \quad (11.55)$$

$$P_{iron} \approx \frac{V_1^2}{R_o} \approx \frac{(V_2')^2}{R_o} \quad (11.56)$$

Using Equation 11.52, the efficiency of the transformer can be computed as

$$\eta = \frac{P_{out}}{P_{input}} = \frac{V_2' \, I_2' \cos\theta_2}{V_2' \, I_2' \cos\theta_2 + P_{cu} + P_{iron}} = \frac{V_1 \, I_1 \cos\theta_1 - P_{cu} - P_{iron}}{V_1 \, I_1 \cos\theta_1} \quad (11.57)$$

where
θ_1 is the power factor angle of I_1
θ_2 is the power factor angle of I_2 and the angle of the load impedance

Example 11.13

Compute the efficiency of the transformer in Example 11.12.

Solution

$$P_{cu} = (I_2')^2 \, R_{eq} = 17.7^2 \times 1 = 313.29 \text{ W}$$

$$P_{iron} = \frac{V_1^2}{R_o} = \frac{1000^2}{1000} = 1 \text{ kW}$$

The output power is

$$P_{out} = V_2' \, I_2' \cos\theta_2 = 885 \times 17.7 \times \cos 30 = 13.57 \text{ kW}$$

The efficiency of the transformer is

$$\eta = \frac{P_{out}}{P_{input}} = \frac{V_2' \, I_2' \cos\theta_2}{V_2' \, I_2' \cos\theta_2 + P_{cu} + P_{iron}} = \frac{13,570}{13,570 + 313.29 + 1,000} = 91.2\%$$

11.5.3 Voltage Regulation

The voltage regulation VR of the transformer is defined as

$$\mathrm{VR} = \frac{|V_{\mathrm{no\ load}}| - |V_{\mathrm{full\ load}}|}{|V_{\mathrm{full\ load}}|} \tag{11.58}$$

where

$|V_{\mathrm{no\ load}}|$ is the magnitude of the open circuit voltage measured at the load terminals

$|V_{\mathrm{full\ load}}|$ is the magnitude of the voltage at the load terminals when the rated current is delivered to the load

The VR represents the change in the load voltage from no load to full load. It indicates the voltage reduction due to the various parameters of the transformer.

Consider the equivalent circuit in Figure 11.25. At no load (open circuit), $I_2' = 0$. Hence, the voltage measured at the load terminals is

$$V_{\mathrm{no\ load}} = V_1 \tag{11.59}$$

At full load

$$V_{\mathrm{full\ load}} = V_2' \tag{11.60}$$

Hence, the voltage regulation of the transformer is

$$\mathrm{VR} = \frac{V_1 - V_2'}{V_2'} \tag{11.61}$$

Example 11.14

Compute the voltage regulation of the transformer in Example 11.12. Assume that the secondary winding of the transformer carries the full load current when $0.5 \angle 30°\ \Omega$ is connected across the secondary terminals.

Solution

$$\mathrm{VR} = \frac{V_1 - V_2'}{V_2'} = \frac{1000 - 885}{885} = 13\%$$

EXERCISES

1. A single-phase transformer has a turns ratio of 10,000/5,000. A direct current voltage of 30 V is applied to the primary winding. Compute the voltage of the secondary winding.
2. A single-phase transformer is rated at 2 kVA, 240/120 V. The transformer is fully loaded by an inductive load of 0.8 power factor lagging. Compute the following:
   (a) Real power delivered to the load
   (b) Load current
3. A single-phase transformer is rated at 10 kVA, 220/110 V.
   (a) Compute the rated current of each winding.

(b) If a 2 Ω load resistance is connected across the 110 V winding, what are the currents in the high-voltage and low-voltage windings?

(c) What is the equivalent load resistance referred to the 220 V side?

4. Three single-phase transformers are connected in a wye/delta configuration. Each single-phase transformer has an identical number of turns in the primary and secondary windings. If a line-to-line voltage of 480 V is applied on the wye windings, compute the line-to-line voltage on the delta windings.

5. Three single-phase transformers, each rated at 10 kVA, 400/300 V, are connected as a wye Y–delta configuration. Compute the following:

(a) Rated power of the transformer bank

(b) Line-to-line voltage ratio of the transformer bank

6. A single-phase transformer has a voltage regulation of 5%. The input voltage of the transformer is 120 V, and the turns ratio N_1/N_2 is 2. Compute the voltage across the load V_2.

7. Three single-phase transformers are connected as a transformer bank rated at 18 MVA, 13.8 kV (Δ)/120 kV(Y). One side of the transformer bank is connected to a 120 kV transmission line, and the other side is connected to a three-phase load of 12 MVA at 0.8 lagging power factor. Compute the following:

(a) Turns ratio of the transformer bank

(b) Line current at the 120 kV side

8. A single-phase 10 kVA, 2300/230 V two-winding transformer is connected as an autotransformer to step up the voltage from 2300 to 2530 V.

(a) Draw the schematic diagram of the autotransformer showing the winding connections and all voltages and currents at full load.

(b) Find the kilovolt ampere rating of the autotransformer. Do not allow the currents of windings to exceed their rated values.

9. A single-phase, 240/120 V transformer has the following parameters:

$$R_1 = 1 \, \Omega; \quad R_2 = 0.5 \, \Omega; \quad X_1 = 6 \, \Omega; \quad X_2 = 2 \, \Omega; \quad R_o = 500 \, \Omega; \quad X_o = 1.5 \, k\Omega$$

A load of 10 Ω at 0.8 power factor lagging is connected across the low-voltage terminals of the transformer. The voltage measured across the load side is 110 V. Compute the following:

(a) Load voltage referred to the primary side

(b) Currents of the primary and secondary windings

(c) Source voltage

(d) Voltage regulation

(e) Load power

(f) Efficiency of the transformer

10. Three identical single-phase transformers, each rated at 100 kVA, 7 kV/3.5 kV, are connected as a three-phase transformer bank. The high-voltage side of the transformer is connected in delta and the low-voltage side is connected in wye. A 200 kVA, wye-connected load is attached across the secondary winding. Compute the following:

(a) Ratio of the line-to-line voltages

(b) Line current on both sides of the transformer

11. The transformer shown in Figure 11.6 consists of one primary winding and two secondary windings. The numbers of turns of the windings are $N_1 = 10,000$, $N_2 = 5,000$, and $N_3 = 1,000$. A voltage source of 120 V is applied to the primary winding. The load of winding N_2 consumes 600 W and 300 VAr inductive power. The load of winding N_3 consumes 24 W and 36 VAr capacitive power. Compute the following:

(a) Voltages of the secondary windings

(b) Currents of all windings

12. A single-phase transformer has three windings. The primary winding N_1 carries a current of $I_1 = 10$ A. One of the secondary windings has 3000 turns and carries 2 A, and the other secondary winding has 6000 turns and carries 1 A. All currents are in phase. Compute the number of turns in the primary winding.

13. Three single-phase transformers are used to form a three-phase transformer bank rated 13.8 kV (delta)/120 kV(wye). One side of the transformer bank is connected to a 120 kV transmission line, and the other side is connected to a three-phase load of 12 MVA at 0.8 power factor lagging. Determine the turns ratio of each transformer and the line current of the bank at the 120 kV side.

12 Electric Machines

Most electric machines are dual-action electromechanical converters. The machines that convert electrical energy into mechanical are called *motors*, and the machines that convert mechanical energy into electrical are called *generators*. Electric motors are probably the most used power devices anywhere. You can find them in almost every equipment with mechanical movements from children's toys to space crafts. They play incredible roles in our daily life by performing numerous vital tasks in various applications. Refrigerators, washers, dryers, stoves, air conditioners, hair dryers, computers, printers, clocks, electric toothbrushes, electric shavers, and fans are some of the household devices that use motors. In the industrial and commercial sectors, motors are used in numerous applications such as transportation vehicles, elevators, forklifts, blowers, robots, actuators, electric and hybrid cars, machine tooling, paper mills, cooking machines, medical tools, assembly lines, and conveyor belts. The computer hard disk, for example, has at least two motors and the computer printer has at least four motors. National Aeronautics and Space Administration (NASA)'s Mars exploration rover built by the Jet Propulsion Laboratory (JPL) has about 200 motors; most are used for actuation, sensing, sampling, and control. Many of them are for one-time use only during the cruising phase and the entry, descent, and landing (EDL) phase.

In total, about 65% of the electrical energy in the United States is consumed by electric motors, and over 99% of the energy produced by utilities worldwide is produced by electrical generators. In stand-alone systems such as aircrafts, ships, and automobiles, the generators are the main source of electric power for these mobile systems.

Electrical machines come in various sizes and power levels. Their weights range from micrograms for motors installed inside silicon chips to 7700 ton for the generator built by the consortium from Hewitt, Siemens, and General Electric for China's Yangtze River Three Gorges Hydraulic Power Plant. The power capacity of the machines is also of a wide range, from microwatts to more than 2 GW.

12.1 ROTATING MAGNETIC FIELD

A multiphase alternating current (ac) source produces a rotating magnetic field inside ac motors, which causes the shaft of the motor to spin. Before we explain how an ac machine rotates, we need to understand the concept of rotating magnetic fields. Consider the conceptual diagram of the stator of a three-phase machine shown in Figure 12.1. The figure shows three windings mounted symmetrically inside a hollow metal tube called stator, which are shifted by 120° from each other. Each winding goes along the length of the stator on one side and then returns from the other side. If we apply a three-phase balanced voltage across the terminals of the windings, the currents of the windings create balanced three-phase magnetic fields inside the tube, as shown in Figure 12.2.

Figure 12.3 shows a more convenient representation of the stator windings. The circles embedded inside the stator represent the windings of the motor; a–a′ represent the winding of phase a, b–b′ represent the winding of phase b, and c–c′ represent the winding of phase c. The crosses and dots inside the circles represent current entering and leaving the windings, respectively.

Because of the mechanical arrangement of the stator windings, the flux of each phase, according to the right hand rule, travels along its axis as shown in Figure 12.3. This flux is inside the tube, thus it is called *airgap flux*.

Now, let us consider any three consecutive time instances (e.g., t_1, t_2, and t_3 shown in Figure 12.2). At t_1, the angle is 60°, and the magnitude of the flux of each phase is

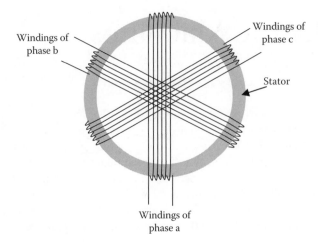

FIGURE 12.1   Three-phase windings mounted on a stator.

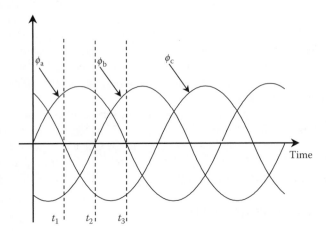

FIGURE 12.2   Airgap flux of the three phases.

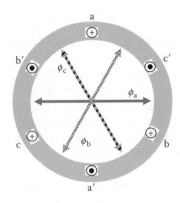

FIGURE 12.3   Loci of airgap fluxes.

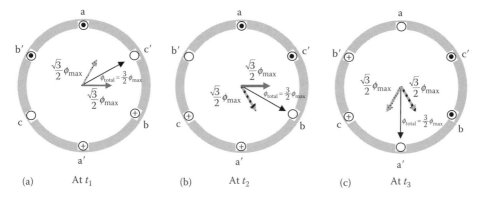

FIGURE 12.4 Rotating airgap flux.

$$\phi_a = -\phi_b = \frac{\sqrt{3}}{2}\phi_{max}$$
$$\phi_c = 0 \tag{12.1}$$

The fluxes at t_1 are mapped along their corresponding axes in Figure 12.4a. Note the direction of the current in each winding as it determines the direction of its flux according to the right hand rule. The total airgap flux is the phasor sum of all fluxes present in the airgap. Hence, at t_1

$$\overline{\phi}_{total}(t_1) = \overline{\phi}_a(t_1) + \overline{\phi}_b(t_1) + \overline{\phi}_c(t_1) = \frac{\sqrt{3}}{2}\phi_{max}\angle 0° + \frac{\sqrt{3}}{2}\phi_{max}\angle 60° + 0$$
$$\overline{\phi}_{total}(t_1) = \frac{3}{2}\phi_{max}\angle 30° \tag{12.2}$$

Similarly, at t_2, the angle is 120°, and the magnitudes of the fluxes of the three phases are

$$\phi_a = -\phi_c = \frac{\sqrt{3}}{2}\phi_{max}$$
$$\phi_b = 0 \tag{12.3}$$

These fluxes are shown in Figure 12.4b, where the total airgap flux at t_2 is

$$\overline{\phi}_{total}(t_2) = \overline{\phi}_a(t_2) + \overline{\phi}_b(t_2) + \overline{\phi}_c(t_2) = \frac{\sqrt{3}}{2}\phi_{max}\angle 0° + 0 + \frac{\sqrt{3}}{2}\phi_{max}\angle -60°$$
$$\overline{\phi}_{total}(t_2) = \frac{3}{2}\phi_{max}\angle -30° \tag{12.4}$$

Finally, at t_3, the angle of the waveform is 180°, and the magnitudes of the fluxes of the three phases are

$$\phi_a = 0$$
$$\phi_b = -\phi_c = \frac{\sqrt{3}}{2}\phi_{max} \tag{12.5}$$

These fluxes are shown in Figure 12.4c. In this case, the total airgap flux at t_3 is

$$\overline{\phi}_{total}(t_3) = \overline{\phi}_a(t_3) + \overline{\phi}_b(t_3) + \overline{\phi}_c(t_3) = 0 + \frac{\sqrt{3}}{2}\phi_{max}\angle -120° + \frac{\sqrt{3}}{2}\phi_{max}\angle -60°$$
$$\overline{\phi}_{total}(t_3) = \frac{3}{2}\phi_{max}\angle -90° \tag{12.6}$$

By examining Equations 12.2, 12.4, and 12.6, we can conclude the following:

- Magnitude of the total flux in the airgap is constant and equal to 1.5 ϕ_{max}.
- Angle of the total airgap flux changes with time. The flux in the above case is rotating in the clockwise direction. This rotating flux is one of the main reasons for the development of three-phase systems.
- Total flux in the airgap completes one revolution in every ac cycle. Hence, the mechanical speed of the total flux in the airgap n_s is one revolution per ac cycle.

$$n_s = f \text{ rev/s} \tag{12.7}$$

where f is the frequency of the ac supply (Hz). The mechanical speed is often measured in revolution per minute (rpm).

Hence,

$$n_s = 60f \text{ rpm} \tag{12.8}$$

The speed n_s is known as *synchronous speed* because its value is determined by the frequency of the supply voltage; that is, the speed is synchronized with the supply frequency.

The machine in Figure 12.3 is considered two pole because every phase has one winding that creates one north pole and one south pole. If each phase is composed of two windings (i.e., a_1–a_1' and a_2–a_2') arranged symmetrically along the inner circumference of the stator as shown in Figure 12.5, the machine is considered to have four poles. In this arrangement, the mechanical angle between the phases is 60° instead of 120° for the two-pole machine. If you repeat the analyses in Equations 12.1 through 12.6, you will find that the flux moves 180° (mechanical angle) for each complete ac cycle. Hence, we can write a general expression for the mechanical speed of the airgap flux as

$$n_s = \frac{60f}{pp} = 120\frac{f}{p} \text{ rpm} \tag{12.9}$$

where
   pp is the number of pole pairs
   p is the number of poles ($p = 2$ pp)

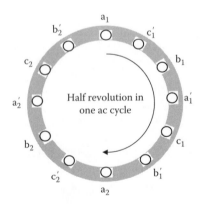

FIGURE 12.5   Four-pole arrangement.

EXAMPLE 12.1

Compute the synchronous speed for 2-, 4-, 6-, 8-, and 10-pole machines operating at 50 and 60 Hz.

Solution

With the direct substitution in Equation 12.9, you can get the data tabulated below.

Number of Poles	Synchronous Speed (rpm)	
	50 Hz	60 Hz
2	3000	3600
4	1500	1800
6	1000	1200
8	750	900
10	600	720

Note that the synchronous speed for the 50 Hz system is slower than that for the 60 Hz system.

12.2 ROTATING INDUCTION MOTOR

About 65% of the electric energy in the United States is consumed by electric motors. In the industrial sector alone, about 75% of the total energy is consumed by motors. Some of the unfamiliar uses of electric motors are shown in Figure 12.6.

Over 90% of the energy consumed by electric motors is consumed by induction motors. This is because they are rugged, reliable, easy to maintain, and relatively inexpensive. The induction motor is composed of one stator and one rotor; a small induction motor is shown in Figure 12.7. The stator has three-phase windings, which are excited by a three-phase supply. The rotor, as the name implies, is the rotating part of the motor. It is mounted inside the stator and is supported by two sets of ball bearings; one on each end of the rotor. The mechanical load of the motor is attached to the shaft of the rotor. The rotor circuit consists of windings that are shorted either permanently as part of the rotor structure or externally through a system of slip rings and brushes. The rotor with an external short, called a *slip-ring rotor*, has three rings mounted on the rotor shaft inside the motor structure as shown in Figure 12.7. These slip rings are electrically isolated from one another, but each is connected to one terminal of the three-phase windings of the rotor; most rotor windings are connected in wye. Carbon brushes mounted on the stator structure are placed to continuously touch the rotating slip rings, thus achieving connectivity between the rotor windings and any external device. The rotor circuit with an internal short, called *squirrel cage rotor*, consists of slanted wire bars shorted on both ends of the rotor, as shown in Figure 12.8.

12.2.1 ROTATION OF INDUCTION MOTOR

The rotation of the induction motor can be explained using Faraday's law and Lorentz equation. When a conductor carries a current in a uniform magnetic field, the electromechanical relationships can be represented by the following equation:

$$e = Bl\,\Delta v$$
$$\text{Force} = Bli$$

(12.10)

FIGURE 12.6 (See color insert following page 300.)   Induction machines are the prime movers of these vehicles. (a) Mass rover. (Image courtesy of NASA.) (b) Underwater unmanned robot. (c) Electric propulsion ferry. (d) Hybrid electric vehicle. (e) City transportation. (f) City electric bus. (g) Maglev train. (Image courtesy of Transrapid International.) (h) Electric train. (i) Monorail.

where

　　e is the voltage across the conductor
　　B is the flux density
　　l is the length of the conductor
　　Δv is the relative speed between the conductor and the field
　　i is the current of the conductor
　　"force" is the mechanical force exerted on the conductor

If we generalize Equation 12.10 for rotating fields, we can rewrite them in the following forms:

$$e = f(\phi, \Delta n) \tag{12.11}$$

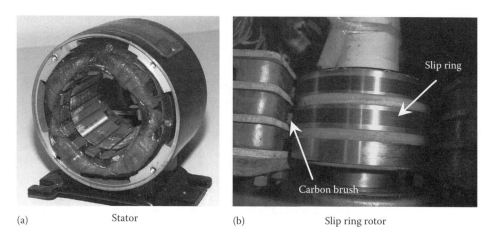

(a)   Stator   (b)   Slip ring rotor

FIGURE 12.7 (See color insert following page 300.)   Main parts of an induction machine. (a) Stator and (b) slip-ring rotor.

$$T = f(\phi, i) \tag{12.12}$$

where
  $f(.)$ is the functional relationship
  T is the torque of the conductors
  Δn is the difference in angular speed between the conductor and the flux in the airgap

   To understand the mechanism by which the rotor is spinning, let us assume that the rotor is at standstill. When a three-phase voltage is applied across the stator windings, a rotating flux at synchronous speed n_s is produced in the airgap. The speed difference Δn is the difference between the rotor speed n and the speed of the airgap flux n_s.

$$\Delta n = n_s - n \tag{12.13}$$

Since the rotor is stationary, Δn is equal to the synchronous speed n_s. Hence, the rotating field induces a voltage e in the rotor windings as given in Equation 12.11. Since the rotor windings are shorted, the induced voltage causes a current i to flow in the rotor windings, producing the Lorentz torque T in Equation 12.12. This torque spins the rotor.

FIGURE 12.8   Squirrel cage rotor.

As you may conclude from Equations 12.11 and 12.12, the induction motor cannot spin at the synchronous speed. This is because Δn in this case is zero resulting in no induced voltage across the rotor windings, so no current is flowing in the windings and therefore no torque is developed to spin the motor.

So what is the steady-state speed of the induction motor? The answer depends on the load torque. The third law of motion developed by Isaac Newton in 1686 states that "whenever one body exerts a force on another, the second exerts a force on the first that is equal in magnitude, opposite in direction and has the same line of action." For the induction motor, this theory means that the motor develops a torque equal in magnitude and opposite in direction to the load torque. Hence, the steady-state operation is achieved when the motor, on a continuous basis, provides the torque needed by the load. Assuming that the flux is unchanged, this developed torque needs certain magnitude of rotor current i as shown in Equation 12.12. The magnitude of this rotor current requires a certain value of induced voltage e in the rotor windings that is equal to the rotor current i multiplied by the rotor impedance. This voltage in turn requires a certain speed difference Δn as given in Equation 12.11. Hence, the steady-state speed of the rotor must always be slightly less than the synchronous speed to maintain the balance between the load torque and the motor's developed torque. Δn is larger for heavy load torques and smaller for light load torques.

The per unit value of the speed difference is known as the slip S.

$$S = \frac{\Delta n}{n_s} = \frac{n_s - n}{n_s} \tag{12.14}$$

Keep in mind that the unit of n is revolution per minute. If we use the angular speed ω (rad/s) instead of n, the slip of the motor is

$$S = \frac{\Delta \omega}{\omega_s} = \frac{\omega_s - \omega}{\omega_s} \tag{12.15}$$

where

$$\omega = \frac{2\pi}{60} n \tag{12.16}$$

The slip at starting, when the rotor speed is zero, is equal to one. At no load, when the motor speed is very close to the synchronous speed, the slip is close to zero. During a normal steady-state operation, the slip is small and often less than 0.1.

Example 12.2

A two-pole, 60 Hz induction motor operates at a slip of 0.02. Compute its rotor speed.

Solution

First, let us compute the synchronous speed.

$$n_s = 120 \frac{f}{p} = 120 \frac{60}{2} = 3600 \text{ rpm}$$

Use Equation 12.14 to compute the speed of the rotor.

$$S = \frac{n_s - n}{n_s}; \quad n = n_s(1 - S) = 3600(1 - 0.02) = 3528 \text{ rpm}$$

EXAMPLE 12.3

The steady-state speed of a 60 Hz induction motor is 1150 rpm. Compute the number of poles of the induction motor and the slip.

Solution

As stated earlier, during the steady-state operation, the slip of the motor is just below the synchronous speed. Using the table in Example 12.1, you will find that the synchronous speed just above 1150 is 1200. Hence, the motor is a six-pole machine.

$$S = \frac{n_s - n}{n_s} = \frac{1200 - 1150}{1200} = 0.0417 \text{ or } 4.17\%$$

12.2.2 EQUIVALENT CIRCUIT OF INDUCTION MOTOR

Since the induction motor is considered a balanced three-phase device, a single-phase equivalent circuit is adequate to model the machine. The stator of the machine consists of a set of copper windings mounted on an iron core. This is almost the same as the primary circuit of the transformer. The stator winding can be represented by a resistance R_1 and inductive reactance X_1. Since the stator copper windings are imbedded in the stator's iron core, the core can be represented by a parallel combination of a core resistance R_c and a core inductive reactance X_c. By using these parameters, it can be observed that the model for the stator circuit shown in Figure 12.9 is very similar to the model for the primary circuit of the transformer. The sum of the currents in R_c and X_c is called *magnetizing* (or core) current I_c. E_1 in the figure is equal to the source voltage V minus the drop across the copper impedance.

$$\overline{E}_1 = \overline{V} - \overline{I}_1(R_1 + jX_1) \tag{12.17}$$

The equivalent circuit of the rotor is also similar to the equivalent circuit for the secondary circuit of the transformer, but only at standstill (where $n = 0$) and when the secondary winding is shorted (because the rotor windings are shorted). The rotor can be represented by its winding resistance R_2 and inductive reactance X_2 as shown in Figure 12.10. E_2 is the induced voltage across the rotor winding at standstill. If you assume that the number of turns of the stator winding is N_1 and that for the rotor winding is N_2, then

$$\frac{E_2}{E_1} = \frac{N_2}{N_1} \tag{12.18}$$

The inductive reactance X_2 of the rotor winding is

$$X_2 = 2\pi f L_2 \tag{12.19}$$

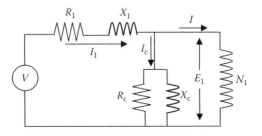

FIGURE 12.9 Equivalent circuit of the stator.

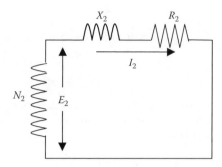

FIGURE 12.10 Equivalent circuit of the rotor at standstill.

where

  f is the frequency of the supply voltage
  L_2 is the inductance of the rotor winding

When the rotor spins, two variables change:

  1. Induced voltage across the rotor winding
  2. Frequency of the rotor current (or rotor voltage)

As shown in Equation 12.11, the induced voltage across the rotor winding is proportional to the speed difference Δn. Hence, the induced voltage E_2 is at standstill when $n = 0$

$$E_2 \sim \Delta n = n_{\mathrm{s}} \tag{12.20}$$

At any other speed, the induced voltage across the rotor winding E_{r} is

$$E_{\mathrm{r}} \sim \Delta n = n_{\mathrm{s}} - n \tag{12.21}$$

Dividing Equation 12.21 by Equation 12.20, we get the relationship between the induced voltage of the rotor at any speed and the induced voltage at standstill.

$$\frac{E_{\mathrm{r}}}{E_2} = \frac{n_{\mathrm{s}} - n}{n_{\mathrm{s}}} = S \tag{12.22}$$

Hence, the voltage across the rotor winding at any speed E_{r} is equal to the rotor voltage at standstill E_2 multiplied by the slip S. In other words, the induced voltage across the rotor windings is always equal to

$$E_{\mathrm{r}} = SE_2 = \frac{N_2}{N_1} SE_1 \tag{12.23}$$

The frequency of the rotor current is directly proportional to the rate at which the magnetic field cuts the rotor winding. At standstill, the airgap flux cuts the rotor windings at a rate proportional to the synchronous speed n_{s}. Hence, the frequency of the rotor current at standstill is f_{s} and is proportional to the synchronous speed. This makes f_{s} equal to the frequency of the stator voltage f. If the rotor is spinning at a speed n, the flux cuts the rotor windings at a rate proportional to Δn. Hence, the frequency of the rotor current at any speed f_{r} is proportional to Δn.

$$\begin{aligned} f = f_{\mathrm{s}} &\sim n_{\mathrm{s}} \\ f_{\mathrm{r}} &\sim \Delta n \end{aligned} \tag{12.24}$$

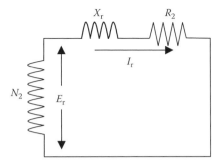

FIGURE 12.11   Equivalent circuit of the rotor at any speed.

Hence, the frequency of the rotor current at any speed is

$$\frac{f_r}{f} = \frac{\Delta n}{n_s} = S$$

$$f_r = S f_s = S f \tag{12.25}$$

The equivalent circuit of the rotor at any speed is shown in Figure 12.11. Because the rotor voltage and rotor frequency are changing with speed, it is hard to use the equivalent circuit in Figure 12.11. Hence, we need to modify the circuit to make it easier to analyze at any speed. The inductive reactance of the rotor X_r is different from X_2 in Equation 12.19 because the rotor frequency at any speed is f_r. Hence,

$$X_r = 2\pi f_r L_2 \tag{12.26}$$

Substituting the value of f_r in Equation 12.25 into Equation 12.26 yields

$$X_r = 2\pi f_r L_2 = 2\pi (Sf) L_2 \tag{12.27}$$

Using the value of X_2 in Equation 12.19 yields

$$X_r = 2\pi (Sf) L_2 = S\, X_2 \tag{12.28}$$

The rotor current I_r for the circuit in Figure 12.11 is

$$\overline{I}_r = \frac{\overline{E}_r}{R_2 + jX_r} = \frac{S\,\overline{E}_2}{R_2 + jSX_2} = \frac{\overline{E}_2}{\frac{R_2}{S} + jX_2} \tag{12.29}$$

Using Equation 12.29, the equivalent circuit of the rotor can be modified as shown in Figure 12.12. Now, we can add the stator circuit to complete the model for the motor as shown in Figure 12.13. This circuit can further be modified as shown in Figure 12.14 by referring the rotor circuit to the stator using the turns ratio. This is the same process used to refer the secondary circuit of the transformer to the primary side. Thus, the resistance R_2' and inductive reactance X_2' of the rotor winding referred to the stator circuit are computed as follows:

$$R_2' = R_2 \left(\frac{N_1}{N_2}\right)^2 \tag{12.30}$$

$$X_2' = X_2 \left(\frac{N_1}{N_2}\right)^2 \tag{12.31}$$

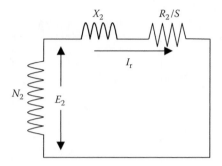

FIGURE 12.12   Modified equivalent circuit of the rotor at any speed.

where N_1 and N_2 are the number of turns of the stator and rotor windings, respectively. Also, the rotor current referred to the stator circuit I_2' can be computed as

$$I_2' = I_r \left(\frac{N_2}{N_1} \right) \tag{12.32}$$

The resistance R_2'/S has two components; one of them is the resistance of the rotor winding R_2'.

$$\frac{R_2'}{S} = R_2' + \frac{R_2'}{S}(1 - S) \tag{12.33}$$

The resistive element $R_2'/S/(1 - S)$ is an electrical representation of the mechanical load of the motor and is therefore known as the load resistance. The parsing of R_2'/S leads to the equivalent circuit in Figure 12.15a. We can further modify the equivalent circuit by assuming that the core current I_c is much smaller than I_1. Hence, $I_1 \approx I_2'$ and we can thus assume that the impedances of the stator and rotor windings are in series as shown in Figure 12.15b.

R_{eq} and X_{eq} in Figure 12.15c are defined as

$$\begin{aligned} R_{eq} &= R_1 + R_2' \\ X_{eq} &= X_1 + X_2' \end{aligned} \tag{12.34}$$

12.2.3 Power Analysis

Figure 12.16 shows the power flow of the induction motor. The motor receives input power P_{in} from the electric source

$$P_{in} = 3VI_1 \cos \theta_1 \tag{12.35}$$

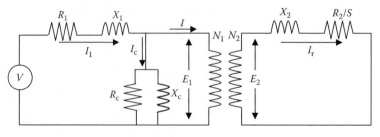

FIGURE 12.13   Equivalent circuit of an induction motor.

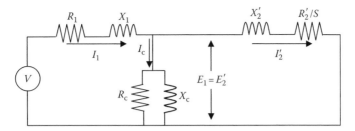

FIGURE 12.14 Equivalent circuit of an induction motor referred to the stator.

where

P_{in} is the input power of the motor from the three phases of the source

V is the phase voltage of the source

I_1 is the phase current of the stator

θ_1 is the phase angle of the current (the angle between the phase voltage and phase current)

Part of the input power is wasted in the stator circuit in the form of copper loss inside the windings P_{cu1} and core loss P_c. These losses can be computed using the equivalent circuit in Figure 12.15a.

$$P_{cu1} = 3I_1^2R_1$$

$$P_c \approx 3\frac{V^2}{R_c}$$

(12.36)

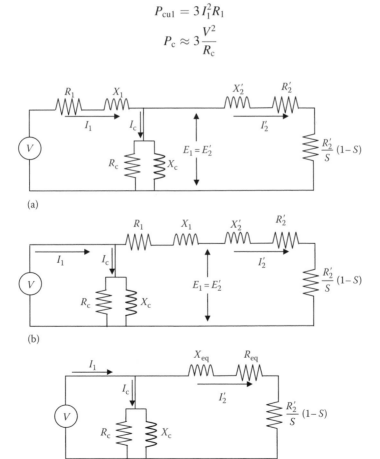

(a)

(b)

(c)

FIGURE 12.15 More equivalent circuits of the induction motor.

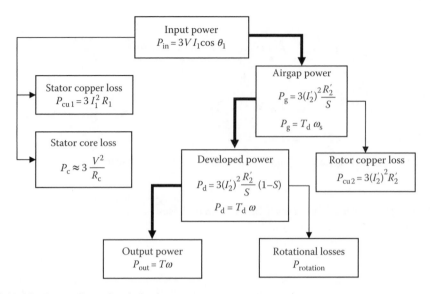

FIGURE 12.16   Power flow of an induction motor.

The rest of the power is transmitted to the rotor by the airgap flux, thus it is called the *airgap power* P_g, which is consumed in R'_2/S.

$$P_g = 3\,(I'_2)^2\,\frac{R'_2}{S} \qquad (12.37)$$

The airgap power is also a form of mechanical power since it involves the mechanical rotation of the airgap flux. Therefore, P_g can also be written in the mechanical form.

$$P_g = T_d\,\omega_s \qquad (12.38)$$

where
   T_d is the developed torque of the airgap flux
   ω_s is the synchronous speed of the flux

T_d is the torque produced by all three phases. When the airgap power enters the rotor circuit, part of it is wasted in the copper winding of the rotor P_{cu2}.

$$P_{cu2} = 3\,(I'_2)^2\,R'_2 \qquad (12.39)$$

The rest of the power is the developed power P_d consumed by $R'_2(1 - S)/S$

$$P_d = 3\,(I'_2)^2\,\frac{R'_2}{S}\,(1 - S) \qquad (12.40)$$

This developed power can be represented in the mechanical form

$$P_d = T_d\,\omega \qquad (12.41)$$

where ω is the speed of the rotor (not the synchronous speed). The developed power P_d is not completely converted to useful power (shaft power) as part of it is wasted in the form of rotational losses such as friction and windage. The output power of the motor is equal to the output torque on the motor shaft T (not the developed torque) multiplied by the rotor speed.

$$P_{out} = T\,\omega \qquad (12.42)$$

Note that the relationships between the airgap power, developed power, and rotor copper loss are

$$P_{cu2} = SP_g$$
$$P_d = (1 - S)P_g \qquad (12.43)$$

EXAMPLE 12.4

A 60 Hz, three-phase, Y-connected induction motor produces 100 hp at 1150 rpm. The friction and windage losses of the motor are 1.2 kW; the stator copper and core losses are 2.1 kW. Compute the input power and the efficiency of the motor.

Solution

The mechanical unit for the output power is horsepower (hp), which can be converted into kilowatt by the conversion factor given in the appendix.

$$P_{out} = \frac{100}{1.34} = 74.63 \text{ kW}$$

To compute the input power, we can use the power diagram in Figure 12.16. We can work the problem from the output toward the input. First, calculate the developed power

$$P_d = P_{out} + P_{rotation} = 74.63 + 1.2 = 75.83 \text{ kW}$$

To compute the airgap power, we need to know the rotor copper loss, which is not explicitly provided. Another alternative is to compute the slip and use the relationship in Equation 12.43 to compute the airgap power. To calculate the slip, we need the synchronous speed and the actual speed of the motor. The actual speed is given, but not the synchronous speed. However, from the principle of operation, the speed of the induction motor is just below the synchronous speed. Using the table in Example 12.1, we can find that the number of poles of this machine is six, and its synchronous speed is 1200 rpm. Hence,

$$S = \frac{n_s - n}{n_s} = \frac{1200 - 1150}{1200} = 0.0417$$

Using Equation 12.43, we can compute the airgap power.

$$P_g = \frac{P_d}{1 - S} = \frac{75.83}{0.9583} = 79.13 \text{ kW}$$

Now use the diagram in Figure 12.16 to compute the input power.

$$P_{in} = P_g + P_{cu1} + P_c = 79.13 + 2.1 = 81.23 \text{ kW}$$

The motor efficiency η is

$$\eta = \frac{P_{out}}{P_{in}} = \frac{74.63}{81.23} = 91.87\%$$

12.2.4 SPEED–TORQUE RELATIONSHIP

By using the mechanical expression for the developed power in Figure 12.16, we can compute the developed torque of the induction motor.

$$T_d = \frac{P_d}{\omega} = \frac{3(I_2')^2 R_2'(1-S)}{S\omega} \qquad (12.44)$$

From the equivalent circuit in Figure 12.15c, the rotor current is

$$I_2' = \frac{V}{\sqrt{\left(R_1 + \frac{R_2'}{S}\right)^2 + X_{eq}^2}} \qquad (12.45)$$

Substituting Equation 12.45 into Equation 12.44 leads to

$$T_d = \frac{3V^2 R_2'(1-S)}{S\omega \left[\left(R_1 + \frac{R_2'}{S}\right)^2 + X_{eq}^2\right]} \qquad (12.46)$$

From Equation 12.15, the speed of the motor ω can be written in terms of the synchronous speed and slip

$$\omega = \omega_s(1-S) \qquad (12.47)$$

Substituting Equation 12.47 into Equation 12.46, we can obtain Equation 12.48 for the developed torque

$$T_d = \frac{3V^2 R_2'}{S\omega_s \left[\left(R_1 + \frac{R_2'}{S}\right)^2 + X_{eq}^2\right]} \qquad (12.48)$$

Keep in mind that V is the phase voltage, and the torque is developed by all three phases. Equation 12.48 gives torque as a function of the slip, which can be modified to represent the torque as a function of speed.

$$T_d = \frac{3V^2 R_2'}{(\omega_s - \omega)\left[\left(R_1 + \frac{\omega_s R_2'}{(\omega_s - \omega)}\right)^2 + X_{eq}^2\right]} \qquad (12.49)$$

The speed–torque or slip–torque characteristic of the induction motor based on Equations 12.48 and 12.49 is shown in Figure 12.17. The characteristic is nonlinear and has several key operating points.

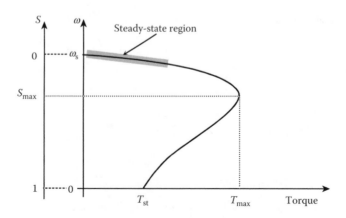

FIGURE 12.17 Speed–torque characteristic of an induction motor.

When the motor starts (at zero speed), the motor develops a starting torque T_{st} that spins the motor. When the motor accelerates while starting, the developed torque increases until it reaches its maximum value T_{max} at the slip S_{max}. The speed of the motor continues to increase until it reaches the steady-state operating region (the shaded area) where the speed of the motor is constant and the developed torque of the motor is equal to the load torque.

EXAMPLE 12.5

A 480 V, 60 Hz, three-phase, four-pole induction motor has the following parameters:

$$R_1 = 0.3 \; \Omega; \quad R_2' = 0.2 \; \Omega; \quad X_{eq} = 2.0 \; \Omega$$

At full load, the motor speed is 1760 rpm. Calculate the following:

(a) Slip of the motor
(b) Developed torque of the motor at full load
(c) Developed power in horsepower
(d) Rotor current
(e) Copper losses of the motor

Solution

(a) The synchronous speed of the motor can be computed using Equation 12.9.

$$n_s = 120\frac{f}{p} = 120\frac{60}{4} = 1800 \text{ rpm}$$

$$\omega_s = \frac{2\pi}{60}n_s = \frac{2\pi}{60}1800 = 188.5 \text{ rad/s}$$

Using Equation 12.15, we can compute the slip of the motor.

$$S = \frac{n_s - n}{n_s} = \frac{1800 - 1760}{1800} = 0.0222$$

(b) Equation 12.48 or Equation 12.49 can be used to compute the load torque.

$$T_d = \frac{3V^2 R_2'}{S\omega_s\left[\left(R_1 + \frac{R_2'}{S}\right)^2 + X_{eq}^2\right]} = \frac{3\left(\frac{480}{\sqrt{3}}\right)^2 0.2}{0.0222 \times 188.5\left[\left(0.3 + \frac{0.2}{0.0222}\right)^2 + 4\right]} = 121.46 \text{ N m}$$

(c) The angular speed of the motor is

$$\omega = \frac{2\pi}{60}n = \frac{2\pi}{60}\ 1760 = 184.3 \text{ rad/s}$$

The developed power of the motor is

$$P_d = T_d\omega = 121.46 \times 184.3 = 22.386 \text{ kW}$$

The developed power in (hp) unit is

$$P_d = 22.386 \times 1.34 = 30 \text{ hp}$$

(d) The rotor current can be computed using Equation 12.45.

$$I_2' = \frac{V}{\sqrt{\left(R_1 + \frac{R_2'}{S}\right)^2 + X_{eq}^2}} = \frac{480/\sqrt{3}}{\sqrt{\left(0.3 + \frac{0.2}{0.0222}\right)^2 + 4}} = 29.1 \text{ A}$$

(e) The copper losses of the motor are in the stator and rotor windings, that is,

$$P_{cu} = P_{cu1} + P_{cu2} = 3(I_2')^2 R_{eq} = 3 * (29.1)^2 \ (0.3 + 0.2) = 1.27 \text{ kW}$$

12.2.5 STARTING TORQUE AND STARTING CURRENT

At starting, the speed of the motor is zero and $S = 1$. Substituting these values either in Equation 12.48 or in Equation 12.49 yields the starting torque T_{st}.

$$T_{st} = \frac{3V^2 R_2'}{\omega_s \left[(R_1 + R_2')^2 + X_{eq}^2\right]} = \frac{3V^2 R_2'}{\omega_s Z_{eq}^2} \tag{12.50}$$

where

$$Z_{eq} = \sqrt{(R_1 + R_2')^2 + X_{eq}^2} = \sqrt{R_{eq}^2 + X_{eq}^2} \tag{12.51}$$

The starting current I_{st} of the motor can be computed using Equation 12.45, when $S = 1$.

$$I_{st}' = \frac{V}{\sqrt{(R_1 + R_2')^2 + X_{eq}^2}} = \frac{V}{Z_{eq}} \tag{12.52}$$

EXAMPLE 12.6

For the motor in Example 12.5, compute the following:

(a) Starting torque
(b) Starting current
(c) Ratio of the starting torque to the full load torque
(d) Ratio of the starting current to the full load current

Solution

(a) The starting torque can be computed using Equation 12.50.

$$T_{st} = \frac{3V^2 R_2'}{\omega_s Z_{eq}^2} = \frac{480^2 \times 0.2}{188.5(0.5^2 + 4)} = 57.52 \text{ N m}$$

(b) Using Equation 12.52, we can compute the starting current.

$$I_{st}' = \frac{V}{Z_{eq}} = \frac{480/\sqrt{3}}{\sqrt{0.5^2 + 4}} = 134.42 \text{ A}$$

(c) The full load torque computed in Example 12.5 is 121.46 N m. The ratio of the starting torque to the full load torque is

$$\frac{T_{st}}{T_d} = \frac{57.52}{121.46} = 0.4736$$

As seen, the starting torque is less than half the full load torque. Heavily loaded induction motors may not start with low starting torques. In the following sections, we shall discuss some techniques to increase the starting torque.

(d) The full load current computed in Example 12.5 is 29.1 A. The ratio of the starting current to the full load current is

$$\frac{I_{st}}{I_2'} = \frac{134.42}{29.1} = 4.62$$

The starting current of this motor is more than four times the rated current. This excessive current could damage the motor. In the following sections, we shall discuss methods by which the starting current can be reduced.

12.2.6 MAXIMUM TORQUE

The maximum torque of the induction motor can be computed by setting the derivative of the torque equation with respect to the slip to zero. The slip that satisfies Equation 12.53 is the slip at the maximum torque S_{max}.

$$\frac{dT_d}{dS} = 0; \quad S \Rightarrow S_{max}$$

$$\left(R_1^2 + X_{eq}^2\right)S_{max}^2 - \left(R_2'\right)^2 = 0 \tag{12.53}$$

Hence,

$$S_{max} = \frac{R_2'}{\sqrt{R_1^2 + X_{eq}^2}} \tag{12.54}$$

Substituting the value of S_{max} into Equation 12.48 leads to the maximum torque of the motor.

$$T_{max} = \frac{3V^2}{2\omega_s \left(R_1 + \sqrt{R_1^2 + X_{eq}^2}\right)} \tag{12.55}$$

EXAMPLE 12.7

For the motor in Example 12.5, compute the following:

(a) Slip at maximum torque
(b) Speed at maximum torque
(c) Maximum torque
(d) Ratio of the maximum torque to the full load torque
(e) Current at the maximum torque

Solution

(a)
$$S_{max} = \frac{R_2'}{\sqrt{R_1^2 + X_{eq}^2}} = \frac{0.2}{\sqrt{0.3^2 + 4}} = 0.0989$$

(b) Using Equation 12.47, we can compute the speed at maximum torque.

$$n = n_s(1 - S_{max}) = 1800(1 - 0.0989) = 1621.98 \text{ rpm}$$

(c) Using Equation 12.55, we can compute the maximum torque.

$$T_{max} = \frac{3\,V^2}{2\omega_s\left[R_1 + \sqrt{R_1^2 + X_{eq}^2}\right]} = \frac{480^2}{2 \times 188.5\left[0.3 + \sqrt{0.2^2 + 4}\right]} = 264.57 \text{ N m}$$

(d) The full load torque computed in Example 12.5 is 121.46 N m. The ratio of the maximum torque to the full load torque is

$$\frac{T_{max}}{T_d} = \frac{264.57}{121.46} = 2.18$$

The maximum torque is larger than the full load torque.

(e) Rotor current at S_{max} can be computed using Equation 12.45.

$$I_2' = \frac{V}{\sqrt{\left(R_1 + \frac{R_2'}{S_{max}}\right)^2 + X_{eq}^2}} = \frac{480/\sqrt{3}}{\sqrt{\left(0.3 + \frac{0.2}{0.0989}\right)^2 + 4}} = 90.42 \text{ A}$$

The current at the maximum torque is smaller than the starting current, but is larger than the full load current.

12.2.7 STARTING METHODS

The induction motor has two problems during starting: the starting current is large and the starting torque is small. The large starting current can damage the rotor windings due to the excessive thermal heat inside the windings. The small starting torque may not be enough to start the motor under heavy loading conditions. As shown in Equation 12.52, the starting current can be reduced if we reduce the terminal voltage of the motor or increase its equivalent impedance Z_{eq}. These two methods are explored below.

12.2.7.1 Voltage Reduction

If an ac/ac converter is used to drive the induction motor as shown in Figure 12.18, the voltage across the motor at starting can be reduced, thus reducing the starting current. However, the reduction of the stator voltage can also reduce the starting torque and the maximum torque as shown in Figure 12.19 and explained by Equations 12.50 and 12.55. A 20% reduction in the stator voltage reduces the starting torque and the maximum torque by 36% each.

EXAMPLE 12.8

For the motor in Example 12.5, compute the following:

(a) Terminal voltage that limits the starting current to twice the full load current
(b) Reduction in the starting torque
(c) Reduction in the maximum torque

Solution

(a) The full load current computed in Example 12.5 is 29.1 A. Hence, the starting current is limited to 58.2 A.

$$I'_{st} = \frac{V_{st}}{Z_{eq}} = \frac{V_{st}}{\sqrt{0.5^2 + 4}} = 58.2 \text{ A}$$

where V_{st} is the reduced phase voltage at starting.

$$V_{st} = 119.98 \text{ V}$$

The line-to-line voltage at starting is $\sqrt{3} \times 119.98 = 207.8$ V, which is about 43% of the rated voltage.

(b) The starting torque can be computed using Equation 12.50.

$$T_{st} = \frac{3V_{st}^2 R'_2}{\omega_s Z_{eq}^2} = \frac{207.8^2 \times 0.2}{188.5(0.5^2 + 4)} = 10.78 \text{ N m}$$

This starting torque is about 19% of the starting torque computed in Example 12.6 for the full voltage. Hence, the reduction in the starting torque is about 81%.

(c) Using Equation 12.55, we can compute the maximum torque for the new voltage.

$$T_{max} = \frac{3V_{st}^2}{2\omega_s \left[R_1 + \sqrt{R_1^2 + X_{eq}^2}\right]} = \frac{207.8^2}{2 \times 188.5\left[0.3 + \sqrt{0.2^2 + 4}\right]} = 49.58 \text{ N m}$$

This maximum torque is also about 19% of the maximum torque computed in Example 12.7 for the full voltage. The reduction is about 81%.

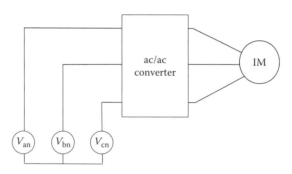

FIGURE 12.18   Voltage control of induction motor.

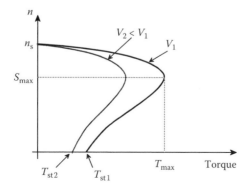

FIGURE 12.19   Speed–torque characteristics at different voltage levels.

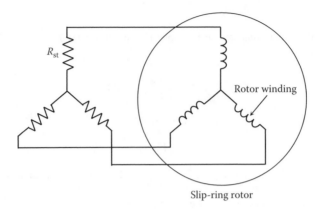

Slip-ring rotor

FIGURE 12.20 Starting of an induction motor by inserting a resistance in the rotor circuit.

12.2.7.2 Insertion of Resistance

As seen in Example 12.8, the voltage reduction at starting reduces the starting current, which is desirable. However, the method leads to undesirable reduction in the starting torque. Therefore, the motor may produce less torque at starting than that needed by the load. To address this problem, we should use other methods that can increase the starting torque while reducing the starting current. One of these methods is based on increasing the rotor resistance. For the squirrel cage rotor, the windings can be designed from material that exhibit skin effects at the frequency of the source voltage, thus increasing its resistance at starting only. When the rotor speed is close to the full load speed, the rotor frequency is very low, thus no skin effect is present. For the slip-ring motor, the rotor windings are accessible from the stator through the slip-ring mechanism shown in Figure 12.7. Thus, an external resistance can be added to the rotor circuit at starting as shown in Figure 12.20. In either type of motors, when the rotor resistance increases, the starting current decreases as shown in Equation 12.52. But how do the starting and maximum torques change? Upon the first examination of Equation 12.50, it is hard to tell whether the increase in the rotor resistance would increase the starting torque. However, because most machines have their $R_{eq}^2 \ll X_{eq}^2$, we can approximate the starting torque equation as

$$T_{st} = \frac{3V^2 R_2'}{\omega_s \left[\left(R_1 + R_2' \right)^2 + X_{eq}^2 \right]} \approx \frac{3V^2 R_2'}{\omega_s X_{eq}^2} \qquad (12.56)$$

Now, it is obvious that the increase in the rotor resistance would result in an increase in the starting torque. This is another desirable feature.

The maximum torque, as shown in Equation 12.55, is unaffected by the increase in the rotor resistance. However, the slip at maximum torque S_{max} in Equation 12.54 increases when the rotor resistance is increased. This means that the maximum torque occurs at lower speeds. In fact, if enough resistance is added to the rotor circuit so that $S_{max} = 1$, the starting torque is equal to the maximum torque of the motor as shown in Figure 12.21.

Example 12.9

For the motor in Example 12.5, compute the following:

(a) Resistance that should be added to the rotor circuit to achieve the maximum torque at starting
(b) Starting current

Solution

(a) To compute the value of the inserted resistance in the rotor circuit to achieve the maximum torque at starting, we can set the slip at maximum torque in Equation 12.54 to 1.

$$S_{max} = \frac{R'_2 + R'_{st}}{\sqrt{R_1^2 + X_{eq}^2}} = 1$$

Hence,

$$R'_{st} = \sqrt{R_1^2 + X_{eq}^2} - R'_2 = \sqrt{0.3^2 + 4} - 0.2 = 1.822 \ \Omega$$

This is the starting resistance referred to the stator circuit. The actual rotor resistance can be computed if we know the turns ratio of the motor windings.

(b) The starting current in Equation 12.52 can be modified to include the starting resistance in the rotor circuit.

$$I'_{st} = \frac{V}{\sqrt{\left(R_1 + \left(R'_2 + R'_{st}\right)\right)^2 + X_{eq}^2}} = \frac{\frac{480}{\sqrt{3}}}{\sqrt{(0.3 + (0.2 + 1.822))^2 + 4}} = 90.43 \text{ A}$$

Comparing this starting current to that computed in Example 12.6, you will find that the starting current is decreased because of the insertion of the starting resistance.

12.3 LINEAR INDUCTION MOTOR

Linear induction motors (LIMs) are very effective drive mechanisms for transportation and actuation systems. High-power LIMs are used in rapid transportations, baggage handling systems, conveyors, crane drives, theme park rides, and flexible manufacturing systems. Low-power ones are used in robotics, gate controls, guided trajectories (e.g., aluminum can propulsion), and stage and curtain movements. NASA envisions the use of such motors in launching spacecrafts in the future.

LIMs come in two general designs: wheeled linear induction motor (WLIM) and magnetically levitated linear induction motor (Maglev). The Maglev technology, which was developed in 1934 by Hermann Kemper, is becoming popular for high-speed transportation. Some of the LIM applications are shown in Figure 12.22.

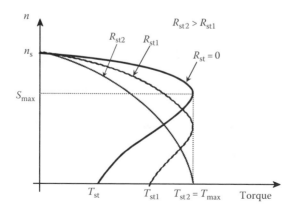

FIGURE 12.21 Speed–torque characteristics when a resistance is added to the rotor circuit.

(a)   (b)

(c)

FIGURE 12.22 (See color insert following page 300.) Examples of LIM applications. (a) Roller coaster. (b) Maglev launcher. (Image courtesy of NASA Marshall Space Flight Center.). (c) Maglev rapid transportation. (Image courtesy of Transrapid International.)

12.3.1 WHEELED LINEAR INDUCTION MOTOR

The LIM is similar to the rotating induction motor except that the LIM has a flat structure instead of the cylindrical structure of the rotating motor. Consider the rotating induction motor in Figure 12.23a. If you imagine that we can cut the motor along the dashed line and flatten the machine,

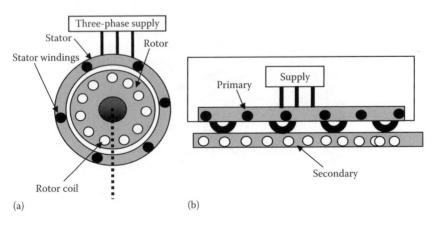

(a)   (b)

FIGURE 12.23 Wheeled LIM.

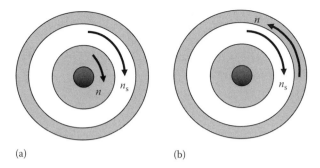

FIGURE 12.24   Relative speed of a rotating induction motor. (a) Stationary stator and (b) stationary rotor.

we will get the WLIM in Figure 12.23b, which can now be used to propel a train. In this case, the rotor is the track of the train, and the stator is the train's engine. The stator of the LIM is called the primary circuit, and the rotor is called the secondary circuit. In rotating motors, the separation between the rotor and stator is maintained by the ball bearings of the rotor's shaft. In WLIMs, the separation is maintained by the wheels of the train.

For the rotating induction motors in Figure 12.24a, when the stator is anchored and the rotor is allowed to rotate freely, the speed of the rotor is in the same direction as the speed of the magnetic field. However, if you hold the rotor stationary and allow the stator to spin, the stator rotates in the direction opposite to the synchronous speed of the field as shown in Figure 12.24b. Similarly, for the LIM, the primary circuit moves opposite to the synchronous speed of the field.

The synchronous speed of the rotating field of the induction motor is given in Equation 12.9. The linear synchronous speed v_s of this rotating field at any point located along the inner surface of the stator is

$$v_s = \omega_s r = \left(\frac{2\pi}{60}n_s\right)r = \left(\frac{2\pi}{60}\frac{120f}{p}\right)r = 2\left(\frac{2\pi r}{p}\right)f = 2\tau_p f \qquad (12.57)$$

where
   r is the inner radius of the stator (m) (Figure 12.25)
   n_s is the synchronous speed of the field in rpm
   ω_s is synchronous speed of the field in rad/s
   τ_p is the pole pitch

The pole pitch is the circumference of the inner circle of the stator divided by a number of poles. This is the separation between a–a′, b–b′, or c–c′ as shown in Figure 12.25.

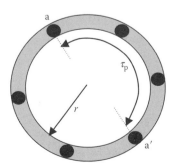

FIGURE 12.25   Pole pitch of a motor.

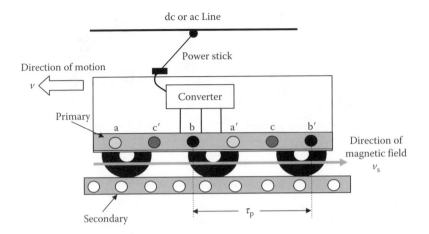

FIGURE 12.26 Motion of an LIM.

Let us consider the train that is powered by the WLIM in Figure 12.26. The primary windings are mounted under the floor of the train's compartment, called *bogie*. The secondary windings consist of metal alloy bars embedded along the track's guideway and are perpendicular to the track, called *reaction plates*. The length of the secondary circuit is the length of the track itself. The train is powered by direct current (dc) or ac power line alongside the track or above it. The train taps its energy from the power line through brushes that are always in contact with the line. A converter is mounted on the vehicle to convert the waveform of the power line into balanced multiphase waveforms with variable voltage and frequency.

The three-phase current of the LIM produces a magnetic field traveling in the direction shown in Figure 12.26. The speed of this magnetic field is given in Equation 12.57. The train itself moves opposite to the direction of the magnetic field as explained earlier. The pole pitch of the LIM is the distance between the opposite poles of any winding; that is, the distance between a and a'.

The equivalent circuit of the LIM is basically the same as that for the rotating induction motor in Figure 12.15. One key difference is that the resistance of the secondary winding of the LIM is made larger than the rotor resistance of the rotating motor. This way, the motor develops a large torque at starting.

In linear motion analysis, force is used instead of torque. The developed force of the LIM F_d is also known as the *thrust* of the motor. It can be computed from the developed torque in Equation 12.48, but is modified for a linear motion where angular velocities are replaced by linear speeds.

$$F_d = \frac{3V^2 R_2'}{Sv_s\left[\left(R_1 + \frac{R_2'}{S}\right)^2 + X_{eq}^2\right]} \tag{12.58}$$

where
v is the actual speed of the train
v_s is the synchronous speed of the magnetic field

The slip S is defined as

$$S = \frac{v_s - v}{v_s} \tag{12.59}$$

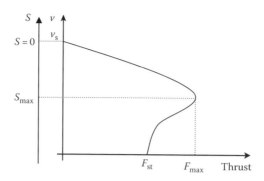

FIGURE 12.27 Speed–thrust of an LIM.

In linear motion, the developed power in the mechanical form is the force multiplied by the speed.

$$P_d = F_d v \qquad (12.60)$$

Substituting Equation 12.58 into Equation 12.60, we can compute the thrust power of the LIM.

$$P_d = F_d v = \frac{3V^2 R_2' v}{Sv_s\left[\left(R_1 + \frac{R_2'}{S}\right)^2 + X_{eq}^2\right]} \qquad (12.61)$$

The speed–thrust characteristic of the LIM is shown in Figure 12.27. The developed thrust must compensate for three main components: the friction between the wheels and the track, the drag force, and the acceleration/deceleration force. For steady-state speed, the third component is zero.

The friction force $F_{friction}$ is a function of the normal force F_{normal} on the track.

$$F_{friction} = \mu F_{normal} \qquad (12.62)$$

where μ is the coefficient of friction between the wheels and the track. For a leveled track, the normal force is the weight of the vehicle.

$$F_{normal} = mg \qquad (12.63)$$

where
m is the mass of the moving vehicle
g is the acceleration of gravity

For an inclined track, the normal force is the component of the weight that is perpendicular to the track.

$$F_{normal} = mg \cos \theta \qquad (12.64)$$

where θ is the inclination angle of the track. Example 12.12 deals with an inclined track.

The air drag force, which slows down the moving vehicle, is due to many factors such as the frontal area of the vehicle, the shape of the vehicle, the velocity of the vehicle, and the air density. The air drag force F_{air} can be computed by

$$F_{air} = 0.5\delta v^2 A C_d \qquad (12.65)$$

where
    δ is the air density (about 1 kg/m^3)
    v is the velocity of the vehicle (m/s)
    A is the frontal area of the vehicle that is perpendicular to the air flow (m^2)
    C_d is the coefficient of drag that depends on the shape of the frontal area of the vehicle; C_d can
        be about 1 for train, 0.4 for roller coaster, and 0.5 for a typical passenger car

EXAMPLE 12.10

A train with WLIM is traveling at 100 km/h. The frontal area of the train is 20 m^2 and the coefficient of its drag is 0.8. The friction coefficient between the wheels of the train and the track is 0.05, and the weight of the train is 100,000 kg. Compute the thrust of the motor at steady state. Also, compute the developed power of the motor in horsepower.

Solution

At steady state, the induction motor produces enough thrust to compensate for the friction force plus the drag force. The frictional force is computed based on the normal force of the vehicle and the coefficient of friction μ.

$$F_{normal} = mg = 10^5 \times 9.8 = 980 \text{ kN}$$
$$F_{friction} = \mu F_{normal} = 0.05 \times 980 = 49 \text{ kN}$$

The drag force can be computed using Equation 12.65.

$$F_{air} = 0.5 \delta v^2 A C_d = 0.5 \times 1 \times \left(100 \times \frac{1000}{3600} \right)^2 \times 20 \times 0.8 = 6.174 \text{ kN}$$

The developed force is the sum of the frictional force and the drag force.

$$F_d = F_{friction} + F_{air} = 49 + 6.174 = 55.174 \text{ kN}$$

The developed power of the LIM is

$$P_d = F_d v = 55.174 \times 27.78 = 1532.7 \text{ kW}$$
$$P_d = 1532.7 \times 1.34 = 2054 \text{ hp}$$

EXAMPLE 12.11

A vehicle with WLIM has a mass of 4000 kg. The friction coefficient between the wheels of the vehicle and the track is 0.05. The pole pitch of the vehicle is 3 m. At steady state, the slip of the motor is 0.1 when the frequency of the primary windings is 10 Hz. Assume that the drag force at steady state is 5000 N. Compute the developed power of the motor.

Solution

At steady state, there is no acceleration, so the induction motor produces enough thrust to compensate for the friction force plus the drag force.

$$F_{normal} = mg = 4000 \times 9.8 = 39.2 \text{ kN}$$
$$F_{friction} = \mu F_{normal} = 0.05 \times 39.2 = 1.96 \text{ kN}$$

To compute the developed power, we need the speed of the vehicle. The synchronous speed of the magnetic field is

$$v_s = 2\tau_p f = 2 \times 3 \times 10 = 60 \text{ m/s}$$

Hence, the speed of the vehicle is

$$v = v_s(1 - S) = 60(1 - 0.1) = 54 \text{ m/s}$$

The developed power is

$$P_d = F_d v = (F_{\text{friction}} + F_{\text{air}})v = (1960 + 5000) \times 54 = 375.84 \text{ kW}$$

The equations for the maximum developed force (maximum thrust) and the slip at maximum force for the LIM are the same as Equations 12.55 and 12.54 for the rotating motor.

$$S_{\text{max}} = \frac{R_2'}{\sqrt{R_1^2 + X_{\text{eq}}^2}} \tag{12.66}$$

$$F_{\text{max}} = \frac{3V^2}{2v_s \left(R_1 + \sqrt{R_1^2 + X_{\text{eq}}^2} \right)} \tag{12.67}$$

For the LIM, the rotor resistance is higher than that for the rotating motor. Hence, S_{max} for the LIM is large, occurring near the starting region. This way, the LIM develops enough thrust at starting to speed up the train even at heavy loading conditions. Therefore, a reasonable approximation is to assume that the speed–thrust characteristic of the motor is linear between the operating slip S and S_{max}; that is,

$$S = \frac{F_d}{F_{\text{max}}} S_{\text{max}} \tag{12.68}$$

EXAMPLE 12.12

A theme ride vehicle is moving uphill at a constant speed. The slope of the hill is $10°$. The mass of the vehicle plus the riders is 2000 kg. The vehicle is powered by a WLIM. The pole pitch of the motor is 0.5 m. The friction coefficient of the surface is 0.01. The primary circuit resistance R_1 is 0.5 Ω, and the secondary resistance referred to the primary circuit R_2' is 1.0 Ω. The equivalent inductance of the primary and secondary circuits is 0.01 H. The line-to-line voltage of the primary windings of the motor is 480 V, and its frequency is 15 Hz. Assume that the drag force is 500 N and compute the speed of the vehicle.

Solution

The system can be described by Figure 12.28. The weight of the vehicle is resolved into two components: one normal to the road surface F_{normal} and the other F_g parallel to the road surface pulling the vehicle down due to gravity. F_{normal} produces the frictional force F_{friction}.

$$F_d = F_g + F_{\text{friction}} + F_{\text{air}}$$
$$F_{\text{normal}} = mg \cos \theta = 2000 \times 9.8 \times \cos 10 = 1.93 \times 10^4 \text{ N}$$
$$F_{\text{friction}} = \mu F_{\text{normal}} = 0.01 \times 1.93 \times 10^4 = 193 \text{ N}$$
$$F_g = mg \sin \theta = 2000 \times 9.8 \times \sin 10 = 3.403 \times 10^3 \text{ N}$$
$$F_d = F_g + F_{\text{friction}} + F_{\text{air}} = 4.097 \text{ kN}$$

The synchronous speed of the motor is

$$v_s = 2\tau_p f = 2 \times 0.5 \times 15 = 15 \text{ m/s}$$

The motor speed can be computed using the relationship in Equation 12.68. But first, let us compute the maximum force and the slip at maximum force.

$$S_{max} = \frac{R_2'}{\sqrt{R_1^2 + X_{eq}^2}} = \frac{1.0}{\sqrt{0.5^2 + (2\pi \times 15 \times 0.01)^2}} = 0.9373$$

$$F_{max} = \frac{3V^2}{2v_s\left[R_1 + \sqrt{R_1^2 + X_{eq}^2}\right]} = \frac{480^2}{2 \times 15\left[0.5 + \sqrt{0.5^2 + (2\pi \times 15 \times 0.01)^2}\right]} = 4.901 \text{ kN}$$

Now, we have the information needed to compute the slip of the motor.

$$S = \frac{F_d}{F_{max}} S_{max} = \frac{4.097}{4.901} 0.9373 = 0.784$$

Hence, the speed of the vehicle is

$$v = v_s(1 - S) = 15(1 - 0.784) = 3.24 \text{ m/s or } 11.66 \text{ km/h}$$

12.3.2 MAGNETICALLY LEVITATED INDUCTION MOTOR

One of the main limitations of the WLIM is the friction between the wheels and the track. Besides the extra power consumption, the friction limits the maximum speed of the vehicle as well as its maximum acceleration and deceleration.

The Maglev technology allows the vehicle to magnetically levitate with virtually no friction between the bogie of the vehicle (undercarriage) and the track. The technology is based on having magnetic poles on the track that are similar to the magnetic poles on the bogie of the vehicle (both are north or south). These poles repel each other and the bogie then levitates. Hence, a Maglev train can achieve very high speeds and very smooth rides because the vehicle essentially flies. In addition, the Maglev trains can climb steep hills that regular trains cannot reach. With the current technology, a Maglev train can speed up to 600 km/h, which is much faster than any other ground transportation system. The travel time of a medium trip (1000 km) on a Maglev train is essentially equal to the travel time of the airplane when you include the check-in time at the airport as well as the taxing time of the airplane.

There are two types of Maglev systems: levitation and propulsion. In both systems, magnetic force is developed along the track to repel the bogie of the vehicle. This is done in various ways;

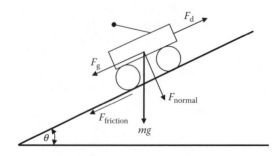

FIGURE 12.28 Force vectors of a vehicle moving uphill.

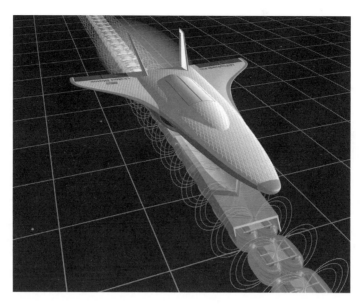

FIGURE 12.29 (See color insert following page 300.)   Concept of Maglev space launcher. (Image courtesy of NASA Marshall Space Flight Center.)

one of them is to install magnetic coils along the track as shown in Figure 12.29. These coils are excited by a separate source when the vehicle approaches them, thus keeping the vehicle levitated along the track. This system, however, is suitable for short tracks as the excitation of long tracks is expensive and difficult to implement. Another alternative is to induce voltage in the coils of the track by the vehicle itself. When the vehicle moves over a coil in the track, the electric magnets installed in the bogie are switched on to induce a voltage on the coil of the track. If the track coil is wounded in the direction that creates poles with similar polarity to the magnetic poles at the bottom of the bogie's magnets, a repulsive force is created that levitates the vehicle.

For the propulsion Maglev, the magnetic force of the track coils propels the vehicle along the track in addition to lifting the bogie. To understand this, assume that you have two magnets on top of each other with their similar poles facing each other. The top magnet will levitate due to the magnetic lifting force. Now, fix the bottom magnet to a table and keep the top magnet at constant vertical distance but free to move horizontally. In this case, a repulsion force will move the top magnet horizontally. Hence, we have achieved levitation and propulsion at the same time. The Maglev propulsion is created by switching the magnets of the bogie as they leave track coils.

Besides ground transportation, the U.S. Sandia National Laboratories developed the Maglev technology for high-power, high-thrust military applications. The Maglev is also developed for satellite launchers where the spacecraft accelerates to 1000 km/h at the end of the track causing it to take off like an airplane. While airborne, the spacecraft uses conventional rocket engines to reach its orbit.

12.4 SYNCHRONOUS GENERATOR

The synchronous generator is the most used machine for generating electricity. All power plants use synchronous generators to convert the mechanical power of the turbine into electrical power. Synchronous generators can have enormous capacity; a single generator can be built to produce over 2 GW of electric power. The large ones are used in nuclear power plants and major hydro plants such as China's Yangtze River Three Gorges Hydraulic Power Plant.

Figure 12.30 shows the stator of one of the synchronous generators in the Grand Coulee Hydroelectric Power Plant on the Columbia River in Washington. To put the size into perspective,

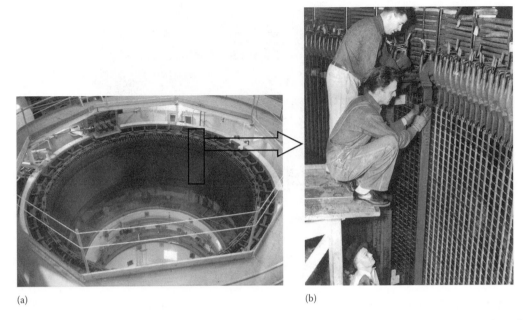

(a)   (b)

FIGURE 12.30 Stator of a synchronous generator in Grand Coulee Dam. (a) Stator and (b) closeup view of stator. (Image courtesy of the U.S. Bureau of Reclamation.)

Figure 12.30a shows part of the stator during the construction of the generator. The area in Figure 12.30b is almost the same as the boxed area in Figure 12.30a. Figure 12.31 shows the rotor of the same synchronous generator.

The main components of the synchronous generator are shown in Figure 12.32. The stator of the synchronous machine is similar to the stator of the induction motor; it consists of three-phase windings mounted symmetrically inside the stator core. The stator windings are also known as *armature windings*. The winding of the rotor of the synchronous machine is excited by an external dc source through a slip-ring system similar to the one shown in Figure 12.33. The rotor winding, which is also known as *field winding* or *excitation winding*, a produces a stationary flux with respect

FIGURE 12.31 Rotor of a synchronous generator in Grand Coulee Dam. (Image courtesy of the U.S. Bureau of Reclamation.)

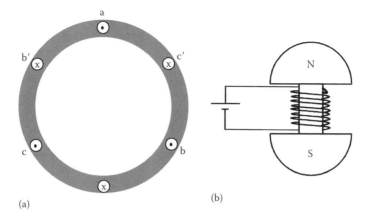

FIGURE 12.32 Basic components of a synchronous generator. (a) Stator and (b) rotor.

to the rotor; that is, the rotor is an electric magnet. The rotor is assembled inside the stator as shown in Figure 12.34a. Two sets of ball bearings are used on both ends of the rotor to allow the rotor to spin freely inside the stator.

The rotor of the synchronous generator is connected to a prime mover such as a hydro or thermal turbine. When the turbine spins the rotor of the synchronous generator, the magnetic field cuts the stator windings, thus inducing a sinusoidal voltage across the stator windings. Since the three stator windings are equally spaced from each other, the induced voltages across the phase windings are shifted by 120° from each other as shown in Figure 12.34b. The frequency of the induced voltage is dependent on the speed of the rotor. The relationship between the speed of the magnetic field and the frequency of the induced voltage is given in Equation 12.9. Hence, to maintain the frequency of the induced voltage at 60 Hz, the rotor speed must be n_s, where

$$n_s = 120\frac{f}{p} = \frac{7200}{p} \tag{12.69}$$

FIGURE 12.33 (See color insert following page 300.) Slip-ring arrangement.

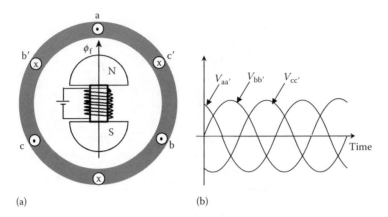

(a)   (b)

FIGURE 12.34   Three-phase voltage waveforms due to the clockwise rotation of the rotor magnet.

Any slight variation in this synchronous speed will result in a frequency variation for the induced voltage. This can cause a stability problem if the machine is connected to a fixed frequency power grid as discussed in Chapter 14. Therefore, the speed of the turbine must be delicately controlled so that the rotor is always spinning at the synchronous speed.

When the field circuit of the synchronous generator is excited and the rotor is spinning, a balanced three-phase voltage is induced in the stator windings. If an external three-phase load is connected across the stator windings as shown in Figure 12.35, three-phase currents will flow into the load. The current of the load is known as *armature current*. If the load is balanced, the armature currents are balanced, and a single-phase representation is all we need to model the generator.

The airgap of the synchronous generator has two fields as shown in Figure 12.36.

1. Excitation field ϕ_f, which is stationary with respect to the rotor. When a turbine spins the rotor at the synchronous speed, ϕ_f rotates with respect to the stator at the synchronous speed.
2. Stator flux ϕ_s due to the three-phase armature currents. This flux, which is only present when the generator is loaded, rotates at the synchronous speed inside the airgap similar to the rotating field inside the induction motor.

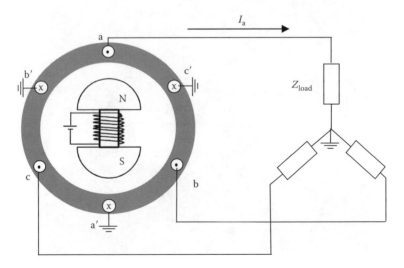

FIGURE 12.35   Loaded synchronous generator.

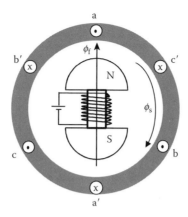

FIGURE 12.36 Total airgap flux inside a synchronous generator.

Hence, the net magnetic field in the airgap ϕ_g is the phasor sum of both fields.

$$\overline{\phi}_g = \overline{\phi}_f + \overline{\phi}_s \tag{12.70}$$

The equivalent circuit of the synchronous generator can be developed by using the schematic of the single-phase representation of the generator in Figure 12.37. The figure shows the rotor circuit excited by a dc source V_f, which produces an excitation current I_f. The excitation current produces the field ϕ_f. The armature current I_a produces the rotating field ϕ_s, which is known as *armature reaction*.

Assume first that the armature is unloaded, that is, $I_a = 0$ and $\phi_s = 0$. In this case, the induced voltage across the armature winding e_f is a function of ϕ_f only.

$$e_f = N \frac{d\phi_f}{dt} \tag{12.71}$$

where N is the number of turns in the armature winding. e_f is directly proportional to the excitation current I_f. The rms value of e_f is E_f, which is known as the *equivalent excitation voltage*.

Now assume that an electric load is connected across the armature windings. In this case, the induced voltage across the winding e_g is a function of the total airgap flux ϕ_g.

$$e_g = N \frac{d\phi_g}{dt} \tag{12.72}$$

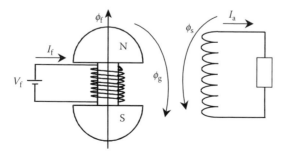

FIGURE 12.37 Representation of a synchronous generator.

The rms component of e_g is E_g. This induced voltage is not equal to the voltage across the load V_t because of the reactance X and resistance R of the armature winding. Hence,

$$\overline{E}_g = \overline{V}_t + \overline{I}_a(R + jX) \tag{12.73}$$

where
  V_t is the phase value of the terminal voltage across the load
  I_a is the armature current

The equivalent circuit represented by Equation 12.73 is shown in Figure 12.38.

Note that the magnitude of E_g is a function of ϕ_f and ϕ_s. Hence, E_g changes its value with any change in the load current. Therefore, it is difficult to analyze the synchronous generator at all loading conditions by the above model. However, we can modify the equivalent circuit by representing E_g in terms of E_f, which is independent of the load current. This can be done by representing the instantaneous voltage e_g in Equation 12.72 as

$$\overline{e}_g = N\frac{d\overline{\phi}_g}{dt} = N\frac{d\overline{\phi}_f}{dt} + N\frac{d\overline{\phi}_s}{dt} \tag{12.74}$$

The magnitude of the term $N(d\overline{\phi}_f/dt)$ is equal to e_f, which is the induced voltage across the stator winding due to flux ϕ_f. The magnitude of the second term $N(d\overline{\phi}_s/dt)$ is the voltage across the stator windings due to the armature currents that produce the flux ϕ_s. Hence, the second term represents a voltage drop across the stator windings. Since Faraday's law gives a positive sign for the induced voltage and a negative sign for the drop voltage, Equation 12.74 can be rewritten as

$$\overline{e}_g = N\frac{d\overline{\phi}_f}{dt} + N\frac{d\overline{\phi}_s}{dt} = \overline{e}_f - \overline{e}_s \tag{12.75}$$

where e_s is the voltage drop across the armature winding due to the armature current. The magnitude of e_s is then

$$e_s = N\frac{d\phi_s}{dt} = L_a\frac{di_a}{dt} \tag{12.76}$$

where
  i_a is the instantaneous current of the armature winding (the current that produces ϕ_s)
  L_a is the equivalent inductance

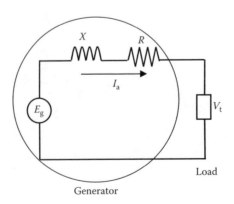

FIGURE 12.38 Equivalent circuit of a synchronous generator.

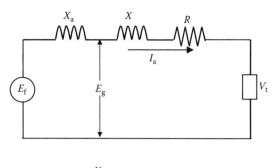

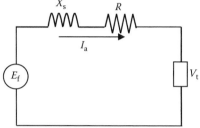

FIGURE 12.39 Equivalent circuit of a synchronous generator.

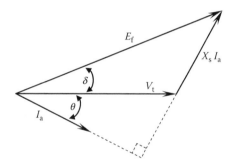

FIGURE 12.40 Phasor diagram of a synchronous generator.

Based on Equation 12.76, we can conclude that e_s is a voltage drop across an inductor L_a. In terms of rms, Equation 12.74 can be rewritten as

$$\overline{E}_g = \overline{E}_f - \overline{E}_s = \overline{E}_f - \overline{I}_a \, \overline{X}_a \tag{12.77}$$

where $X_a = \omega L_a$. Using Equation 12.77, we can modify the equivalent circuit in Figure 12.38 as shown in Figure 12.39. X_s in Figure 12.39 is known as *synchronous reactance*, and is equal to $X + X_a$.

The model for the synchronous generator can be further simplified by ignoring the resistance of the armature windings. This is justified for large machines with windings made of large cross-section wires. In this case, the synchronous generator model can be represented by

$$\overline{E}_f = \overline{V}_t + \overline{I}_a \, \overline{X}_s \tag{12.78}$$

The phasor diagram of Equation 12.78 is shown in Figure 12.40. Note that E_f leads V_t.

Example 12.13

A 60 Hz, four-pole synchronous generator of a thermal power plant has a synchronous reactance of 5 Ω. The stator windings are connected in wye, and its line-to-line voltage is 15 kV. The line current of the generator is 1 kA at 0.9 power factor lagging. Compute the following:

(a) Equivalent field voltage E_f
(b) Real power delivered to the power grid
(c) Reactive power delivered to the power grid

Solution

(a) Equivalent field voltage can be computed using Equation 12.78 assuming the reference to be the phase terminal voltage.

$$\overline{E}_f = \overline{V}_t + \overline{I}_a \overline{X}_s = \frac{15}{\sqrt{3}} \angle 0° + (1 \angle -\cos^{-1} 0.9) \times 5 \angle 90° = 11.74 \angle 22.54° \text{ kV}$$

(b) $$P = \sqrt{3} V_{t-ll}\, I_a \cos\theta = \sqrt{3} \times 15 \times 1 \times 0.9 = 23.38 \text{ MW}$$

where V_{t-ll} is the line-to-line value of the terminal voltage.

(c) $$Q = \sqrt{3} V_{t-ll}\, I_a \sin\theta = \sqrt{3} \times 15 \times 1 \times \sin(\cos^{-1} 0.9) = 11.32 \text{ MVAr.}$$

12.4.1 Synchronous Generator Connected to Infinite Bus

The infinite bus, by definition, is a large power grid consisting of a number of large generators. The voltage and frequency of the infinite bus are constant and cannot be changed regardless of any action made by any one generator. The infinite bus can absorb any amount of active or reactive powers from any single generator.

When a synchronous generator is connected to an infinite bus as shown in Figure 12.41, the voltage and frequency at the infinite bus are constant and cannot be changed due to any change in the excitation current I_f or the speed of the generator. This is much like having two objects tied together. If the mass of one object is much larger than the mass of the other, any attempted movement by the small object may not affect the large object. The infinite bus in this scenario is the large object and the synchronous generator is the small one.

When the generator is connected directly to an infinite bus, the terminal voltage of the machine V_t (Figure 12.39) is equal to the infinite bus voltage V_o. Since V_o is constant, V_t is also constant. Keep in mind that E_f is a function of the field current I_f. Hence, E_f is controlled by the operator of the generator. The phasor diagram of the synchronous machine connected to an infinite bus is shown in Figure 12.42. Note that the only difference between this phasor diagram and the one in Figure 12.40 is that the terminal voltage of the generator is equal to the infinite bus voltage.

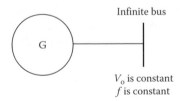

FIGURE 12.41 Synchronous generator connected to an infinite bus.

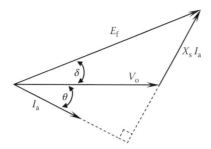

FIGURE 12.42 Phasor diagram of a synchronous generator connected to an infinite bus.

12.4.1.1 Real Power of Synchronous Generator

The real power delivered to the infinite bus by the synchronous generator is

$$P = 3V_t I_a \cos \theta = 3V_o I_a \cos \theta = 3E_f I_a \cos (\theta + \delta) \tag{12.79}$$

where
   θ is the phase angle of the current
   δ is the angle between the infinite bus voltage and the equivalent field voltage E_f. δ is known
      as the *power angle*

In Equation 12.79, V_t, V_o, and E_f are phase quantities. Examining the phasor diagram in Figure 12.42 leads to the following relationship:

$$I_a X_s \cos \theta = E_f \sin \delta \tag{12.80}$$

Hence,

$$I_a \cos \theta = \frac{E_f \sin \delta}{X_s} \tag{12.81}$$

Substituting Equation 12.81 into Equation 12.79 yields the equation for the real power

$$P = \frac{3V_o E_f}{X_s} \sin \delta \tag{12.82}$$

Equation 12.82 is called the *power-angle equation*. It relates the output power of the generator to the power angle δ as well as other parameters and variables. Equation 12.82 shows that the maximum power that can be generated by the machine P_{max}, which is known as the *pullover power*, occurs when $\delta = 90°$.

$$P_{max} = \frac{3V_o E_f}{X_s} \tag{12.83}$$

Figure 12.43 shows the power-angle curve representing Equation 12.82. Let us assume that the internal losses of the generator are insignificant. In this case, the input mechanical power P_m to the generator is equal to the electrical power P delivered to the infinite bus. The mechanical power P_m is controlled by the operator at the power plant. Hence, the point at which the mechanical power

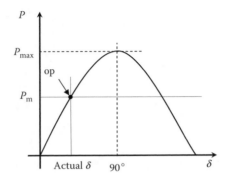

FIGURE 12.43   Real power as a function of the power angle.

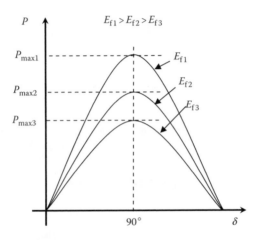

FIGURE 12.44   Pullover power due to changes in excitation current.

equals the electrical power is the operating point op of the generator. The power angle at this point is the actual angle between the infinite bus voltage V_o and the equivalent field voltage E_f.

As seen in Equation 12.83, the maximum power P_{max} of the system increases when E_f increases. However, the power delivered to the infinite bus will not increase unless the input mechanical power to the generator increases. Therefore, increasing P_{max} increases the capacity of the generator to deliver more power. Changing E_f is done by changing the excitation current I_f, which is continuously controlled at the power plant. Figure 12.44 shows various values of pullover powers due to different E_f.

EXAMPLE 12.14

A synchronous generator is connected directly to an infinite bus. The voltage of the infinite bus is 23 kV. The excitation of the generator is adjusted until the equivalent field voltage E_f is 25 kV. The synchronous reactance of the machine is 2 Ω.

1. If the output mechanical power of the turbine is 100 MW, compute the power angle.
2. Compute the pullover power.
3. If the excitation of the generator is decreased by 10%, compute the power angle and the pullover power.

Solution

(a) The electric power of the generator is equal to the mechanical power of the turbine. By direct substitution in Equation 12.82, we can compute the power angle. However, keep in mind that all voltages in the equation are phase quantities and all powers are for the three phases. Also, synchronous generators are almost always connected in wye.

$$P = \frac{3V_o E_f}{X_s} \sin \delta$$

$$100 = \frac{3 \frac{23}{\sqrt{3}} \frac{25}{\sqrt{3}}}{2} \sin \delta$$

$$\delta = 20.35°$$

(b) The pullover power can be computed using Equation 12.83.

$$P_{max} = \frac{3V_o E_f}{X_s} = \frac{23 \times 25}{2} = 287.5 \text{ MW}$$

(c) Assume a linear relationship between the excitation current and the equivalent field voltage. When the excitation is decreased by 10%, the equivalent field voltage decreases by 10% as well.

$$P = \frac{3V_o E_f}{X_s} \sin \delta$$

$$100 = \frac{23 \times (0.9 \times 25)}{2} \sin \delta$$

$$\delta = 22.74°$$

$$P_{max} = \frac{3V_o E_f}{X_s} = \frac{23 \times (0.9 \times 25)}{2} = 258.75 \text{ MW}$$

12.4.1.2 Reactive Power of Synchronous Generator

The reactive power delivered by the synchronous generator to the infinite bus Q_t can be computed from the phasor diagram in Figure 12.42.

$$Q_t = 3V_o I_a \sin \theta \tag{12.84}$$

Using trigonometric calculations, the following relationship can be obtained:

$$I_a X_s \sin \theta = E_f \cos \delta - V_o$$
$$I_a \sin \theta = \frac{E_f \cos \delta - V_o}{X_s} \tag{12.85}$$

Substituting Equation 12.85 into Equation 12.84 yields

$$Q_t = 3V_o I_a \sin \theta = \frac{3V_o}{X_s}(E_f \cos \delta - V_o) \tag{12.86}$$

Equation 12.86 shows that the reactive power delivered to the infinite bus is dependent on the magnitude of E_f. The reactive power can be positive, negative, or zero depending on the level of excitation of the synchronous generator.

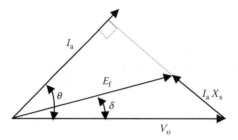

FIGURE 12.45   Underexcited synchronous machine.

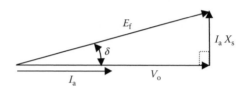

FIGURE 12.46   Synchronous machine with exact excitation.

1. *Overexcited generator*: This is when the excitation is adjusted so that $E_f \cos \delta > V_o$. The reactive power in this case is positive, meaning the generator is delivering reactive power to the infinite bus. The phasor diagram in this case is shown in Figure 12.42. Note that the current lags the voltage of the infinite bus.
2. *Underexcited generator*: If the excitation is adjusted so that $E_f \cos \delta < V_o$, the reactive power is negative; that is, the generator is receiving reactive power from the infinite bus. The phasor diagram in this case is shown in Figure 12.45 where the current is leading the infinite bus voltage.
3. *Exactly excited generator*: If the excitation is adjusted so that $E_f \cos \delta = V_o$, there is no reactive power at the terminals of the generator. The phasor diagram is shown in Figure 12.46. The current in this case is in phase with the infinite bus voltage.

Example 12.15

For the synchronous machine in Example 12.14, compute the equivalent excitation voltage E_f required to deliver 100 MW of real power and no reactive power to the infinite bus.

Solution

The terminal voltage of the machine is unchanged since it is connected to an infinite bus. However, we cannot assume that the power angle is equal to that computed in Example 12.14, because the equivalent excitation voltage must change to modify the reactive power. As seen in Figure 12.44, the change in the excitation alone leads to a change in the power angle.

Our problem has two unknowns: the equivalent excitation voltage and the power angle. To solve these two unknowns, we need two equations, namely the real and reactive power equations.

$$P = \frac{3V_o\, E_f}{X_s}\sin\delta$$

$$100 = \frac{3\,\frac{23}{\sqrt{3}}\,E_f}{2}\sin\delta$$

$$E_f \sin\delta = 5.02 \text{ kV}$$

For zero reactive power at the generator terminals, the generator must be exactly excited. Hence,

$$E_f \cos \delta = V_o = \frac{23}{\sqrt{3}} = 13.28 \text{ kV}$$

Hence,

$$\frac{E_f \sin \delta}{E_f \cos \delta} = \tan \delta = \frac{5.02}{13.28} = 0.38$$

$$\delta = 20.7°$$

Now, we can solve for the phase value of the equivalent excitation voltage.

$$E_f = \frac{5.02}{\sin \delta} = 14.2 \text{ kV}$$

The line-to-line equivalent excitation voltage is $14.2\sqrt{3} = 24.59$ kV.

12.4.2 SYNCHRONOUS GENERATOR CONNECTED TO INFINITE BUS THROUGH A TRANSMISSION LINE

Most power plants are located in remote areas where the energy resources are available or easily transportable, and the possible pollution from the power plants are kept away from population centers. The energy generated by these power plants must then be transmitted to the load centers (cities, factories, etc.) by a system of transmission lines. Figure 12.47 shows a one line diagram of a simple system where the generator is connected to an infinite bus through a transmission line. The terminal voltage of the generator is V_t and the voltage at the infinite bus is V_o. The transmission line is a high-voltage wire that has a resistance and an inductive reactance X_l. However, the resistance is often much smaller than the inductive reactance and is normally ignored.

The equivalent circuit of the system is shown in Figure 12.48. The power angle δ in this case is defined as the angle between the infinite bus voltage V_o and the equivalent field voltage E_f. The basic equations of the system are

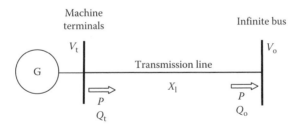

FIGURE 12.47 Synchronous generator connected to an infinite bus through a transmission line.

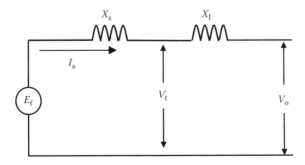

FIGURE 12.48 Equivalent circuit of the system in Figure 12.47.

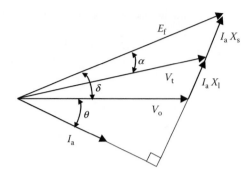

FIGURE 12.49   Phasor diagram of Equation 12.87.

$$\overline{E}_f = \overline{V}_t + \overline{I}_a \overline{X}_s$$
$$\overline{V}_t = \overline{V}_o + \overline{I}_a \overline{X}_l \qquad (12.87)$$
$$\overline{E}_f = \overline{V}_o + \overline{I}_a \overline{X}_l + \overline{I}_a \overline{X}_s$$

Equation 12.87 can be interpreted by the phasor diagram in Figure 12.49, where the current is assumed to be lagging the infinite bus voltage. Equation 12.87 can also be simplified as

$$\overline{E}_f = \overline{V}_o + \overline{I}_a(\overline{X}_s + \overline{X}_l) = \overline{V}_o + \overline{I}_a \overline{X} \qquad (12.88)$$

where $\overline{X} = \overline{X}_s + \overline{X}_l$. The simplified phasor diagram is shown in Figure 12.50.

The real and reactive power equations can be developed in a process similar to that in Sections 12.4.1.1 and 12.4.1.2. Because we are ignoring the resistance of the transmission line, the real power at the infinite bus is the same as the real power at the machine's terminals.

$$P = 3V_o I_a \cos\theta = 3E_f I_a \cos(\theta + \delta) \qquad (12.89)$$

where

$$I_a \cos\theta = \frac{E_f \sin\delta}{X} \qquad (12.90)$$

Hence, the equation for the real power is

$$P = \frac{3V_o E_f}{X} \sin\delta \qquad (12.91)$$

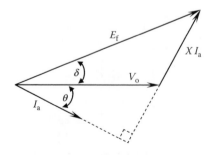

FIGURE 12.50   Simplified phasor diagram of Equation 12.87.

Equations 12.91 and 12.82 are different in two aspects: The inductive reactance in Equation 12.82 is the synchronous reactance of the machine, while the inductive reactance in Equation 12.91 is the sum of the synchronous reactance and the inductive reactance of the transmission line. In addition, the power angle in Equation 12.82 is between the terminal voltage and the equivalent field voltage. In Equation 12.91, the power angle is between the infinite bus voltage and the equivalent field voltage.

Similar to the process in Section 12.4.1.2, the reactive power delivered to infinite bus Q_o can be derived as

$$Q_o = 3V_o I_a \sin \theta = \frac{3V_o}{X}(E_f \cos \delta - V_o) \tag{12.92}$$

As seen in Figure 12.47, the reactive power at the infinite bus is not the same as the reactive power at the terminals of the generator. The difference is the reactive power consumed by the inductive reactance of the transmission line. The reactive power at the generator's terminal Q_t can be computed as

$$Q_t = 3V_t I_a \sin (\theta + \delta - \alpha) \tag{12.93}$$

where

$$I_a X_s \sin (\theta + \delta - \alpha) = E_f \cos \alpha - V_t \tag{12.94}$$

Hence, the reactive power at the terminals of the generator is

$$Q_t = \frac{3V_t}{X_s}(E_f \cos \alpha - V_t) \tag{12.95}$$

EXAMPLE 12.16

A synchronous generator is connected to an infinite bus through a transmission line. The infinite bus voltage is 23 kV, the inductive reactance of the transmission line is 1 Ω, and the synchronous reactance of the machine is 2 Ω. When the excitation of the generator is adjusted so that the line-to-line equivalent field voltage is 28 kV, the generator delivers 100 MW of real power to the infinite bus. Compute the following:

(a) Terminal voltage of the machine
(b) Reactive power delivered to the infinite bus
(c) Reactive power consumed by the transmission line
(d) Reactive power at the terminals of the generator

Solution

(a) Terminal voltage of the generator can be computed if we know the current in the transmission line. The current is a function of the phasor difference between the infinite bus voltage and the equivalent field voltage. From the power equation we can compute the angle of the equivalent field voltage.

$$P = \frac{3V_o E_f}{X} \sin \delta$$

$$100 = \frac{23 \times 28}{(2+1)} \sin \delta$$

$$\delta = 27.76°$$

The current in the transmission line can be computed using the model in Figure 12.48.

$$\bar{I}_a = \frac{\overline{E}_f - \overline{V}_o}{\overline{X}} = \frac{\frac{28}{\sqrt{3}}\angle 27.76° - \frac{23}{\sqrt{3}}\angle 0°}{3\angle 90°} = 2.53\angle -7.75° \text{ KA}$$

$$\overline{V}_t = \overline{V}_o + \bar{I}_a\overline{X}_l = \frac{23}{\sqrt{3}}\angle 0° + (2.53\angle -7.75°)\,1\angle 90° = 13.85\angle 10.44° \text{ kV}$$

The line-to-line terminal voltage is $\sqrt{3} \times 13.85 = 24$ kV.

(b) $$Q_o = \frac{3V_o}{X}(E_f\cos\delta - V_o) = \frac{\sqrt{3}\times 23}{3}\left(\frac{28}{\sqrt{3}}\cos(27.76°) - \frac{23}{\sqrt{3}}\right) = 13.62 \text{ MVAr.}$$

(c) The reactive power consumed by the transmission line is

$$Q_l = 3I_a^2 X_l = 3 \times 2.53^2 \times 1 = 19.25 \text{ MVAr}$$

(d) The reactive power at the terminals of the generator can be computed by two methods. The first is to sum the reactive powers at the infinite bus and the reactive power consumed by the load.

$$Q_t = Q_o + Q_l = 32.87 \text{ MVAr}$$

The other method is to use Equation 12.95. But first, we need to calculate the angle α using the phasor diagram in Figure 12.49.

$$\alpha = \delta - \angle\overline{V}_t = 27.76° - 10.44° = 17.32°$$

$$Q_t = \frac{3V_t}{X_s}(E_f\cos\alpha - V_t) = \frac{3\times 13.85}{2}\left(\frac{28}{\sqrt{3}}\cos 17.32° - 13.85\right) = 32.87 \text{ MVAr}$$

EXAMPLE 12.17

A synchronous generator of a power plant is connected to an infinite bus through a transmission line. The infinite bus voltage is 23 kV, the inductive reactance of the transmission line is 1 Ω, and the synchronous reactance of the machine is 2 Ω. The reactive power measured at the machine's terminals is zero, and the machine delivers 1 kA to the infinite bus. Compute the following:

(a) Equivalent field voltage
(b) Real power delivered to the infinite bus
(c) Real power at the terminals of the machine

Solution

(a) Since there is no reactive power at the terminals of the generator, the current I_a is in phase with the terminal voltage V_t. Using the equivalent circuit in Figure 12.48, the phasor diagram can be constructed as shown in Figure 12.51.
The equivalent field voltage can be computed using Equation 12.87.

$$\overline{E}_f = \overline{V}_t + \bar{I}_a\,\overline{X}_l$$

Since I_a is in phase with V_t, the voltage drop $I_a X_l$ is leading V_t by 90°. Hence, we can use the Pythagorean Theorem to compute the terminal voltage.

$$V_t = \sqrt{V_o^2 - (I_a X_l)^2} = \sqrt{\left(\frac{23}{\sqrt{3}}\right)^2 - (1\times 1)^2} = 13.24 \text{ kV}$$

Similarly,

$$E_f = \sqrt{V_t^2 + (I_a\,X_s)^2} = \sqrt{(13.24)^2 + (1 \times 2)^2} = 13.4 \text{ kV}$$

The line-to-line equivalent excitation voltage is $\sqrt{3} \times 13.4 = 23.2$ kV.
(a) The real power delivered to the infinite bus is

$$P = \frac{3V_o\,E_f}{X}\sin\delta$$

where

$$\delta = \sin^{-1}\left(\frac{I_a\,X_l}{V_o}\right) + \sin^{-1}\left(\frac{I_a\,X_s}{E_f}\right) = \sin^{-1}\left(\frac{1 \times 1}{23/\sqrt{3}}\right) + \sin^{-1}\left(\frac{1 \times 2}{13.4}\right) = 12.91°$$

$$P = \frac{3V_o\,E_f}{X}\sin\delta = \frac{23 \times 23.2}{3}\sin 12.91° = 39.7 \text{ MW}$$

(b) Since we ignored the resistance of the transmission line, the real power at the terminals of the machine is the same as the real power at the infinite bus.

12.4.3 INCREASE TRANSMISSION CAPACITY

If we wish to increase the output electrical power of the synchronous machine, we must increase the mechanical power of the turbine. In addition, the transmission system must be able to transmit the extra power. The power system operates securely when the capacity of the transmission system is higher than the generated electric power. The capacity of the power system can be viewed as the maximum power that can be transmitted by the system. In Equation 12.91, the system capacity is the maximum power.

$$P_{max} = \frac{3V_o\,E_f}{X} \tag{12.96}$$

The power plant cannot generate more than the transmission capacity P_{max}. However, it is possible to increase the transmission capacity if we increase E_f or reduce the total inductive reactance X. There are two common methods to reduce X: (1) inserting a capacitor in series with the transmission line and (2) using parallel transmission lines.

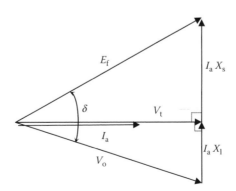

FIGURE 12.51   Phasor diagram of the system in Example 12.17.

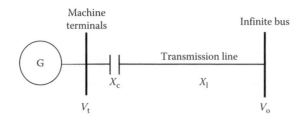

FIGURE 12.52 Capacitor in series with the transmission line.

12.4.3.1 Increasing Transmission Capacity by Using Series Capacitor

A capacitor can be connected in series with the transmission line, as shown in Figure 12.52, to reduce the total inductive reactance of the system. Recall that without the capacitor, $X = X_s + X_l$.

Now with the capacitor, the total reactance is

$$\overline{X} = \overline{X}_s + \overline{X}_l + \overline{X}_c = jX_s + jX_l - jX_c = j(X_s + X_l - X_c)$$
$$X = X_s + X_l - X_c$$

(12.97)

where X_c is the capacitive reactance of the capacitor.

EXAMPLE 12.18

A synchronous generator is connected to an infinite bus through a transmission line. The infinite bus voltage is 230 kV and the equivalent field voltage of the machine is 210 kV. The transmission line's inductive reactance is 10 Ω and the synchronous reactance of the machine is 2 Ω.

(a) Compute the capacity of the system.
(b) If a capacitor is connected in series with the transmission line to increase the transmission capacity by 25%, compute its reactance.

Solution

1. System capacity is

$$P_{max} = \frac{3V_o\,E_f}{X} = \frac{230 \times 210}{(10 + 2)} = 4.025 \text{ GW}$$

2. System capacity increases by 25% using a capacitor.

$$1.25 \times 4025 = \frac{3V_o\,E_f}{X_{new}} = \frac{230 \times 210}{(10 + 2 - X_c)}$$

Hence,

$$X_c = 2.4 \ \Omega$$

This is a large value of capacitive reactance, almost 1.1 μF. Since it carries large currents at high voltages, the capacitor is very large in size and can only be installed in major substations.

12.4.3.2 Increasing Transmission Capacity by Using Parallel Lines

Parallel lines are often used to increase the transmission capacity of the system. It is intuitive knowledge that two lines can transmit more power than any one line. Figure 12.53 shows a schematic of a system with two parallel lines (TL1 and TL2), and Figure 12.54 shows a photo of a transmission tower with two parallel circuits (each circuit consists of three-phase lines).

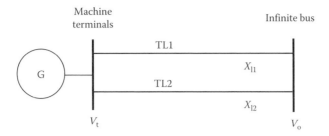

FIGURE 12.53 Two transmission lines in parallel.

FIGURE 12.54 Tower with two transmission lines (the line is a three-phase circuit).

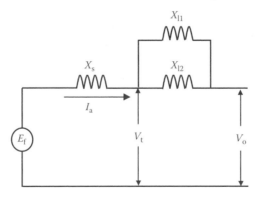

FIGURE 12.55 Equivalent circuit of a power system with two parallel lines.

The equivalent circuit of the system is shown in Figure 12.55. In this system, the total reactance between the infinite bus and the equivalent field voltage is

$$X = X_s + \frac{X_{l1}X_{l2}}{X_{l1} + X_{l2}} \tag{12.98}$$

EXAMPLE 12.19

A synchronous generator is connected to an infinite bus through a transmission line. The infinite bus voltage is 230 kV and the equivalent field voltage of the machine is 210 kV. The transmission line's inductive reactance is 10 Ω and the synchronous reactance of the machine is 2 Ω.

  (a) Compute the capacity of the system.
  (b) Compute the capacity of the system if a second transmission line with a 10 Ω inductive reactance is connected in parallel with the first line.

Solution

  (a) The system capacity is

$$P_{max} = \frac{3V_o E_f}{X} = \frac{230 \times 210}{(10 + 2)} = 4.025 \text{ GW}$$

  (b) After the second line is added, the system capacity is

$$P_{max} = \frac{3V_o E_f}{X_{new}} = \frac{3V_o E_f}{\left(X_s + \frac{X_{l1}X_{l2}}{X_{l1}+X_{l2}}\right)} = \frac{230 \times 210}{2 + \frac{10}{2}} = 6.9 \text{ GW}$$

The power capacity of the system increases by about 71% when the two parallel lines are used. Can you tell why the capacity is not doubled? Also, repeat the example for three parallel lines.

12.5 SYNCHRONOUS MOTOR

A synchronous machine is a dual action device; it can be used as a generator or motor. As a motor the synchronous machine is very popular in applications that demand constant and precise speeds such as electric clocks, movie cameras, tractions, uniform actuations, gate and governor controls, constant feed industrial processes, and many others. Besides driving mechanical loads, the synchronous motor is also used to compensate for the reactive power of industrial loads.

   For small size synchronous motors, the rotor is often made of ferrite permanent magnet material. This is the most economic design for a fractional horsepower motor that does not experience repeated surges in stator currents. If repeatedly occurring, the inrush currents can demagnetize the rotor.

   In latest designs, the rotor is made of rare earth permanent magnet material, such as the Samarium–Cobalt, to produce stronger magnetic fields. The strong rare earth permanent magnet material can be used in motors as large as 100 hp. One great advantage of this rare earth motor is its high power/volume ratio that makes it one of the smallest machines. Furthermore, unlike the ferrite material, the rare earth permanent magnet cannot be easily demagnetized, so it can be used for applications that require heavy currents or inrush currents during starting and braking. Because of these advantages, the rare earth permanent magnet synchronous motor is used to actuate new commercial airplanes using fly-by-wire designs. It is also used in ground transportation; the hybrid electric vehicles in Figure 12.56 use this type of motor.

   For large motors, the rotor is made of an electric magnet that is excited externally by a separate dc source. This is exactly the same rotor used for the synchronous generator; a photograph of its slip-ring arrangement is shown in Figure 12.33.

   The three-phase currents of the stator produce a rotating magnetic field at the synchronous speed n_s, exactly like the induction machine. The magnetic field of the rotor, which is stationary with

FIGURE 12.56 (See color insert following page 300.) Hybrid electric vehicle using a permanent magnet synchronous motor.

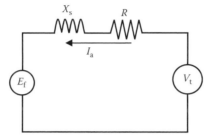

FIGURE 12.57   Equivalent circuit of a synchronous motor.

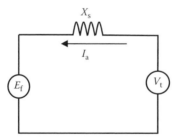

FIGURE 12.58   Simplified equivalent circuit of a synchronous motor.

respect to the rotor, aligns itself with the rotating field, thus spinning the rotor at the synchronous speed. This is similar to having two magnets; if one is moving, the other follows at the same speed. The equivalent circuit of the synchronous generator (Figure 12.39) is also applicable to the synchronous motor (Figure 12.57). The only differences are (1) the stator is excited by a three-phase source and (2) the armature current of the motor flows in the reverse direction when compared with the armature current of the generator. The equivalent circuit of the synchronous motor can be further simplified by ignoring the resistance of the armature winding. This is justified for large machines where the stator windings are made of large cross-section wires. The simplified equivalent circuit of the synchronous motor is shown in Figure 12.58.

The equation of the synchronous motor can be written as

$$\overline{V}_t = \overline{E}_f + \overline{I}_a\,\overline{X}_s \qquad\qquad (12.99)$$

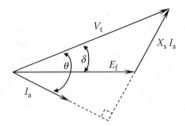

FIGURE 12.59 Phasor diagram of a synchronous motor.

The phasor diagram of Equation 12.99 is shown in Figure 12.59. Note that for the synchronous motor, E_f lags V_t; this is opposite to the synchronous generator. Also, note that the power factor angle θ is the angle between the armature current and the terminal voltage of the motor.

EXAMPLE 12.20

A 30 hp, four-pole, 480 V, 60 Hz, three-phase, wye-connected synchronous motor has a synchronous reactance of 2 Ω. At full load, the armature current of the motor is 30 A and lags the terminal voltage. Ignore all losses and compute the following:

(a) Torque of the motor
(b) Equivalent field voltage
(c) Reactive power consumed by the motor

Solution

(a) We can compute the torque of the motor if we know the output power and the speed of the motor. The output power at full load is 30 hp.

$$P_{out} = \frac{30}{1.34} = 22.39 \text{ kW}$$

$$n_s = 120\frac{f}{p} = 120\frac{60}{4} = 1800 \text{ rmp}$$

$$T = \frac{P_{out}}{\omega_s} = \frac{22390}{\frac{2\pi}{60} \times 1800} = 118.78 \text{ N m}$$

(b) Using the circuit in Figure 12.58, we can compute the equivalent field voltage. But first, we need to compute the phase angle of the current. Since the machine losses are ignored, the output power is equal to the input power. Hence,

$$P_{in} = P_{out} = \sqrt{3}\, VI_a \cos\theta$$

$$22390 = \sqrt{3}\, 480 \times 30 \times \cos\theta$$

$$\theta = 26.14°$$

Using Equation 12.99, we can compute the equivalent field voltage. Select the terminal voltage to be the reference phasor.

$$\overline{E}_f = \overline{V}_t - \overline{I}_a\overline{X}_s = \frac{480}{\sqrt{3}}\angle 0° - (30\angle{-26.14°}) \times 2\angle 90° = 256.4\angle{-12.13°} \text{ V}.$$

The line-to-line value of E_f is 444 V.

(c) The reactive power consumed by the motor is the input reactive power.

$$Q_{in} = \sqrt{3}\, VI_a \sin\theta = \sqrt{3}\, 480 \times 30 \times \sin(26.14) = 11 \text{ kVAr}$$

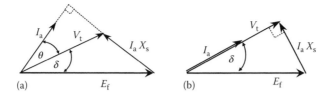

FIGURE 12.60 Phasor diagram of a synchronous motor. (a) Leading current and (b) unity power factor.

12.5.1 Reactive Power Control

In Equation 12.99, the magnitudes of V_t are often fixed since it is connected to the power system. However, E_f can be controlled by adjusting the dc field current in the rotor circuit. Hence, the armature current is dependent on the value of E_f.

There are three possible scenarios for the current in Equation 12.99: lagging, leading, and inphase with the terminal voltage. The three possible phasor diagrams are shown in Figures 12.59 and 12.60. In Figure 12.59, E_f is adjusted so that $E_f \cos \delta < V_t$. In this case, the current I_a lags V_t, and the power factor measured at the terminals of the motor is lagging.

In Figure 12.60a, E_f is increased so that $E_f \cos \delta > V_t$. In this case, the current leads the terminal voltage and the power factor measured at the terminals of the motor is leading.

In Figure 12.60b, E_f is adjusted so that $E_f \cos \delta = V_t$. In this case, the current is in phase with the terminal voltage and the power factor measured at the terminals of the motor is unity.

The reactive power Q at the terminals of the motor can be computed as

$$Q = 3 V_t I_a \sin \theta \qquad (12.100)$$

where V_t is a phase quantity. By examining the phasor diagrams of the synchronous motor, we can show that

$$I_a X_s \sin \theta = E_f \cos \delta - V_t \qquad (12.101)$$

Substituting the current in Equation 12.101 into 12.100 yields

$$Q = \frac{3 V_t}{X_s} [E_f \cos \delta - V_t] \qquad (12.102)$$

The reactive power at the terminals of the motor is leading when the magnitude of Q in Equation 12.102 is positive. This means that the motor is delivering reactive power to the ac source. When Q is negative, the reactive power is lagging, and the motor consumes reactive power.

Example 12.21

A 5 kV, 60 Hz synchronous motor has a synchronous reactance of 4 Ω. The motor runs unloaded and its equivalent field voltage is adjusted to 6 kV. Ignore all losses and compute the reactive power at the terminals of the motor.

Solution

Since the motor is unloaded, the output real power is zero; thus, the power angle is also zero. Hence, the current I_a is either leading or lagging V_t by 90°. However, since the power angle is zero and $E_f > V_t$, the current must be leading the terminal voltage, as shown in the phasor diagram in Figure 12.61.

Using Equation 12.102, we can compute the reactive power.

$$Q = \frac{3V_t}{X_s}[E_f \cos \delta - V_t] = \frac{3 \times \frac{5000}{\sqrt{3}}}{4}\left[\frac{6000}{\sqrt{3}} - \frac{5000}{\sqrt{3}}\right] = 1.25 \text{ MVAr}$$

The positive value of the reactive power means that the synchronous motor is delivering reactive power to the ac source. This is similar to the capacitor that delivers reactive power to the system. Therefore, the machine in this mode is often called *synchronous condenser*.

12.5.2 REAL POWER

The input power to the synchronous motor from the ac source is

$$P = 3V_t\, I_a\, \cos \theta \tag{12.103}$$

where V_t is a phase quantity. By examining any phasor diagram in Figures 12.59 or 12.60, we can show that

$$I_a X_s \cos \theta = E_f \sin \delta \tag{12.104}$$

Substituting I_a of Equation 12.104 into Equation 12.103 yields

$$P = 3\frac{V_t E_f}{X_s} \sin \delta \tag{12.105}$$

Since the synchronous machine rotates at synchronous speed, we can write the developed torque equation as

$$T = \frac{P}{\omega_s} = \frac{3V_t E_f}{\omega_s X_s} \sin \delta \tag{12.106}$$

Figure 12.62 shows the torque curve representing Equation 12.106. The maximum torque of the motor occurs at $\delta = 90°$.

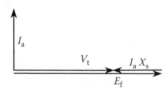

FIGURE 12.61   Phasor diagram of a synchronous motor running at no load.

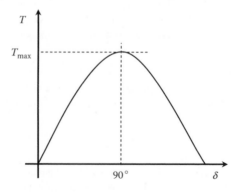

FIGURE 12.62   Torque curve of a synchronous motor.

EXAMPLE 12.22

A 3 kV, 60 Hz, six-pole synchronous motor is driving a load of 6 kNm. The synchronous reactance of the motor is 5 Ω. The equivalent field voltage as a function of the excitation current is $E_f = 40I_f$. Compute the following:

(a) Power consumed by the load
(b) Minimum excitation current of the machine
(c) Reactive power at the motor terminals

Solution

(a) Power of the motor is

$$P = T\,\omega_s$$

where

$$\omega_s = \frac{2\pi}{60}\left(120\frac{60}{6}\right) = 125.66 \text{ rad/s}$$

$$P = T\,\omega_s = 6000 \times 125.66 = 754 \text{ kW}$$

(b) Minimum excitation current is the value that makes the maximum torque of the motor in Equation 12.106 equal the load torque. Any excitation current less than this value will produce torque less than that needed by the load.

$$T_1 = T_{max} = \frac{3V_t\,E_{f\,min}}{\omega_s\,X_s}$$

$$6000 = \frac{3\frac{3000}{\sqrt{3}}E_{f\,min}}{125.66 \times 5}$$

$$E_{f\,min} = 725.5 \text{ V}$$

Hence, the minimum excitation current is

$$E_{f\,min} = 40\,I_{f\,min}$$

$$I_{f\,min} = \frac{E_{f\,min}}{40} = 18.14 \text{ A}$$

(c) The reactive power at the terminals of the motor is

$$Q = \frac{3V_t}{X_s}[E_f\,\cos\delta - V_t]$$

Since the machine is running under minimum excitation, the power angle is 90°.

$$Q = \frac{-3V_t^2}{X_s} = -1.8 \text{ MVAr}$$

The negative sign of the reactive power means that the machine is consuming reactive power from the source.

12.6 DC MOTOR

The dc machine is one of the oldest electrical devices. The dc machine can operate as a motor or a generator without any change in its design. However, it is rarely used as a dc generator because of the wide use of power electronic converters.

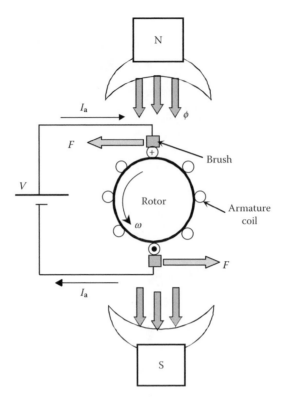

FIGURE 12.63   Main components of a dc motor.

The dc motor consists of a stator (called *field*) and a rotor (called *armature*). The stator is either a permanent magnet or an electric magnet. The winding of this electric magnet is excited by a dc source. Thus, the magnetic field is unidirectional from the north pole to the south pole as shown in Figure 12.63. A photo of the electric field winding in the stator circuit is shown in Figure 12.64.

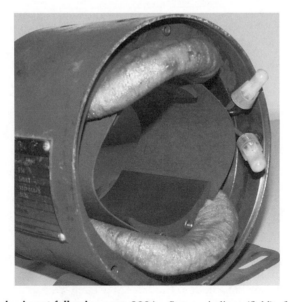

FIGURE 12.64 (See color insert following page 300.)   Stator windings (field) of a dc motor.

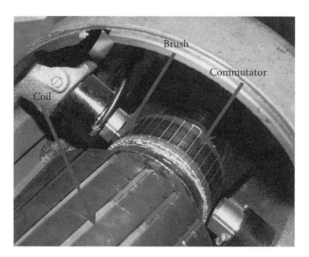

FIGURE 12.65 (See color insert following page 300.)   Rotor of a dc motor.

The rotor of the dc motor is more complex than the rotor of the other motors. It consists of independent coils connected together in cascaded form. The terminals of each coil are connected to two copper strips (segments) mounted on the rotor shaft on opposite sides as shown in Figure 12.65. The copper segments of all coils form what is called a *commutator*. The commutator rotates with the shaft. On the stator two brushes are mounted on opposite sides of the commutator touching the terminals of one coil. The brush system consists of a spring and a carbon piece called brush as shown in Figure 12.66. The spring pushes the brush against the commutator to ensue connectivity with a coil. Because the position of the brush is fixed on the stator and the commutator is rotating, the carbon brushes make contact with the coil that moves under the two brushes. This way, the coils are switched on when they are under the brushes, and switched off when they move away from the brushes.

To explain how the motor rotates, consider Fleming's left hand rule that explains the motion of electric motors. The rule states that if the y-axis is the direction of the magnetic field and the x-axis is the direction of current inside a conductor, then the z-axis is the direction of the force

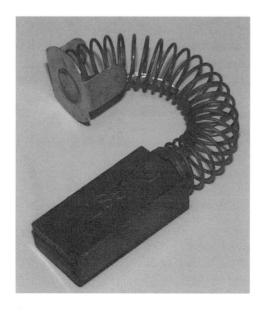

FIGURE 12.66   Carbon brush.

exerted on the conductor. Let us apply this rule for the conductor under the top brush in Figure 12.63. If the armature is excited by an external source whose voltage is V, the current I_a flows through the top brush to the top terminal of the coil and exits from the lower terminal of the same coil to the lower brush and back to the source. The current of the top terminal of the conductor enters the conductor. Since the magnetic field is moving from the top pole to the bottom pole, the force exerted on the top conductor F is to the left and the one on the lower conductor is to the right. These two forces create a torque causing the rotor to rotate in the counterclockwise direction. Once this coil moves away from the brushes, the next one arrives under the brushes and the process is repeated. Thus, the motor is continuously rotating.

Direct current (dc) motors are popular because (1) they are simple to control by just adjusting the armature voltage or the field current; (2) they can deliver high starting torque, which makes them suitable for heavy loads such as elevators, forklifts, and transportations; and (3) their performance is almost linear making them simple to stabilize.

12.6.1 THEORY OF OPERATION OF DC MOTOR

We can rewrite Equations 12.11 and 12.12 in the following forms:

$$E_a = f(\phi, \Delta n) = K\phi\omega \tag{12.107}$$

$$T = f(\phi, i_a) = K\phi I_a \tag{12.108}$$

where
   E_a is the voltage across the armature coil
   I_a is the armature current
   ω is the angular speed of the rotor
   T is the torque of the rotor
   ϕ is the magnetic field
   K is a constant that accounts for a number of design parameters such as the number of turns of
      the coil and the length of the coil

The equivalent circuit of the dc motor is shown in Figure 12.67. The left circuit is the field circuit of the stator. If the stator is a permanent magnet, we do not need to consider it. However, most high-power motors have an electric magnetic field circuit in the stator. The field circuit is excited by a dc source. So all we have to do is to represent the field circuit by its field voltage V_f, which causes a field current I_f to pass through the field windings. Since it is a dc circuit, the field winding is represented by its resistance R_f alone. The right side represents the armature circuit (rotor). The circle with two brushes represents the voltage across the armature coil E_a, R_a is the resistance of the coil, V is the voltage of the source connected to the armature circuit, and I_a is the armature current.

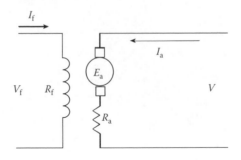

FIGURE 12.67 Representation of a dc motor.

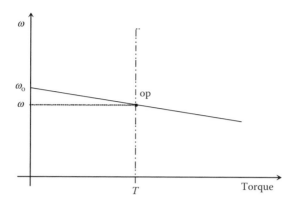

FIGURE 12.68 Speed–torque characteristics of a dc motor.

The equation of the armature circuit is

$$V = E_a + I_a R_a \qquad (12.109)$$

Substituting the value of E_a and I_a in Equations 12.107 and 12.108 into Equation 12.109 yields

$$V = K\phi\omega + \frac{T}{K\phi}R_a \qquad (12.110)$$

Hence,

$$\omega = \frac{V}{K\phi} - \frac{R_a}{(K\phi)^2}T \qquad (12.111)$$

Equation 12.111 can be used to sketch the speed–torque characteristics in Figure 12.68. At no load when the torque of the load $T = 0$, the speed of the motor is called the no-load speed ω_o.

$$\omega_o = \frac{V}{K\phi} \qquad (12.112)$$

The actual speed of a loaded motor, as given in Equation 12.111, depends on the magnitude of the load torque, the value of the magnetic field, and the resistance of the armature winding.

EXAMPLE 12.23

A 200 V dc motor has an armature resistance of 1 Ω and an equivalent field voltage $K\phi = 3$ Vs. If the motor is driving a full load of 30 N m, compute the speed of the motor.

Solution

Direct substitution in Equation 12.111 yields

$$\omega = \frac{V}{K\phi} - \frac{R_a}{(K\phi)^2}T = \frac{200}{3} - \frac{1}{9}30 = 63.33 \text{ rad/s}$$

or

$$n = \frac{\omega}{2\pi} = \frac{63.33}{2\pi} = 10.1 \text{ rev/s or } 606 \text{ rev/min}$$

12.6.2 STARTING OF DC MOTOR

The current of the dc motor can be computed using Equation 12.109.

$$I_a = \frac{V - E_a}{R_a} = \frac{V - K\phi\omega}{R_a} \tag{12.113}$$

At starting, the speed of the motor is zero, hence the starting current I_{st} is

$$I_{st} = \frac{V}{R_a} \tag{12.114}$$

This starting current is excessive as seen in Example 12.24.

EXAMPLE 12.24

For the motor in Example 12.23, compute the armature current at full load and the current at starting.

Solution

The full load current can be computed using Equation 12.108.

$$I_a = \frac{T}{K\phi} = \frac{30}{3} = 10 \text{ A}$$

The starting current from Equation 12.114 is

$$I_{st} = \frac{V}{R_a} = \frac{200}{1} = 200 \text{ A}$$

The starting current is 20 times greater than the full load current.

The high starting current is a major problem for dc motors. The large currents at starting can damage the armature winding due to the excessive heat they generate, especially if the motor is large and the inertia of the load is high, causing the motor to be slow at starting. From Equation 12.114, the starting current can be reduced by reducing the voltage at starting or inserting a resistance in the armature circuit at starting, as shown in Figure 12.69.

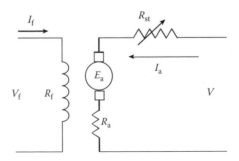

FIGURE 12.69   Starting of a dc motor using starting resistance.

Example 12.25

Compute the voltage at starting for the motor in Example 12.23 that leads to a starting current equal to three times the rated current.

Solution

Use Equation 12.114 to compute the starting voltage V_{st}.

$$V_{st} = I_{st}R_a = (3 \times 10) \times 1 = 30 \text{ V}$$

The voltage should be reduced to 30 V at starting.

Example 12.26

Compute the resistance that should be inserted in series with the armature of the motor in Example 12.23 to reduce the starting current to three times the rated current.

Solution

Use Equation 12.114 to compute the starting resistance R_{st}.

$$R_a + R_{st} = \frac{V}{I_{st}}$$

$$R_{st} = \frac{V}{I_{st}} - R_a = \frac{200}{30} - 1 = 5.67 \ \Omega$$

12.6.3 Speed Control of dc Motor

If you examine Equation 12.111, you find that the speed of the dc motor can be controlled by increasing the armature resistance, by changing the voltage of the armature circuit, and by changing the field ϕ. The field can be changed by adjusting the field current I_f. The first method requires the insertion of a resistance in the armature circuit, which reduces the efficiency of the motor.

Adjustment of armature voltage is a very common method of speed control as it is highly efficient and simple to implement.

Example 12.27

For the motor in Example 12.23, compute the armature voltage that reduces the speed of the motor to 400 rpm.

Solution

Use Equation 12.110 to compute the new value of the voltage.

$$V = K\phi\omega + \frac{T}{K\phi}R_a = 3 \times 2\pi \frac{400}{60} + \frac{30}{3} \times 1 = 135.7 \text{ V}$$

Keep in mind that the voltage of the armature must be below the rated value of the motor. If the voltage exceeds the rated value, the armature winding insulation could be damaged rendering the motor unusable. Hence, the armature voltage control reduces the speed of the motor below its full load speed.

To increase the speed of the motor higher than its full load speed, we can reduce the field voltage V_f. Doing so reduces the field current. Since the magnetic field ϕ is directly proportional to the field current I_f, the reduction of the field current reduces the field by the same percentage.

EXAMPLE 12.28

For the motor in Example 12.23, compute the field voltage that increases the speed of the motor to 700 rpm. Assume that the field voltage at the full load speed is 300 V.

Solution

The magnetic field is proportional to the field current, and the field current is linearly proportional to the field voltage. Hence,

$$K\phi = CV_f$$

where C is a constant of proportionality; it can be computed using the data for the full load speed.

$$C = \frac{K\phi}{V_f} = \frac{3}{300} = 0.01 \text{ s}$$

Now compute $K\phi$ for the new speed.

$$V = K\phi\omega + \frac{T}{K\phi}R_a$$

$$K\phi V - (K\phi)^2\omega - TR_a = 0$$

$$200 \times K\phi - (K\phi)^2\left(2\pi\frac{700}{60}\right) - 30 \times 1 = 0$$

The solution of the above equation leads to

$$K\phi = 2.57$$

The new value of the field voltage that rotates the motor at 700 rpm is

$$V_f = \frac{K\phi}{C} = \frac{2.57}{0.01} = 257 \text{ V}$$

12.7   STEPPER MOTOR

One major drawback of the motors discussed so far is their lack of simple and precise position control capability. However, the stepper motor is an excellent option for position control devices because it can control the position of the rotor and can hold the motor at this position for as long as needed. These features are essential in several applications that demand fine movements such as hard disk drives, printers, plotters, scanners, fax machines, medical equipment, laser guiding systems, robots, and actuators. The stepper motors can achieve fine position control with a step resolution of 1°.

Stepper motors come in three basic types: (1) variable reluctance, (2) permanent magnet, and (3) hybrid. The stator of these three types is similar to that of the induction or synchronous machine. However, for a stepper motor, the phases could be any number depending on the design preference and required resolution. Figure 12.70 shows a stator with four phases. The stator windings are not embedded inside the cylindrical stator, but are wrapped around salient poles extended inside the airgap.

When any of the stator windings (a–a′, b–b′, c–c′, or d–d′) is excited, the flux of the winding moves along the axis between the opposing poles of the same phase (e.g., a and a′) as shown in Figure 12.70b. The direction of the flux is dependent on the polarity of the applied voltage. Thus, the flow of the current determines the direction of the flux in the airgap.

The phases of the stator are switched by a power electronic converter; the most common one is the dc/ac converter such as the one shown in Figure 12.71. Each transistor switches one phase, and

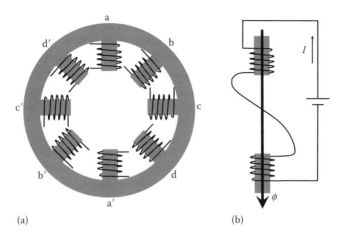

(a)   (b)

FIGURE 12.70   Stator of the stepper motor.

all phases are connected in a star configuration with the common neutral point providing the return path for all currents. The switching pattern of the transistors is a design parameter depending on the resolution of the step angle. The typical switching pattern in case one phase is switched at a time is given at the bottom of Figure 12.71. The period of the switching cycle τ is the sum of the conduction period t_c of all phases.

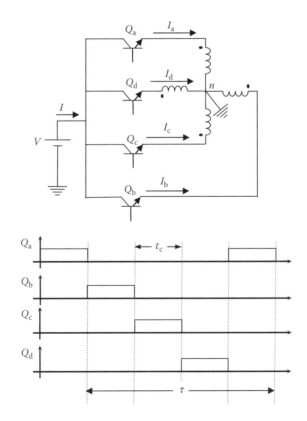

FIGURE 12.71   Switching of stepper motor winding.

12.7.1 Variable Reluctance Stepper Motor

This is the oldest type of stepper motors. Unlike synchronous or induction motors, the rotor of the variable reluctance stepper motor (VRSM) is neither a magnet nor has any windings, but is made of a multi-tooth iron material. Stepper motors are described by their number of phases in the stator windings and the number of poles (or the number of teeth) in their rotor. Figure 12.72 shows a four-phase, six-teeth stepper motor.

When a tooth is aligned with the axis of a stator phase, the airgap distance between the tooth and the phase is called *tooth airgap*. Similarly, when a groove is aligned with the axis of a stator phase, the airgap distance between the groove and the phase is called *groove airgap*. The difference between the two airgaps develops a torque that aligns the rotor. To explain this, let us consider the force of attraction between an energized phase and the rotor, which is governed by the electromagnetic force equation.

$$F = \frac{\mu_o}{4\pi d} I^2 \qquad (12.115)$$

where

μ_o is the absolute permeability (1.257×10^{-6} H/m)
d is the airgap distance between the energized phase of the stator and the rotor
I is the current of the stator winding

Since the tooth airgap is smaller than the groove airgap, the force of the attraction of the tooth is higher than that of the groove. Hence, a tooth is always aligned with the energized pole (in Figure 12.72, phase a–a′ is energized and teeth 1 and 4 are aligned along the axis of phase a). This attraction force is known as *detent force*.

If we disconnect phase a and energize phase b, the nearest teeth (2 and 5) move under poles b–b′, thus rotating one step in the counterclockwise direction. The movement angle is called *step angle δ*, which is a function of the number of phases in the stator and the number of teeth in the rotor.

$$\delta = \frac{360}{N_{ph}\, N_t} \qquad (12.116)$$

where

N_{ph} is the number of stator phases
N_t is the number of teeth in the rotor

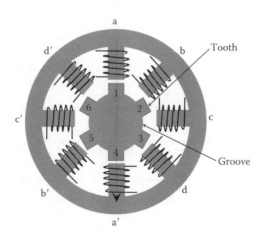

FIGURE 12.72 Variable reluctance stepper motor.

For the variable reluctance stepper, the rotor pole pair pp consists of one tooth and one groove. Thus, the relationship between the number of rotor poles p and number of rotor teeth is

$$pp = N_t$$
$$p = 2N_t \qquad (12.117)$$

Equation 12.116 can be written as a function of the rotor poles.

$$\delta = \frac{720}{N_{ph}\, p} \qquad (12.118)$$

In the four-phase, six-teeth stepper motor (Figure 12.72), the number of rotor poles is 12 and the step angle is 15°. A finer step angle can be achieved if we increase the number of stator phases, the number of teeth, or energize more than one phase at a time.

The stepper motors are often rated by their number of steps per revolution (spr) s, which can be as high as 400 spr.

$$s = \frac{360}{\delta} = \frac{N_{ph}\, p}{2} \qquad (12.119)$$

To rotate the motor continuously, the switching of the phases must also be continuous. The motor speed n can be calculated in a process similar to that in Equation 12.9

$$n = \frac{120f}{p} = \frac{120}{\tau p} \qquad (12.120)$$

where τ is the switching period (s). The unit of n in Equation 12.120 is revolution per minute.

The rotating stepper motor can also be rated by its *step rate* s_r, which is the number of steps per second (sps). Hence,

$$s_r = \frac{sn}{60} = \frac{N_{ph}}{\tau} \qquad (12.121)$$

EXAMPLE 12.29

A six-phase VRSM has a step angle of 10°. Compute the following:

(a) Number of teeth in the rotor.
(b) If the conduction period of each transistor is 2 ms, compute the continuous speed of the motor.

Solution

(a)
$$\delta = \frac{360}{N_{ph} N_t}$$

$$N_t = \frac{360}{10 N_{ph}} = \frac{360}{60} = 6 \text{ teeth}$$

(b)
$$\tau = t_c \times N_{ph} = 2 \times 6 = 12 \text{ ms}$$

$$n = \frac{120}{\tau p} = \frac{120}{12 \times 12} = 833.33 \text{ rpm}$$

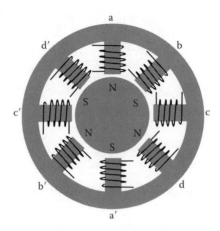

FIGURE 12.73 Permanent magnet stepper motor.

12.7.2 PERMANENT MAGNET STEPPER MOTOR

Because its rotor has no magnet, the variable reluctance stepper develops little torque. Therefore, it is used only in applications such as disk drives, printers, or low torque actuations. For higher torque applications, the permanent magnet stepper is often selected. The rotor of the permanent magnet stepper motor (PMSM) is cylindrical with several permanent magnets as shown in Figure 12.73. The magnet provides better holding force under the energized stator than the VRSM.

One common characteristic of permanent magnet motors is that they tend to cog as you turn the rotor manually without energizing the stator. This is because the rotor magnetic poles are attracted to the salient iron poles of the stator. This attraction force is the detent force.

Equations 12.118 through 12.121 for the VRPM are also applicable to the PMSM. The only difference is that the PMSM has poles instead of teeth.

EXAMPLE 12.30

A two-phase, four-pole PMSM is driven by a 100 Hz dc/ac converter. Compute the following:

(a) Speed of the motor
(b) Conduction period of each transistor in the converter

Solution

(a)
$$n = \frac{120f}{p} = \frac{120 \times 100}{4} = 3000 \text{ rpm}$$

(b)
$$\tau = t_c \times N_{ph}$$

$$t_c = \frac{\tau}{N_{ph}} = \frac{1}{f\,N_{ph}} = \frac{1}{100 \times 2} = 5 \text{ ms}$$

12.7.3 HYBRID STEPPER MOTOR

The hybrid stepper motor (HSM) provides the best resolution and the highest torque among all other types of stepper motors. The HSM can achieve a resolution of up to 400 spr. The rotor of the HSM is

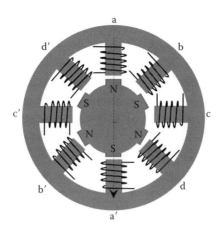

FIGURE 12.74 Hybrid stepper motor.

a multi-tooth rotor like the VRSM, but the teeth are also magnets (Figure 12.74). The attraction force of this motor is composed of two components; the attraction force due to the reluctance and the attraction force of the permanent magnet. Thus, a stronger attraction force is achieved.

12.7.4 HOLDING STATE OF STEPPER MOTOR

When the stepper motor is unloaded, the rotor is perfectly aligned with the energized phase as shown in Figures 12.72 through 12.74. In this case, there is no angle between the axes of the energized phase and the attracted poles of the rotor. If no other switching action is made, the rotor will maintain its holding position.

During the holding state, when an external load torque T_l is applied on the stepper motor, the rotor may no longer be perfectly situated under the axis of the energized phase. Take, for example, the case in Figure 12.75 when phase a–a$'$ is excited and the load torque is in the clockwise direction. The load torque tends to move the rotor in the clockwise direction, thus causing the mechanical displacement angle α in Figure 12.75. The greater the torque applied, the larger the *mechanical displacement angle*. When the load torque exceeds the maximum static torque of the motor, which is called holding torque T_h, the motor enters an unstable state where the rotor can no longer maintain its holding state.

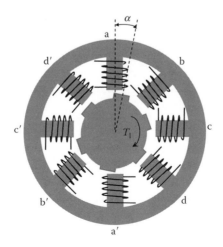

FIGURE 12.75 Displacement angle of a stepper motor.

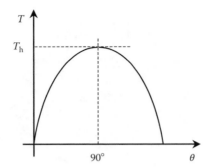

FIGURE 12.76 Torque curve of a stepper motor.

If the stepper motor is powered by one phase at a time, the developed torque of the motor can be computed using Faraday's rule in Equation 12.10. Hence

$$T_d \sim \text{force} \sim I$$
$$T_d = KI \tag{12.122}$$

where K is a constant that depends on the motor geometry and the property of its windings. The unit of K is volt second. I is the current of the winding. When the axes of the stator windings and the attracted poles are not perfectly aligned, the developed torque in the rotor that would align the axes can be written as

$$T_d = KI \sin\theta \tag{12.123}$$

where θ is called *electrical displacement angle*. The electrical displacement angle θ is the same as the mechanical displacement angle δ if the stepper is a two-pole machine. For any other number of poles, the relationship between the mechanical and electrical displacement angles is

$$\alpha = \frac{\theta}{p/2} = \frac{\theta}{N_t} \tag{12.124}$$

Equation 12.123 is plotted in Figure 12.76, and it shows that the maximum holding torque T_h occurs at $\theta = 90°$. Hence,

$$T_h = KI \tag{12.125}$$

EXAMPLE 12.31

A 10-teeth, two-phase VRSM is loaded by a 10 mN m torque. The torque constant of the motor is 3 Vs and the holding torque is 30 mN m. Compute the following:

    a. Current of the stator windings
    b. Mechanical displacement angle
    c. Displacement error
    d. Stator current that reduces the mechanical displacement angle by 50%

Solution

    (a) The holding torque is the maximum torque in Equation 12.123.

$$T_h = KI$$
$$I = \frac{T_h}{K} = \frac{30}{3} = 10 \text{ mA}$$

(b) The mechanical displacement angle of the rotor can be computed using Equation 12.123.

$$T_d = KI \sin(N_t \alpha)$$
$$10 = 30 \sin(10\alpha)$$

Hence,

$$\alpha = 2°$$

(c) The displacement error is the displacement angle with respect to the step angle. The step angle is

$$\delta = \frac{360}{N_{ph} N_t} = \frac{360}{210} = 18°$$

Displacement error $= (2/18) \times 100 = 11.11\%$

(d) For $\alpha = 1°$

$$T_d = KI \sin(N_t \alpha)$$
$$10 = 3I \sin(10)$$

Hence,

$$I = 19.2 \text{ mA}$$

12.7.5 ROTATING STEPPER MOTOR

During the holding state, the stepper motor must develop enough torque to compensate for the load torque. When the motor rotates, the developed torque of the motor must compensate for three components: (1) load torque T_l, (2) friction and drag torque T_f, and (3) acceleration or deceleration torque T_j.

$$T_d = T_l + T_f + T_j \tag{12.126}$$

The acceleration and deceleration torque is also known as *inertia torque*.

$$T_j = J \frac{d\omega}{dt} = \frac{2\pi}{60} J \frac{dn}{dt} \tag{12.127}$$

where J is the inertia of the rotor plus load (g m^2).

EXAMPLE 12.32

A six-phase VRSM with a step angle of 7.5° and a step rate of 200 sps is driving an inertia load. The inertia of the load plus the rotor is 1 mg m^2, and the friction torque is 15 mN m. Compute the following:

(a) Number of teeth of the rotor
(b) Switching period τ
(c) Speed of the motor
(d) Input power at steady state (ignore electrical losses)
(e) Inertia torque to accelerate the motor at 40 rad/s^2

Solution

(a)
$$\delta = \frac{720}{N_{ph}\, p}$$

$$p = \frac{720}{\delta N_{ph}} = \frac{720}{7.5 \times 6} = 16 \text{ poles}$$

Hence, the number of teeth is eight.

(b)
$$s_r = \frac{N_{ph}}{\tau}; \quad \tau = \frac{6}{200} = 30 \text{ ms.}$$

(c)
$$n = \frac{120}{\tau p} = \frac{120}{30 \times 16} = 250 \text{ rpm.}$$

(d) Since the load is just inertia, the developed torque of the motor at steady state compensates only for the friction torque. Hence, the input power is

$$P = T_f\, \omega = 15 \times \frac{2\pi}{60} 250 = 392.7 \text{ mW}$$

(e)
$$T_j = J \frac{d\omega}{dt} = 1 \times 40 = 0.04 \text{ N m.}$$

12.8   SINGLE-PHASE MOTORS

As discussed in Section 12.1, the three-phase source produces a naturally rotating magnetic field in the airgap of the induction and synchronous motors. This is the main advantage of using the three-phase system as it makes the heavy industrial motors start and rotate smoothly. However, for low-power applications, such as household and office equipment, we must rotate the motors without using a three-phase system. This is because the outlets in these places are only single-phase systems. To rotate the single-phase motors, we need to create a rotating magnetic field in the airgap by using various designs and circuits such as the ones discussed in this section.

12.8.1   SPLIT-PHASE MOTOR

A split-phase motor uses two stator windings as shown in Figure 12.77. The main winding is M–M′ and the auxiliary winding is AU–AU′. The auxiliary winding has less turns of smaller cross-section conductors than the primary winding. This makes the resistance of the auxiliary winding much higher than that of the primary winding ($R_{AU} > R_M$). Now consider the simple representation of the

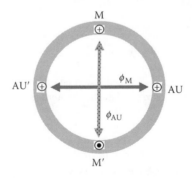

FIGURE 12.77   Winding arrangement of a split-phase motor.

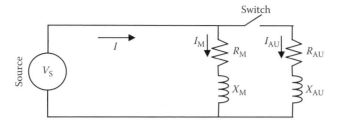

FIGURE 12.78 Representation of windings of a split-phase motor.

motor's windings in Figure 12.78. The figure shows a single-phase source connected to both windings. Each winding is represented by an equivalent resistance and an equivalent inductive reactance. The auxiliary winding is switched on at starting and switched off when the motor reaches its full speed. Assume that the switch is closed and the motor is energized, in this case the currents of the two windings are

$$\bar{I}_M = \frac{\bar{V}_s}{Z_M \angle \theta_M} = \frac{\bar{V}_s}{Z_M} \angle -\theta_M$$

$$\bar{I}_{AU} = \frac{\bar{V}_s}{Z_{AU} \angle \theta_{AU}} = \frac{\bar{V}_s}{Z_{AU}} \angle -\theta_{AU}$$

$$(12.128)$$

where

$$\bar{Z}_M = R_M + jX_M$$

$$\bar{Z}_{AU} = R_{AU} + jX_{AU}$$

$$(12.129)$$

Since $R_{AU} > R_M$ and $X_{AU} < X_M$ due to the smaller number of turns in the auxiliary winding, then

$$\theta_M > \theta_{AU} \qquad (12.130)$$

The difference in these two angles creates a phase shift between the currents inside the two windings of as much as 30°. These two currents produce magnetic fields in the airgap that is rotating as explained in Example 12.33.

EXAMPLE 12.33

The phase shift between the currents of the auxiliary and primary windings of a split-phase induction motor is 30°. Assume that the magnetic fields due to these currents are

$$\phi_M = K \sin(377\,t)$$

$$\phi_{AU} = 0.2K \sin(377\,t + 30)$$

where K is a constant value. Assume that the auxiliary winding is placed 90° from the main winding. Compute the magnetic field in the airgap at $t = 1$, 4, and 7 ms.

Solution

As seen in Figure 12.77, the current of each winding produces its own magnetic field. The total magnetic field in the airgap is the phasor sum of both. Since they are placed 90° from each other, the total field in the airgap is

$$\overline{\phi} = \sqrt{\phi_M^2 + \phi_{AU}^2} \angle \tan^{-1}\left(\frac{\phi_{AU}}{\phi_M}\right)$$

At $t = 1$ ms

$$\phi_M = K\sin(377 \times 0.001) = 0.3681\ K$$

$$\phi_{AU} = 0.2\ K\sin\left(377 \times 0.001 + \frac{\pi}{6}\right) = 0.1567\ K$$

$$\overline{\phi} = \sqrt{\phi_M^2 + \phi_{AU}^2} \angle \tan^{-1}\left(\frac{\phi_{AU}}{\phi_M}\right) = 0.4\ K\angle 23°$$

At $t = 4$ ms

$$\phi_M = K\sin(377 \times 0.004) = 0.998\ K$$

$$\phi_{AU} = 0.2\ K\sin\left(377 \times 0.004 + \frac{\pi}{6}\right) = 0.1791\ K$$

$$\overline{\phi} = \sqrt{\phi_M^2 + \phi_{AU}^2} \angle \tan^{-1}\left(\frac{\phi_{AU}}{\phi_M}\right) = 1.01\ K\angle 10.2°$$

At $t = 8$ ms

$$\phi_M = K\sin(377 \times 0.01) = 0.1253\ K$$

$$\phi_{AU} = 0.2\ K\sin\left(377 \times 0.01 + \frac{\pi}{6}\right) = -0.0775\ K$$

$$\overline{\phi} = \sqrt{\phi_M^2 + \phi_{AU}^2} \angle \tan^{-1}\left(\frac{\phi_{AU}}{\phi_M}\right) = 0.15\ K\angle -32°$$

Note that the magnitude and the angle of the magnetic field in the airgap change with time as shown in Figure 12.79. The change in the angle is always in one direction causing the magnetic field to rotate in that direction, thus rotating the rotor of the motor. Because the magnitude of the rotating field is changing, the motor cannot be used to start heavy loads. Recall that the magnitude of the rotating field in three-phase systems is constant; thus, three-phase motors are more suitable for heavy loads.

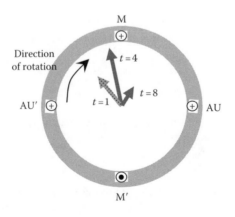

FIGURE 12.79 Rotating magnetic field of a split-phase motor.

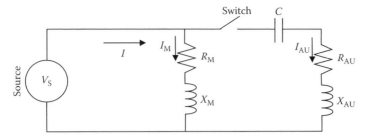

FIGURE 12.80 Representation of a capacitor starting motor.

The auxiliary winding is often switched by a centrifugal or solid-state switch. The centrifugal switch is normally closed and is opened only when the motor reaches its full speed. Thus, it activates the auxiliary windings at starting only. After the switch is opened, the motor continues rotating as the system inertia keeps the rotor spinning while it chases the magnetic field of the main winding.

12.8.2 CAPACITOR STARTING MOTOR

A capacitor starting motor is very similar to the split-phase motor. However, the phase shift between the main and auxiliary windings is achieved by inserting a capacitor in series with the auxiliary winding at starting, as shown in Figure 12.80. The phase shift of the currents in both windings causes the magnetic field in the airgap to rotate much the same way as that explained in Example 12.33. After the motor starts, the capacitor and the auxiliary winding are switched off by a centrifugal switch or a solid-state device.

12.8.3 SHADED-POLE MOTOR

A shaded-pole motor has its stator core split into main and shaded cores as shown in Figure 12.81. A small short-circuited winding composed of a few turns of thick wire is wrapped around the shaded leg of the core. The main winding is energized by a power source, thus producing the main

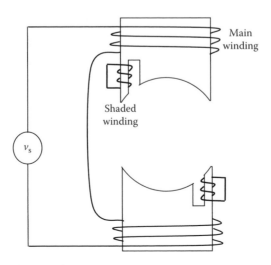

FIGURE 12.81 Structure of a shaded-pole motor.

magnetic field. This magnetic field in the core induces a current in the shaded winding, which creates its own magnetic field. The current in the shaded winding lags that of the main winding. Consequently, the airgap magnetic field of the shaded winding lags the magnetic field of the main winding. Since the shaded winding is displaced from the center axis of the main winding, a rotating magnetic field is produced in the airgap as explained in Example 12.33.

EXERCISES

1. The output power of a four-pole, three-phase induction motor is 100 hp. The motor operates at a slip of 0.02 and efficiency of 90%. Compute the following:
   (a) Input power of the motor (kw)
   (b) Shaft torque (output torque)
2. The synchronous speed of a 60 Hz, three-phase induction motor is 1200 rpm. What is the synchronous speed of the motor if it is used in a 50 Hz system?
3. A 100 hp, 60 Hz, three-phase induction motor has a slip of 0.03 at full load. Compute the efficiency of the motor at full load when the friction and windage losses are 900 W, the stator core loss is 4200 W, and the stator copper loss is 2700 W.
4. A three-phase induction motor is rated at 50 hp, 480 V, 60 Hz, and 1150 rpm. If the motor operates at full load, compute the following:
   (a) Developed torque
   (b) Frequency of the rotor current
5. A three-phase, 480 V, 60 Hz, 12-pole induction motor has the following parameters:

$$R_1 = 1.0 \ \Omega; \quad R_2' = 0.5 \ \Omega; \quad X_{eq} = 10 \ \Omega; \quad X_c = 100 \ \Omega; \quad R_c = 800 \ \Omega$$

   Calculate the following:
   (a) Slip at maximum torque
   (b) Current at maximum torque
   (c) Speed at maximum torque
   (d) Maximum torque
6. A 480 V, three-phase, six-pole, slip-ring, Y-connected induction motor has $R_1 = R_2' = 0.1 \ \Omega$ and $X_{eq} = X_1 + X_2' = 0.5 \ \Omega$. The motor slip at full load is 3%, and its efficiency is 90%. Calculate the following:
   (a) Starting current (You may ignore the magnetizing current.)
   (b) Starting torque
   (c) Maximum torque
   (d) Value of the resistance that should be added to the rotor circuit to reduce the starting current by 50%
   (e) Starting torque of case (d)
   (f) Value of the resistance that should be added to the rotor circuit to increase the starting torque to the maximum value
   (g) Starting current of case (f)
7. A three-phase, four-pole, 480 V, 60 Hz, Y-connected slip-ring induction motor has the following parameters:

$$R_1 = R_2' = 0.5 \ \Omega; \quad X_1 + X_2' = 5 \ \Omega$$

   Compute the following:
   (a) Starting torque
   (b) Inserted rotor resistance that increases the starting torque by a factor of 4

8. A three-phase, 2.2 kV, 60 Hz, Y-connected slip-ring induction motor has the following parameters:

$$R_1 = R_2' = 0.2\ \Omega; \quad X_1 + X_2' = 1.5\ \Omega$$

The motor runs at 570 rpm. Ignore the rotational losses and calculate the full load torque.

9. A 480 V, 60 Hz, three-phase induction motor has the following parameters:

$$R_1 = R_2' = 0.3\ \Omega; \quad X_{eq} = 1.0\ \Omega; \quad X_c = 600\ \Omega$$

At full load, the motor speed is 1120 rpm, the rotational loss is 400 W, and the core loss is 1 kW. Calculate the following:
(a) Motor slip
(b) Developed torque at full load
(c) Developed power in hp
(d) Rotor current
(e) Copper losses
(f) Input power
(g) Reactive power consumed by the motor
(h) Power factor of the motor

10. The speed of a 12-pole, 480 V three-phase induction motor at full load is 560 rpm. The motor has the following parameters;

$$R_1 = 0.1\ \Omega; \quad R_2' = 0.5\ \Omega; \quad X_1 + X_2' = 5\ \Omega$$

(a) Compute the developed torque.
(b) While the torque is unchanged, the voltage is changed to reduce the speed of the motor to 520 rpm. Compute the new voltage.

11. A train is driven by an LIM at 80 km/h when the frequency of the primary windings is 15 Hz. The frontal area of the train is 25 m^2 and its coefficient of drag is 0.9. The friction coefficient between the wheels of the train and the track is 0.1, and the weight of the train is 500,000 kg. The pole pitch of the vehicle is 3 m. Compute the following:
(a) Speed of the thrust force
(b) Slip of the motor
(c) Developed force of the motor at steady state
(d) Developed power in hp

12. A train with LIM is traveling uphill at 80 km/h when the frequency of its primary windings is 15 Hz. The inclination of the hill is 5°. The frontal area of the train is 25 m^2 and its coefficient of drag is 0.9. The friction coefficient between the wheels of the train and the track is 0.1, and the weight of the train is 500,000 kg. The pole pitch of the vehicle is 3 m. Compute the following:
(a) Developed force of the motor at steady state
(b) Developed power in hp

13. A vehicle is powered by an LIM with a pole pitch of 2 m. The primary circuit resistance R_1 is 0.5 Ω, and the secondary resistance referred to the primary circuit R_2' is 1.0 Ω. The equivalent inductance of the primary plus secondary circuits is 0.02 H. The frequency of the primary windings is 10 Hz. Assume that the drag force is 500 N and the friction force is 2 kN. Compute the voltage across the primary windings when the motor travels at 100 km/h.

14. A synchronous generator connected directly to an infinite bus is generating at its maximum real power. The line-to-line voltage of the infinite bus is 480 V and the synchronous reactance of the generator is 5 Ω. Compute the reactive power of the generator.

15. A 1 GW synchronous generator is connected to an infinite bus through a transmission line. The synchronous reactance of the generator is 9 Ω and the inductive reactance of the transmission line is 3 Ω. The infinite bus voltage is 110 kV. If the generator delivers its rated power at unity power factor to the infinite bus, compute the following:
    (a) Terminal voltage of the generator
    (b) Equivalent field voltage
    (c) Real power output at the generator terminals
    (d) Reactive power output at the generator terminals
16. A synchronous generator is connected directly to an infinite bus. The voltage of the infinite bus is 15 kV. The excitation of the generator is adjusted until the equivalent field voltage E_f is 14 kV. The synchronous reactance of the machine is 5 Ω. Compute the following:
    (a) Pullover power (maximum power)
    (b) Equivalent excitation voltage that increases the pullover power by 20%
17. A 100 MVA synchronous generator is connected to a 25 kV infinite bus through two parallel transmission lines. The synchronous reactance of the generator is 2.5 Ω, and the inductive reactance of each transmission line is 2 Ω. The generator delivers 100 MVA to the infinite bus at 0.8 power factor lagging. Suppose a lightning strike causes one of the transmission lines to open. Assume that the mechanical power and excitation of the generator are unchanged. Can the generator still deliver the same amount of power to the infinite bus?
18. A synchronous generator is connected to an infinite bus through a transmission line. The infinite bus voltage is 15 kV and the equivalent field voltage of the machine is 14 kV. The transmission line inductive reactance is 4 Ω, and the synchronous reactance of the machine is 5 Ω.
    (a) Compute the capacity of the system.
    (b) If a 2 Ω capacitor is connected in series with the transmission line, compute the new capacity of the system.
19. A synchronous motor excited by a 5 kV source has a synchronous reactance of 5 Ω. The motor is running without any mechanical load (i.e., the output real power is zero). Ignore all losses and calculate the equivalent field voltage Ef that delivers 3 MVAr to the source. Also, draw the phasor diagram.
20. A six-pole, 60 Hz synchronous motor is connected to an infinite bus of 15 kV through a transmission line. The synchronous reactance of the motor is 5 Ω, and the inductive reactance of the transmission line is 2 Ω. When Ef is adjusted to 16 kV (line-to-line), the reactive power at the motor's terminals is zero.
    (a) Draw the phasor diagram of the system.
    (b) Calculate the terminal voltage of the motor.
    (c) Calculate the developed torque of the motor.
21. A twelve pole, 480 V, 60 Hz synchronous motor has a synchronous reactance of 5 Ω. The field current is adjusted so that the equivalent field voltage Ef is 520 V (line-to-line). Calculate the following:
    (a) Maximum torque
    (b) Power factor at the maximum torque
22. An industrial load is connected across a three-phase, Y-connected source of 4.5 KV (line-to-line). The real power of the load is 160 kW, and its power factor is 0.8 lagging.
    (a) Compute the reactive power of the load.
    (b) A synchronous motor is connected across the load to improve the total power factor to unity. The synchronous motor is running unloaded (no mechanical load). If the synchronous reactance of the motor is 10 Ω, compute the equivalent field voltage of the motor.
23. A dc motor is driving a constant-torque load. If the armature voltage of the motor increases by 20%, calculate the percentage change in no-load speed.
24. A dc motor is driving a constant-torque load. If the field voltage of the motor decreases by 20%, calculate the percentage change in no-load speed.

25. A 150 V dc motor drives a constant-torque load at a speed of 1200 rpm. The armature and field resistances are 2 and 150 Ω, respectively. The motor draws an armature current of 9 A and field current of 1 A. Assume that a resistance is added in the field circuit to reduce the field current by 20%. Calculate the armature current, motor speed, and value of the added resistance.

26. A 200 V dc motor is loaded by a constant torque of 10 N m. The armature resistance of the motor is 2 Ω and the field constant $K\phi = 2.5$ Vs. Calculate the motor speed.

27. A 10-teeth VRSM is loaded by a 10 mN m torque. The phase voltage is adjusted so that the winding current is 20 mA. The torque constant K of the motor is 3 Vs. Compute the following.
    (a) Holding torque of the motor
    (b) Mechanical displacement angle

28. An eight-pole, three-phase PMSM has a torque constant K of 2 Vs. When the load torque is 20 mN m, the mechanical displacement angle is 10°. Compute the following:
    (a) Holding torque
    (b) Current of the stator windings
    (c) Stator current that reduces the mechanical displacement angle to 4°

29. An eight-phase VRSM with a step angle of 4.5° and a step rate of 300 sps is driving an inertia load. The inertia of the load plus the rotor is 4.4 mg m^2, and the friction torque is 13 mN m. Compute the following.
    (a) Number of teeth of the rotor
    (b) Switching period τ
    (c) Speed of the motor
    (d) Input power (ignore electric losses)
    (e) Inertia torque to accelerate the motor at 20 rad/s^2

30. Explain the rotation of the shaded-pole motor using your own words.

31. Write a computer program that shows how the capacitor in the auxiliary winding of a single-phase motor causes the airgap magnetic field to rotate.

32. For the split-phase motor, write a computer program to compute the magnetic field in the airgap.

33. Why do we need to switch off the auxiliary winding of the single-phase motor?

13 Power Quality

The ideal alternating current (ac) waveform has two important features: (1) it contains no harmonic component besides the fundamental frequency (50 or 60 Hz) and (2) its magnitude is constant over a long period. Figure 13.1 shows the shape of an ideal waveform that is symmetrical and cyclic, and whose period is constant. A deviation from any of these characteristics results in what is known as power quality problem.

Unfortunately, the ideal waveform in Figure 13.1 does not exist in the real world. The nonlinear characteristics of most power equipment and the switching of loads tend to distort the voltage and current waveforms. Devices such as transformers and motors can cause distortions due to the saturation of their magnetic cores. Power electronic converters can also cause distortions due to the continuous switching of their components as shown in Chapter 10. Cyclic load switching, such as that in saw mills and rock crushing plants, can cause flickers in voltage. Overloading distribution feeders can cause low-voltage problems at customers' sites.

Most electrical equipment and devices tolerate small distortions in the ac waveforms. However, when the distortions are large enough to affect equipment or be perceived by the public as annoyances, they are power quality problems.

Although the use of modern power electronic converters cause some of the power quality problems due to their switching actions, power electronics can solve a large number of these problems as well. In the early 1980s when personal computers were widely available, they were very susceptible to a wide range of power quality problems. Any dip in voltage, due to the use of heavy loads such as vacuum cleaners at starting, would result in freezing or rebooting computers. Nowadays, the advancement in power electronics allows us to build computers with power supplies that can ride through any voltage dip and most transients.

In this chapter, you will find that most of the power quality problems are created by some customers and can affect other customers on the same grid. This situation creates a dilemma to utilities when it comes to identifying the parties responsible for the power quality problems and who should bear the cost of fixing the problems. Utilities often argue that polluters of the waveforms must bear the costs of ensuring supply quality. Nevertheless, unless there are enforceable standards, utilities will continue to carry the burden of fixing the power quality problems on their grid.

The main power quality problems are related to voltage and frequency. Voltage problems include voltage flickers, voltage sags, voltage swells, stray voltages, voltage transients, and outages. Frequency problems include harmonic distortions and frequency variations. Electromagnetic interferences, whether from nearby energized circuits or from solar flares, are also considered power quality problems.

13.1 VOLTAGE PROBLEMS

Most of the voltage problems can be explained by the circuit in Figure 13.2. The figure shows two loads (load$_1$ and load$_2$) representing two customers connected to the same source via a feeder (cable, transmission line, etc.). The parallel impedance of the two loads can be computed by

$$\frac{1}{\overline{Z}} = \frac{1}{\overline{Z}_1} + \frac{1}{\overline{Z}_2} \tag{13.1}$$

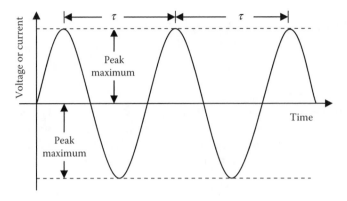

FIGURE 13.1   Ideal ac waveform.

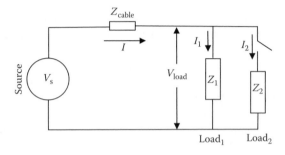

FIGURE 13.2   Loads connected to a voltage source through cable.

The voltage at the load side can be computed by

$$\overline{V}_{\text{load}} = \overline{I}\,\overline{Z} = \left(\frac{\overline{V}_s}{\overline{Z} + \overline{Z}_{\text{cable}}}\right)\overline{Z} = \frac{\overline{V}_s}{1 + \left(\frac{\overline{Z}_{\text{cable}}}{\overline{Z}}\right)} \tag{13.2}$$

When load$_2$ is switched on, the magnitude of the total impedances Z is smaller than the magnitude of any one of the two parallel impedances. Knowing this fact, let us examine Equation 13.2. It shows that when more loads are added, the total impedance Z is reduced, thus the load voltage is reduced. Therefore, the voltage at customer 1 depends on the load impedance of customer 2.

EXAMPLE 13.1

For the system in Figure 13.2, assume that $V_s = 120$ V, $Z_{\text{cable}} = 2\angle 90°$ Ω, $Z_1 = 10\angle 0°$ Ω, and $Z_2 = 5\angle 0°$ Ω. Compute the following:

(a) Load voltage when load$_2$ is switched off
(b) Load voltage when load$_2$ is switched on
(c) Voltage reduction on load$_1$ due to the presence of load$_2$

Solution

(a) Without load$_2$, the line current is

$$\overline{I} = \frac{\overline{V}_s}{\overline{Z}_1 + \overline{Z}_{\text{cable}}} = \frac{120\angle 0°}{10 + j2} = 11.54 - j2.3 \text{ A}$$

The voltage on $load_1$ without $load_2$ is

$$\overline{V}_{load_1} = \overline{Z}_1\,\overline{I} = 10\,(11.5 - j2.3) = 115.4 - j23 = 117.7\angle{-11.3°}\ \text{V}$$

(b) With $load_2$, the line current I is

$$\overline{I} = \frac{\overline{V}_s}{\left(\frac{\overline{Z}_1\overline{Z}_2}{\overline{Z}_1+\overline{Z}_2}\right) + \overline{Z}_{cable}} = \frac{120\angle{0°}}{3.33 + j2} = 31\angle{-31°}\ \text{A}$$

Hence, the load voltage is

$$\overline{V}_{load_2} = \left(\frac{\overline{Z}_1\overline{Z}_2}{\overline{Z}_1 + \overline{Z}_2}\right)I = 3.33(31\angle{-31°}) = 103\angle{-31°}\ \text{V}$$

(c) Voltage reduction on $load_1$ due to $load_2$ is

$$\frac{\Delta V}{V_1} = \frac{V_{load_2} - V_{load_1}}{V_{load_1}} = \frac{103 - 117.7}{117.7} = -12.5\%$$

The voltage on $load_1$ is reduced by 12.5% when $load_2$ is switched on. This is because $load_2$ is twice as large as $load_1$ (or $Z_1 = 2Z_2$). Repeat the example and assume $load_2$ is smaller than $load_1$ ($Z_2 > Z_1$). Can you draw a conclusion?

As seen in Example 13.1, the voltage at the customer side changes with the change in the overall loading of the feeder. However, it is normal for the voltage to change slightly and slowly. People and equipment are not affected by the small and slow changes in voltage. Actually if you measure the voltage in your home, you will find that its value changes slightly several times a day. In severe cases, when the changes are large, rapid, or cyclic, voltage problems could occur such as the following:

- *Fluctuations*: Small changes in voltage between 90% and 110% of the rated value. If the change is slow, every few hours, the fluctuations have no effect on people or equipment.
- *Flickers*: Fast and cyclic changes in voltage that are detected by human eyes. Although the magnitude of the flicker could be within the fluctuation range of the voltage, it is often fast enough allowing human eyes to detect it.
- *Sags*: Voltage below 90% of the rated value for a short period (up to a few seconds).
- *Swells*: Voltage above 110% for a short period (up to a few seconds).
- *Undervoltage (brownout)*: Voltage drop below 90% for at least several minutes.
- *Overvoltage*: Voltage increase above 110% for at least several minutes.
- *Interruptions*: Decrease in voltage below 10% for any period.

Most of the above problems can be caused by similar scenarios but with different time scales and magnitudes. In this chapter, we shall cover the most common problems: flickers and sags. The analyses of the other problems are mostly extensions of the ones we cover.

13.1.1 VOLTAGE FLICKERS

The term "flicker" was originally used to describe visible modulation in light intensity of electric incandescent lamps. In general, flicker is the fluctuation of voltage at frequencies much lower than the power frequency.

One of the most irritating effects of voltage flickers is light flickers (fast variations in light intensity). Light intensity λ is a measure of the brightness of a light source, which is roughly dependent on the power consumed by the lightbulb. For incandescent lamps, which are essentially resistors, their light intensities or output powers are proportional to the square of the applied voltages.

$$\lambda_{\text{incandescent}} \propto P \propto V^2 \tag{13.3}$$

The ballasts in ballasted lamps (arc and fluorescent lamps) tend to make the light intensity proportional to the applied voltage.

$$\lambda_{\text{ballast}} \propto V \tag{13.4}$$

Thus, different lamps respond differently to voltage fluctuations. Incandescent lamps tend to be more sensitive to changes in voltage than ballasted lamps.

EXAMPLE 13.2

A 120 V system experiences a 10% voltage drop. Compute the change in light intensity for incandescent and ballasted lamps.

Solution

For incandescent lamps, the change in light intensity is

$$\Delta\lambda_{\text{incandescent}} = \frac{\lambda_2 - \lambda_1}{\lambda_1}$$

where
  λ_1 is the light intensity at 120 V
  λ_2 is the light intensity at 120–10% V

Hence,

$$\Delta\lambda_{\text{incandescent}} = \frac{V_2^2 - V_1^2}{V_1^2} = \frac{(0.9 \times 120)^2 - 120^2}{120^2} = 19\%$$

For ballasted lamps, the change in light intensity is proportional to the voltage.

$$\Delta\lambda_{\text{ballast}} = \frac{\lambda_2 - \lambda_1}{\lambda_1} = \frac{V_2 - V_1}{V_1} = \frac{(0.9 \times 120) - 120}{120} = 10\%$$

The changes in light intensity can be easily perceived by the human eye, which can detect variations in brightness as small as 1%. However, the human eye cannot normally detect light flickers above 35 Hz. Experimental studies show that the maximum sensitivity to light flickers is in the range of 8–10 Hz.

Besides the visual annoyance, voltage flickers can cause a wide range of problems including computer freeze up (especially older computers), jitter of television pictures, faulty signals in electronic circuits, malfunction of sensitive equipment, and loss of information stored in electronic memories.

There are two types of flickers: cyclic flicker and noncyclic flicker. Cyclic flicker may occur at a regular frequency and may have a steady magnitude. It is caused by periodic heavy load fluctuations such as those in reciprocating compressors, automatic spot welders, arc furnaces, flashing signals, sawmills, group elevators, and cyclic electric hammers. An example of a cyclic flicker is shown in

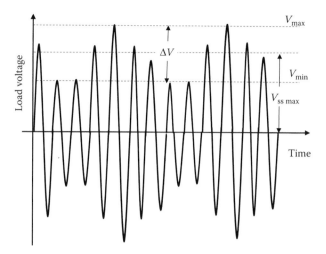

FIGURE 13.3 Cyclic flicker.

Figure 13.3. Noncyclic flicker may occur at random intervals and is caused by occasional heavy load switching.

A simple method to model the flicker in Figure 13.3 is to assume that the voltage is an amplitude-modulated waveform similar to that used in radio broadcasting. In this case, the instantaneous value of the flicker voltage v_f that modulates the load voltage is

$$v_f = V_{ss\,max} \sin(\omega t) \frac{\Delta V}{2} \sin(\omega_f t) \qquad (13.5)$$

where

$V_{ss\,max}$ is the peak value of the steady-state voltage without any flicker
$\Delta V/2$ is the modulation amplitude of the flicker
ω is the frequency of the supply voltage without the flicker
ω_f is the frequency of the flicker

The load voltage is the sum of the load voltage without the flicker v_{ss} and the flicker voltage. This is the well-known amplitude modulation equation.

$$v_{load} = v_{ss} + v_f = V_{ss\,max} \left(1 + \left(\frac{\Delta V}{2} \right) \cos(\omega_f t) \right) \cos(\omega t) \qquad (13.6)$$

The ratio of the modulation amplitude to the peak value of the source voltage without flicker is defined as the flicker factor F.

$$F = \frac{\Delta V/2}{V_{ss\,max}} = \frac{\Delta V}{V_{max} + V_{min}} \qquad (13.7)$$

If the flicker factor is zero, the voltage waveform is without any flicker. The larger the flicker factor, the more severe is the flicker problem.

Another term often used to quantify flicker is voltage fluctuation VF. It is the maximum flicker amplitude ΔV divided by the peak voltage without the flicker.

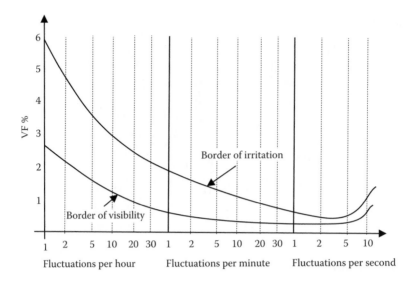

FIGURE 13.4 Effect of flickers on humans.

$$VF = \frac{\Delta V}{V_{ss\,max}} = 2F \qquad (13.8)$$

In the United States, the General Electric Company did an extensive study on human tolerance to light flickers. In this study, a group of volunteers were subjected to different levels of VF at different frequencies. The result of the study is summarized in Figure 13.4. The study shows that the higher the frequency of the light flickers, the lower is the VF that induces irritation to the human eye. The human perception of repeated flicker is usually much more pronounced than the perception of infrequent flickers. However, above 35 Hz, people tend not to be affected by the flicker. Occasional voltage dips, not flickers, even if clearly perceptible, are not objectionable to the majority of individuals.

EXAMPLE 13.3

A system with a steady-state voltage of 120 V is connected to a cyclic load that fluctuates at a rate of 15 cycles/h. Compute the maximum allowable flicker ΔV.

Solution

The maximum allowable flicker is computed at the irritation level, not the level of perception. According to Figure 13.4, VF at the irritation level for 15 cycles/h is about 3%. In this case,

$$\Delta V = VF \times V_{ss\,max} = 0.03 \times 120 \times \sqrt{2} = 5.1 \text{ V}$$

About 5 V variation in voltage can irritate people if the flicker is repeated 15 times/h.

There are a few methods to remedy flicker problems; among them is line compensation. As shown in Example 13.1, the voltage change is due to the large line impedance that feeds the loads. Since this impedance is predominantly inductive reactance, we can reduce the total reactive component of the line impedance by installing a series capacitor along the distribution feeder, as shown in Figure 13.5. In this case, because the total impedance between the source and the load is substantially reduced, the voltage fluctuation is diminished as shown in Example 13.4.

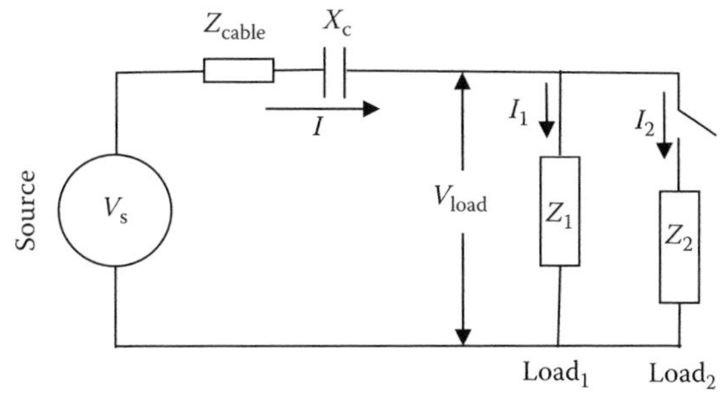

FIGURE 13.5 Loads connected to source by a capacitor compensated cable.

EXAMPLE 13.4

Repeat Example 13.1, but assume that the line is compensated by a series capacitor of 1.8 Ω.

Solution

(a) With load$_2$ switched off, the current of the cable is

$$\bar{I} = \frac{\bar{V}_s}{\bar{Z}_1 + \bar{Z}_{\text{cable}} + \bar{X}_c} = \frac{120\angle 0°}{10 + j2 - j1.8} = 12 - j0.24$$

The steady-state voltage across load$_1$ is

$$\bar{V}_{ss} = \bar{Z}_1 \bar{I} = 120 - j2.4 \approx 120\angle 0° \text{ V}$$

(b) With load$_2$ switched on, the line current is

$$\bar{I} = \frac{\bar{V}_s}{\left(\frac{\bar{Z}_1 \bar{Z}_2}{\bar{Z}_1 + \bar{Z}_2}\right) + \bar{Z}_{\text{cable}} + \bar{X}_c} = \frac{120\angle 0°}{3.33 + j2 - j1.8} = 36\angle -3.44° \text{ A}$$

Hence, the load voltage when load$_2$ is switched on is

$$\bar{V}_{\text{load}} = \left(\frac{\bar{Z}_1 \bar{Z}_2}{\bar{Z}_1 + \bar{Z}_2}\right)\bar{I} = 3.33(36\angle -3.44°) = 119.88\angle -3.44° \text{ V}$$

(c) Voltage reduction of load$_1$ due to the switching of load$_2$ is

$$\frac{\Delta V}{V_{ss}} = \frac{V_{\text{load}} - V_{ss}}{V_{ss}} = \frac{119.88 - 120}{120} = -0.1\%$$

The voltage on load$_1$ is reduced by just 0.1% when load$_2$ is switched on and the line is compensated by the series capacitor. Without the series capacitor, the voltage reduction on load$_1$ is 12.5%, as computed in Example 13.1. According to Figure 13.4, the 0.1% flicker is unnoticeable.

13.1.2 Voltage Sag

Voltage sag (also known as *voltage dip*) is the most common power quality problem. Voltage sag is a short duration of voltage reduction (see Figure 13.6) that is caused by switching on heavy electric loads such as elevators, cranes, pumps, air conditioners, refrigerators, and x-ray equipment. Voltage sags can also be caused by the disconnection of feeders and by faults in power system components. The difference between the sag and the flicker is that the flicker is repetitive or cyclic while the sag is not.

Motors continuously start and stop causing transients in currents (called *inrush current*) due to the saturation of their magnetic cores and the inertia of their mechanical loads. Transformer switching also causes inrush current due to the saturation of its cores. The sudden applied voltage causes the flux in the core to briefly exceed its steady-state value, thus driving the core into saturation (see Section 11.5). The inrush currents can reach values as high as 25 times the full load ratings, and can last between 200 ms and 1 min. Typical starting current characteristics are shown in Figure 13.7. Figure 13.7a shows a symmetrical inrush current without any direct current (dc) component and Figure 13.7b shows the inrush current with a dc component that is often decayed very quickly. In both cases, when the voltage is applied, the inrush current ramps up quickly to its crest value, and then slowly decreases to its steady-state value.

The inrush currents of heavy equipment cause the voltage at the connection point to sag. You have probably noticed that when the vacuum cleaner in your home starts, the light intensity of lamps is reduced. Voltage sags can cause a variety of problems from interrupting controllers or relays to malfunction of sensitive medical equipment. They also cause some types of motors to run at lower speeds or take a longer time to start. Severe sags can cause major disruptions to assembly lines and central air-conditioning systems.

A voltage waveform with a voltage sag typically looks like the one in Figure 13.8 for the symmetrical inrush current without a dc component. The dip in voltage is due to the inrush current around its crest value, and the overshoot is often due to the inductive elements in the equipment.

The symmetrical inrush current in Figure 13.7 can be expressed by the generic equation

$$i = I_{\text{ss max}} \sin(\omega t)\left(1 - e^{\frac{-t}{\tau_1}}\right)\left(1 + k e^{\frac{-t}{\tau_2}}\right) \tag{13.9}$$

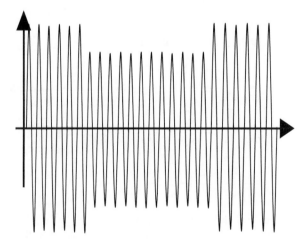

FIGURE 13.6   Voltage sag (dip).

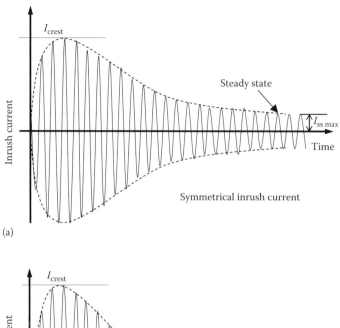

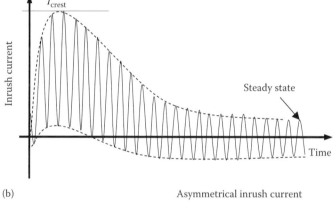

FIGURE 13.7   Typical inrush current waveforms at motor starting or transformer switching. (a) Symmetrical inrush current without a dc component and (b) Asymmetrical inrush current with a dc component.

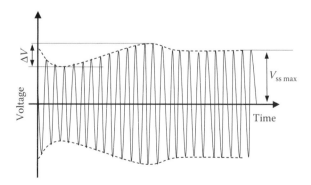

FIGURE 13.8   Voltage sag due to symmetrical inrush current.

where

i is the instantaneous value of the inrush current

τ_1 is the rising time constant

τ_2 is the falling time constant (both τ_1 and τ_2 are functions of the parameters of equipment, especially magnetic circuit and system inertia)

k is a factor that determines the maximum crest of the inrush current I_{crest}

$I_{ss\,max}$ is the peak value of the steady-state current

ω is the angular frequency of the source voltage

If the equipment started is energized through a cable whose resistance is negligibly small and whose inductance is L, the voltage across the equipment v_{load} is

$$v_{load} = v_s - L\left(\frac{di}{dt}\right) \tag{13.10}$$

where v_s is the instantaneous voltage of the source.

$$v_s = V_{max} \sin(\omega t + \theta) \tag{13.11}$$

where θ is the angle of the source voltage with respect to the steady-state current. Solving Equation 13.10 numerically leads to the load voltage in Figure 13.8. To do this, we need to compute the derivative of the current in Equation 13.10, which is obtained by differentiating Equation 13.9:

$$\frac{di}{dt} = I_{ss\,max}\left[\omega\cos(\omega t)\left(1 - e^{\frac{-t}{\tau_1}}\right)\left(1 + ke^{\frac{-t}{\tau_2}}\right) + \sin(\omega t)\left(\frac{e^{\frac{-t}{\tau_1}}}{\tau_1}\right)\left(1 + ke^{\frac{-t}{\tau_2}}\right) - \sin(\omega t)\left(1 - e^{\frac{-t}{\tau_1}}\right)\left(\frac{ke^{\frac{-t}{\tau_2}}}{\tau_2}\right)\right] \tag{13.12}$$

To compute the voltage sag by the above equations is tedious and requires a numerical solution. However, we can predict the approximate value of the sag if we make two reasonable assumptions:

1. Time constants (τ_1 and τ_2) are long enough with respect to the duration of the ac waveform. This would allow us to use root mean square (rms) calculations.
2. Inrush current is totally inductive until it reaches its crest value I_{crest}. This is a reasonable assumption as the inductive elements of the motors and transformers are predominant at starting.

In this case, we can compute the voltage across the load by the rms phasor calculations:

$$\overline{V}_{load} = \overline{V}_s - \overline{X}_L \overline{I} \tag{13.13}$$

where

X_L is the inductive reactance of the line

I is the rms inrush current of the feeder (line)

In Equation 13.13, we ignored the resistance of the cable as it is often much smaller than the inductive reactance.

Since the current is predominantly inductive until it reaches its crest value, the current lags the source voltage by almost 90°. Hence, the source voltage, load voltage, and the drop across the line inductance are all in phase. In this case, we can use the magnitudes only for Equation 13.13.

$$V_{\text{load}} = V_{\text{s}} - X_{\text{L}} I \tag{13.14}$$

The voltage sag VS is defined as the change in voltage ΔV with respect to the steady-state voltage before the sag V_{ss}.

$$\text{VS} = \frac{\Delta V}{V_{\text{ss}}} = \frac{V_{\text{load}} - V_{\text{ss}}}{V_{\text{ss}}} \tag{13.15}$$

EXAMPLE 13.5

A motor has a crest current at starting of 20 A when the source voltage is 120 V (Figure 13.9). The inductive reactance of the cable is 2 Ω. Estimate the voltage sag across the motor's terminals.

Solution

Keep in mind that the crest current is the peak value, not the rms value. Hence, the crest current of the cable in rms is

$$\bar{I} = \frac{\bar{I}_{\text{crest}}}{\sqrt{2}} = \frac{-j20}{\sqrt{2}} = -j14.14 \text{ A}$$

The voltage across the motor at the crest of the starting current is

$$\overline{V}_{\text{load}} = \overline{V}_{\text{s}} - \bar{I}\,\overline{Z}_{\text{cable}} = 120 - (-j14.14)\,(j2) = 91.72\angle-0°\text{ V}$$

Before switching the motor, the voltage at the load side is equal to the source voltage. Hence, the voltage sag VS due to the start-up of the motor is

$$\text{VS} = \frac{\Delta V}{V_{\text{ss}}} = \frac{V_{\text{load}} - V_{\text{ss}}}{V_{\text{ss}}} = \frac{91.72 - 120}{120} = -23.6\%$$

The voltage at the load side has dropped by 23.6%. This high level of voltage drop will be felt by any load connected in parallel with this motor.

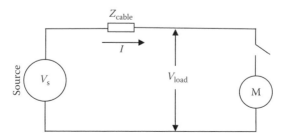

FIGURE 13.9 Motor connected to source via cable.

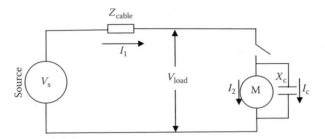

FIGURE 13.10 Motor with a starting capacitor.

Most of the methods used to remedy voltage sags are the same as the ones used for voltage flickers. We have discussed earlier the use of series capacitors to reduce the effect of a voltage flicker in Section 13.1.1. The same method can be used to reduce voltage sags. The capacitors can also be installed in parallel with the load.

Example 13.6

Repeat Example 13.5, but assume that a 6 Ω capacitor is added in parallel with the motor at starting.

Solution

Since the crest inrush current is inductive, we can compute the equivalent reactance X_m of the motor at the crest point of the inrush current when the voltage is 91.72 V.

$$\overline{X}_m = \frac{\overline{V}_{load}}{\overline{I}_{crest}} = \frac{91.72\angle 0°}{-j\left(\frac{20}{\sqrt{2}}\right)} = j6.5 \ \Omega$$

The total load reactance X_{load} is the parallel combination of the motor equivalent inductive reactance X_m and the capacitive reactance of the capacitor X_c.

$$\overline{X}_{load} = \frac{\overline{X}_m \overline{X}_c}{\overline{X}_m + \overline{X}_c} = \frac{j6.5(-j6)}{j0.5} = -j78 \ \Omega$$

The cable current at the crest of the inrush current is

$$\overline{I}_1 = \frac{\overline{V}_s}{\overline{X}_{load} + \overline{Z}_{cable}} = \frac{120\angle 0°}{-j78 + j2} = j1.58 \ \text{A}$$

Now, we can compute the load voltage

$$\overline{V}_{load} = \overline{I}_1 \overline{X}_{load} = j1.58(-j78) = 123.24 \ \text{V}$$

The voltage sag at the start-up of the motor with the capacitor is

$$VS = \frac{\Delta V}{V_{ss}} = \frac{V_{load} - V_{ss}}{V_{ss}} = \frac{123.24 - 120}{120} = 2.7\%$$

The voltage at the load side is actually increased by 2.7% at starting. A smaller increase in voltage can be obtained if the size of the capacitor is reduced. Compare this change in voltage with the 23.6% drop without the capacitor. Can you find the size of the capacitor that reduces the VS to just 1%?

Faults in power systems can also cause voltage sags. The faults could be initiated by a number of scenarios such as falling trees on distribution feeders or failing insulations of power cables. Consider

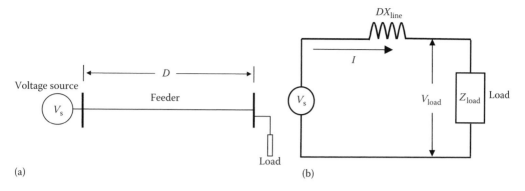

FIGURE 13.11   Representation of a simple power system. (a) One-line diagram and (b) circuit diagram.

the simple system in Figure 13.11. The system in Figure 13.11a consists of a voltage source whose magnitude is assumed constant, a feeder, and a load. The equivalent circuit of the system is shown in Figure 13.11b. V_s is the voltage of the source, X_{line} is the inductive reactance of the feeder per unit length, D is the length of the feeder, and Z_{load} is the impedance of the load. Assume that a fault occurrs at a distance d from the load and the impedance of the fault is Z_{fault}, as shown in Figure 13.12. Using Kirchhoff's equations, we can develop the following relationships:

$$\overline{V}_s = (D - d)\overline{X}_{line}\,\overline{I} + \overline{Z}_{fault}\,\overline{I}_{fault}$$
$$\overline{V}_s = (D - d)\overline{X}_{line}\,\overline{I} + (d\overline{X}_{line} + \overline{Z}_{load})\overline{I}_{load} \qquad (13.16)$$
$$\overline{I} = \overline{I}_{fault} + \overline{I}_{load}$$

Solving the above equations for the load current yields

$$\overline{I}_{load} = \frac{\overline{V}_s}{\overline{Z}_{load} + D\overline{X}_{line} + \left(\dfrac{(D-d)\overline{X}_{line}(\overline{Z}_{load} + d\overline{X}_{line})}{\overline{Z}_{fault}}\right)} \qquad (13.17)$$

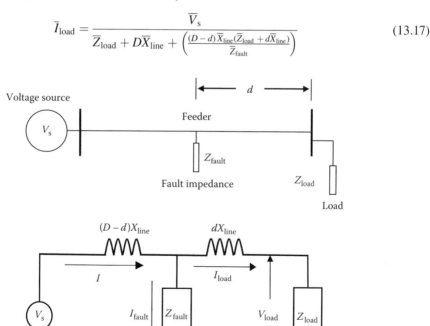

FIGURE 13.12   Representation of a simple power system with a fault between source and load.

Hence, the load voltage is

$$\overline{V}_{\text{load}} = \overline{I}_{\text{load}} \, \overline{Z}_{\text{load}} = \frac{\overline{V}_s \, \overline{Z}_{\text{load}}}{\overline{Z}_{\text{load}} + D\overline{X}_{\text{line}} + \left(\frac{(D-d)\overline{X}_{\text{line}}(\overline{Z}_{\text{load}} + d\overline{X}_{\text{line}})}{\overline{Z}_{\text{fault}}} \right)} \tag{13.18}$$

As seen in Equation 13.18, the load voltage is lower when the fault impedance Z_{fault} is smaller.

EXAMPLE 13.7

For the system in Figure 13.12, the source voltage is 67 kV (phase value), the feeder is 6 km in length, and its inductive reactance is 1.2 Ω/km. The impedance at the load side is $2\angle30°$ kΩ. At 4 km from the source side, a falling tree causes a temporary fault. The impedance of the tree is $100\angle75°$ Ω. Compute the voltage sag at the load side.

Solution

The first step is to compute the steady-state voltage V_{ss} at the load side before the fault. To do this, we can use Equation 13.18 when Z_{fault} is set to infinity.

$$\overline{V}_{\text{ss}} = \frac{\overline{V}_s \, \overline{Z}_{\text{load}}}{\overline{Z}_{\text{load}} + D\overline{X}_{\text{line}}} = \frac{67\angle0°(2000\angle30°)}{2000\angle30° + 6 \times 1.2\angle90°} = 66.88\angle-0.18° \text{ kV}$$

During the fault, the load voltage is

$$\overline{V}_{\text{load}} = \frac{\overline{V}_s \, \overline{Z}_{\text{load}}}{\overline{Z}_{\text{load}} + D\overline{X}_{\text{line}} + \left(\frac{(D-d)\overline{X}_{\text{line}}(\overline{Z}_{\text{load}} + d\overline{X}_{\text{line}})}{\overline{Z}_{\text{fault}}} \right)}$$

where $d = 6 - 4 = 2$ km
 Hence,

$$\overline{V}_{\text{load}} = \frac{67\angle0°(2000\angle30°)}{(2000\angle30°) + (6 \times 1.2\angle90°) + \left(\frac{4(1.2\angle90°)(2000\angle30° + 2 \times 1.2\angle90°)}{100\angle75°} \right)}$$

$$\overline{V}_{\text{load}} = 63.9\angle-0.85° \text{ kV}$$

The voltage sag at the load side during the fault is

$$\text{VS} = \frac{\Delta V}{V_{\text{ss}}} = \frac{V_{\text{load}} - V_{\text{ss}}}{V_{\text{ss}}} = \frac{63.9 - 67}{67} = -4.43\%$$

Notice that we assumed that the generator voltage is unchanged during the fault. In reality, the terminal voltage of the generator is reduced due to the actions of the various voltage regulators in the system as well as the changing internal impedance of the generator. Repeat the solution and assume the source impedance is 10 Ω.

EXAMPLE 13.8

For the system in Figure 13.13, the source voltage is 100 kV (phase value), the feeder reactance X_1 is 8 Ω, and X_2 is 5 Ω. The load impedance is $3\angle20°$ kΩ. A temporary fault occurs at the location in the figure. The fault impedance is $50\angle80°$ Ω. Compute the voltage sag at the load side. Assume the end of the feeder was unloaded before the fault.

Solution

The steady-state voltage V_{ss} across the load before the fault is

$$\overline{V}_{ss} = \frac{\overline{V}_s\,\overline{Z}_{load}}{\overline{Z}_{load} + \overline{X}_1} = \frac{100\angle 0°(3000\angle 20°)}{3000\angle 20° + 8\angle 90°} = 99.9\angle{-0.14}°\ \text{kV}$$

During the fault, you can easily compute the load current by writing the loop and nodal equations of the circuit in the matrix form.

$$\begin{bmatrix} \overline{V} \\ \overline{V} \\ 0 \end{bmatrix} = \begin{bmatrix} \overline{X}_1 & \overline{Z}_{load} & 0 \\ \overline{X}_1 & 0 & \overline{X}_2 + \overline{Z}_{fault} \\ -1 & 1 & 1 \end{bmatrix} \begin{bmatrix} \overline{I} \\ \overline{I}_{load} \\ \overline{I}_{fault} \end{bmatrix}$$

Hence,

$$\begin{bmatrix} \overline{I} \\ \overline{I}_{load} \\ \overline{I}_{fault} \end{bmatrix} = \begin{bmatrix} 8\angle 90° & 3000\angle 20° & 0 \\ 8\angle 90° & 0 & 5\angle 90° + 50\angle 80° \\ -1 & 1 & 1 \end{bmatrix}^{-1} \begin{bmatrix} 100\angle 0° \\ 100\angle 0° \\ 0 \end{bmatrix}$$

The solution of the above equation leads to the load current during the fault

$$\overline{I}_{load} = 27.13 - j10.56\ \text{A}$$

Then, the load voltage during the fault is

$$\overline{V}_{load} = \overline{Z}_{load}\,\overline{I}_{load} = 87.3\angle{-1.3}°\ \text{kV}$$

The voltage sag at the load side during the fault is

$$\text{VS} = \frac{\Delta V}{V_{ss}} = \frac{V_{load} - V_{ss}}{V_{ss}} = \frac{87.3 - 99.9}{99.9} = -12.6\%$$

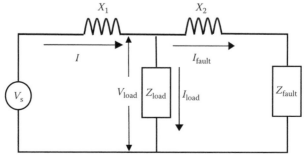

FIGURE 13.13   Representation of a simple power system with load between the fault and the source.

13.2 HARMONIC PROBLEMS

The ac waveform in its ideal shape is purely sinusoidal and its magnitude and frequency are constants. A waveform with any deviation from the ideal sinusoidal shape is considered distorted. An ideal waveform and an example of a distorted and periodic waveform are shown in Figure 13.14.

The reasons for the distortions of power system waveforms are many and include nonlinear loads and the switching of power electronic devices. But before we analyze these common conditions, let us review the basic mathematics of harmonic analysis.

Any periodic waveform g that is not purely sinusoidal has a string of components at different frequencies. These components are called harmonics and can be represented by the well-known Fourier series.

$$g = g_0 + g_1(\omega t) + g_2(2\omega t) + g_3(3\omega t) + g_4(4\omega t) + \cdots \tag{13.19}$$

where

   g is the instantaneous value of the periodic waveform
   g_0 is the dc component in g
   g_1 is the instantaneous value of the sinusoidal component at the fundamental frequency ω
   g_2 is the sinusoidal component at double the fundamental frequency (second harmonic)
   g_3 is the third harmonic component at triple the frequency of the fundamental component, and
      so on

In its general form, the kth harmonic component can be written as

$$g_k = a_k \cos(k\omega t) + b_k \sin(k\omega t) \tag{13.20}$$

where

$$a_k = \frac{1}{\pi} \int_0^{2\pi} g \cos(k\omega t)\, d\omega t$$

$$\tag{13.21}$$

$$b_k = \frac{1}{\pi} \int_0^{2\pi} g \sin(k\omega t)\, d\omega t$$

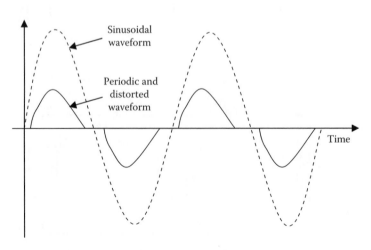

FIGURE 13.14 Sinusoidal and distorted waveforms.

When g is in certain shapes, we can simplify the harmonic analysis. For instance, if g is an even waveform where $g(t) = g(-t)$, the sine term vanishes from Equation 13.20. For odd waveforms, such as the one in Figure 13.14 where $g(t) = -g(-t)$, the cosine term vanishes. For most distorted waveforms in power networks, odd functions are adequate representation. In this case, a current i that is periodic and distorted can be written as

$$i(\omega t) = I_{dc} + I_{1\,max} \sin(\omega t) + I_{2\,max} \sin(2\omega t) + I_{3\,max} \sin(3\omega t) + \cdots \tag{13.22}$$

where
  I_{dc} is a dc component of i
  $I_{1\,max}$ is the peak value of the fundamental frequency component
  $I_{2\,max}$ is the peak value of the second harmonic current, and so on

The periodic current i in Equation 13.22 is hard to analyze in its instantaneous form. However, if we use rms quantities, we can simplify the analysis. The rms current is defined in Chapter 7 as

$$I = \sqrt{\frac{1}{2\pi} \int_0^{2\pi} [i(\omega t)]^2 \; d\omega t} \tag{13.23}$$

Hence, the rms current of a distorted waveform is

$$I = \sqrt{\frac{1}{2\pi} \int_0^{2\pi} [I_{dc} + I_{1\,max} \sin(\omega t) + I_{2\,max} \sin(2\omega t) + I_{3\,max} \sin(3\omega t) + \cdots]^2 \; d\omega t} \tag{13.24}$$

Before we continue with the analysis, you should know that the average value of cross frequency components (two sinusoidal waveforms with different frequencies) is zero, that is,

$$\frac{1}{2\pi} \int_0^{2\pi} \sin(\omega t) \sin(k\omega t) \; d\omega t = 0; \quad \text{if } k \neq 1 \tag{13.25}$$

Using the fact in Equation 13.25, we can ignore all cross frequency terms in Equation 13.24. Thus,

$$I = \sqrt{\frac{1}{2\pi} \left[\int_0^{2\pi} I_{dc}^2 \; d\omega t + \int_0^{2\pi} (I_{1\,max} \sin(\omega t))^2 \; d\omega t + \int_0^{2\pi} (I_{2\,max} \sin(2\omega t))^2 \; d\omega t + \cdots \right]} \tag{13.26}$$

Keep in mind that the rms of any current component is

$$I_k = \sqrt{\frac{1}{2\pi} \left[\int_0^{2\pi} (I_{k\,max} \sin(k\omega t))^2 \; d\omega t \right]} \tag{13.27}$$

Then, the rms current of a distorted waveform is

$$I = \sqrt{I_{dc}^2 + I_1^2 + I_2^2 + I_3^2 + \cdots} \tag{13.28}$$

To quantify the distortion of a periodic function, two indices are often used: individual harmonic distortion (IHD) and total harmonic distortion (THD). They are defined as follows:

$$\text{IHD}_k = \frac{I_k}{I_1}$$

$$\text{THD} = \frac{I_H}{I_1}$$

(13.29)

where I_H is the harmonic current, which is the rms value of all harmonic components except the fundamental.

$$I_H = \sqrt{I_{dc}^2 + I_2^2 + I_3^2 + \cdots}$$

(13.30)

The IHD_k is a measure of the contribution of the individual kth harmonic with respect to the fundamental component of the waveform. The THD is a measure of the total distortion in the waveform.

The THD has several other forms:

$$\text{THD} = \frac{\sqrt{I_{dc}^2 + I_2^2 + I_3^2 + \cdots}}{I_1} = \sqrt{\frac{I_{dc}^2}{I_1^2} + \frac{I_2^2}{I_1^2} + \frac{I_3^2}{I_1^2} + \cdots}$$

$$\text{THD} = \sqrt{\text{IHD}_{dc}^2 + \text{IHD}_2^2 + \text{IHD}_3^2 + \text{IHD}_4^2 + \cdots}$$

(13.31)

EXAMPLE 13.9

A distorted current waveform has four harmonic components represented by their rms values: the fundamental component is 50 A, the third harmonic component is 20 A, the fifth harmonic component is 5 A, and the seventh harmonic component is 2 A. Compute the following:

(a) rms current of the distorted waveform
(b) IHD of the fifth harmonic
(c) THD

Solution

(a) Use Equation 13.28 to compute the rms current of the distorted waveform.

$$I = \sqrt{I_1^2 + I_3^2 + I_5^2 + I_7^2} = \sqrt{50^2 + 20^2 + 5^2 + 2^2} = 54.12 \text{ A}$$

(b) Use Equation 13.29 to compute IHD_5.

$$\text{IHD}_5 = \frac{I_5}{I_1} = \frac{5}{50} = 10\%$$

The above result reveals us that the fifth harmonic current is 10% of the fundamental component current.

(c) Use Equation 13.30 to compute the harmonic current.

$$I_H = \sqrt{20^2 + 5^2 + 2^2} = 20.71 \text{ A}$$

The THD is

$$\text{THD} = \frac{I_H}{I_1} = \frac{20.71}{50} = 41.42\%$$

This waveform contains 41.42% of its magnitude as harmonics.

13.2.1 HARMONIC DISTORTION OF ELECTRIC LOADS

The main question is why harmonics exist in a power system. If the source voltage is purely sinusoidal without any harmonics and the load impedance is linear, the current of the load is also sinusoidal and free of harmonics. However, if the load is nonlinear, the load current is distorted as shown in the following example.

EXAMPLE 13.10

A 120 V (rms) purely sinusoidal voltage source is feeding a nonlinear resistance. Assume that the value of the resistance is dependent on the applied voltage and can be expressed by

$$R = 10 + e^{\frac{v^2}{10^4}} \ \Omega$$

Draw the current waveform.

Solution

The current of the load is

$$i = \frac{v}{R} = \frac{V_{max} \sin(\omega t)}{10 + \exp\left[V_{max}^2 \sin^2(\omega t)/10^4\right]} = \frac{169.7 \sin(\omega t)}{10 + \exp\left[169.7^2 \sin^2(\omega t)/10^4\right]}$$

Using simulation software, you can obtain the current waveform in Figure 13.15.

Notice that the current seems to have two positive peaks and two negative peaks in one cycle. This is a typical shape of a waveform with a large third harmonic component. Draw a fundamental frequency waveform and a third harmonic component on the same graph; then sketch the sum of the two waveforms.

The distortion of the load current can result in distorted voltage across any other load connected to the same service circuit. To explain this important problem, consider the system in Figure 13.16. Assume that the voltage source is purely sinusoidal and the load is switched by a power electronic

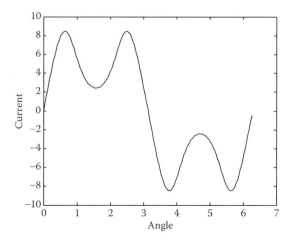

FIGURE 13.15 Current waveform of the nonlinear resistance in Example 13.10.

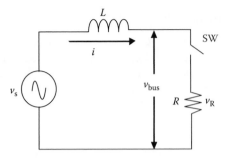

FIGURE 13.16 Representation of a simple system consisting of source, cable, and switching load.

converter represented in the figure by the switch SW. This type of load includes household equipments such as computers, televisions, furnaces, washers, dryers, air conditioners, dimmers, and fans. However, since these household loads are relatively small with respect to the total loads on the utility feeders, their contribution to the THD is negligible. However, industrial loads such as arc furnaces and adjustable speed drives are large enough to produce significant harmonic distortions in the power grid. This results in distorted voltage waveforms at the load end of the feeder.

In Figure 13.16, the source voltage is v_s, the voltage across the resistance is v_R, and the voltage at the load end of the feeder is the bus voltage v_{bus}. Assume that the load is purely resistive and the cable (feeder) is purely inductive. When the switch SW is closed, the loop equation is

$$v_s = iR + L\left(\frac{di}{dt}\right) \tag{13.32}$$

where i is the current of the load. When the switch is open, the current in the circuit is zero. Since the status of the current is dependent on the switching status of SW, the current in Equation 13.32 can be computed much easier if we use the frequency domain. In this case, we can rewrite Equation 13.32 in the Laplace form.

$$V_s(s) = R\,I(s) + sL\,I(s) \tag{13.33}$$

where $V(s)$ and $I(s)$ are the Laplace-transformed voltage and current, respectively. Hence,

$$I(s) = \frac{V_s(s)}{(R + sL)} \tag{13.34}$$

Keep in mind that the above equation is valid only when the switch is closed. When the switch is open, the current is zero. Assume that the voltage source is a purely sinusoidal waveform.

$$v_s = V_{max} \sin(\omega t) \tag{13.35}$$

Also, assume that the switch is closed at angle α and opened at an angle β. In this case, the source voltage in the Laplace form is

$$V_s(s) = \mathcal{L}v_s$$
$$V_s(s) = V_{max}\left[\frac{s \sin\alpha + \omega \cos\alpha}{s^2 + \omega^2}e^{\left(\frac{-\alpha s}{\omega}\right)} - \frac{s \sin\beta + \omega \cos\beta}{s^2 + \omega^2}e^{\left(\frac{-\beta s}{\omega}\right)}\right] \tag{13.36}$$

Substituting Equation 13.36 into 13.34 and inverting the Laplace form back into time domain yield the current at the time the switch SW is closed.

$$i = \frac{V_{max}}{z}\left[\sin(\omega t - \phi) + [\sin(\phi - \alpha)]e^{-\left(\frac{\omega t - \alpha}{\omega \tau}\right)}\right] \qquad (13.37)$$

where z, ϕ, and τ are the total impedance, the phase angle of the impedance, and the time constant of the system impedance, respectively. They are defined as follows:

$$z = \sqrt{R^2 + (\omega L)^2}$$
$$\phi = \tan^{-1}\left(\frac{\omega L}{R}\right) \qquad (13.38)$$
$$\tau = \frac{L}{R}$$

Notice that the current is not purely sinusoidal anymore due to the second term in Equation 13.37. Similarly, the voltage across the load resistance is distorted as shown in Equation 13.39.

$$v_R = Ri = \frac{RV_{max}}{z}\left[\sin(\omega t - \phi) + [\sin(\phi - \alpha)]e^{-\left(\frac{\omega t - \alpha}{\omega \tau}\right)}\right] \qquad (13.39)$$

Now let us find out the bus voltage v_{bus}, which is the voltage at the load end of the feeder. When the switch SW is closed, the bus voltage is equal to the voltage across the load resistance v_R. However, when the switch is opened, the bus voltage is equal to the source voltage v_s.

The waveforms of the source voltage, current, and bus voltage are shown in Figure 13.17. Notice the distorted shape of the bus voltage, which does not resemble the pure sinusoidal waveform of the source voltage. This bus voltage is rich with harmonics. Therefore, when another customer is connected to the same bus, the distorted voltage is expected to be across that customer's load.

Different loads produce different harmonic distortions according to their operating environment or their power electronic designs. Table 13.1 shows a typical range of THD for some common loads. The THD of computers and monitors is very large. However, because their currents are small, their impact on the total system harmonics is relatively small. Ironically, some electrical contractors tend to power all computers in major installations by a dedicated feeder thinking that they protect the computers from any harmonic that might exist if other equipment is allowed to be powered from the same feeder. In fact, these dedicated feeders are highly polluted with harmonics due to the

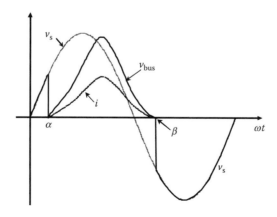

FIGURE 13.17 Waveforms of the circuit in Figure 13.16.

TABLE 13.1
THD of Common Load Currents

Type of Load	Typical THD (%)
Fluorescent lamps	15–25
Adjustable speed drives	30–80
Personal computers	70–120
Monitors	60–120

aggregation of the harmonics of all computers. In such a feeder, the THD is actually very high. A better system is to allow other loads such as heaters and lights to be connected to the same feeder that powers the computers. This important point is further explained in Example 13.11.

EXAMPLE **13.11**

A feeder is powering 100 computers. The total current of all computers can be represented by

$$i = 4 + 50\sin(2\pi 60t) + 30\sin(2\pi 180t) + 10\sin(2\pi 300t) + 5\sin(2\pi 420t) \text{ A}$$

Compute the THD of the feeder. Then, assume that a linear heating load of 100 A (rms) is connected to the same feeder; compute the new THD of the feeder.

Solution

First, compute the rms values of the harmonic components.

$$I_{dc} = 4 \text{ A}$$
$$I_1 = \frac{50}{\sqrt{2}} = 35.36 \text{ A}$$
$$I_3 = \frac{30}{\sqrt{2}} = 21.21 \text{ A}$$
$$I_5 = \frac{10}{\sqrt{2}} = 7.07 \text{ A}$$
$$I_7 = \frac{5}{\sqrt{2}} = 3.54 \text{ A}$$

Use Equation 13.30 to compute the rms value of the harmonic current

$$I_H = \sqrt{4^2 + 21.21^2 + 7.07^2 + 3.54^2} = 23 \text{ A}$$

The THD is then

$$\text{THD} = \frac{I_H}{I_1} = \frac{23}{35.36} = 65\%$$

When the heating load is added, the fundamental component of the total current is the sum of the fundamental components of the computers' current and the current of the heating load. If we assume no phase shift between these two currents, then

$$I_1 = \frac{50}{\sqrt{2}} + 100 = 135.36 \text{ A}$$

The harmonic current when the heating load is added is unchanged as we assume that the heating load is free from harmonics besides the fundamental

$$I_H = \sqrt{4^2 + 21.21^2 + 7.07^2 + 3.54^2} = 23 \text{ A}$$

The THD with the heating load is

$$\text{THD} = \frac{I_H}{I_1} = \frac{23}{135.36} = 17\%$$

Notice that the same feeder now has a lower THD because the heating load increases the fundamental component of the total current.

Harmonic distortions can cause a wide range of problems in power grids as well as to other customers connected to the same feeders. These problems include the following:

- Resonance in power system that could stress the components beyond their designed ratings
- Increasing losses in transmission lines, transformers, and generators
- Damaging to various compensation capacitor banks
- Creating drag forces inside motors that overheat and reducing their torques
- Reducing the overall power factor (pf) of the system and producing inaccurate power measurements
- Increasing the electromagnetic interference to communication networks
- Premature aging of insulators in equipment and cables
- Causing picture jitters, and freezing or rebooting of computers and sensitive equipment

In the remainder of this chapter, we shall discuss some of the most important problems associated with harmonic distortions.

13.2.2 RESONANCE DUE TO HARMONICS

Capacitors are often used in power systems to improve the pf and the voltage profile of the grid. These capacitors together with the inductances that exist in loads and feeders have a resonance frequency that is, by design, far away from the steady-state frequency of the system (50 or 60 Hz). However, if a harmonic component exists with a frequency near the resonance frequency of the system, the current and voltage could oscillate at excessive magnitudes and could damage the power equipment.

Consider the system in Figure 13.18. It consists of voltage source v_s, cable impedance of inductance L_{cable}, and a compensated industrial load. The load consists of an inductive reactance L and a resistance. But for simplicity, we shall assume that the resistance is small with respect to the

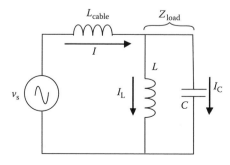

FIGURE 13.18 Compensated load.

inductive reactance at the high frequency of the harmonics. The load is also compensated by a capacitor bank C.

The total load impedance Z_{load} is the parallel combination of the inductive reactance of the load and the compensation capacitor.

$$\overline{Z}_{load} = \frac{j\omega L \left(\frac{1}{j\omega C} \right)}{j\omega L + \left(\frac{1}{j\omega C} \right)} = \frac{j\omega L}{1 - \omega^2 LC} \tag{13.40}$$

The voltage across the load is then

$$\overline{V}_{load} = \overline{Z}_{load} \, \overline{I} \tag{13.41}$$

and the current of the circuit is

$$\overline{I} = \frac{\overline{V}_s}{\overline{Z}_{load} + \overline{Z}_{cable}} \tag{13.42}$$

where

$$\overline{Z}_{cable} = j\omega L_{cable} \tag{13.43}$$

Hence,

$$\overline{V}_{load} = \left(\frac{\overline{Z}_{load}}{\overline{Z}_{load} + \overline{Z}_{cable}} \right) \overline{V}_s \tag{13.44}$$

Substituting the impedances in Equations 13.40 and 13.43 into Equation 13.44 yields

$$\overline{V}_{load} = \left[\frac{L}{(L + L_{cable}) - (LL_{cable}C)\omega^2} \right] \overline{V}_s \tag{13.45}$$

As you see in Equation 13.45, the voltage across the load is a function of the frequency of source voltage v_s. This load voltage can reach a theoretical value of infinity if a frequency ω_o makes the denominator equal zero.

$$(L + L_{cable}) - (LL_{cable} \, C)\,\omega_o^2 = 0 \tag{13.46}$$

At this condition, the frequency is called resonance frequency f_o or ω_o. They are defined as

$$\omega_o = \sqrt{\frac{L + L_{cable}}{LL_{cable} \, C}}$$

$$f_o = \frac{1}{2\pi} \sqrt{\frac{L + L_{cable}}{LL_{cable} \, C}} \tag{13.47}$$

EXAMPLE 13.12

For the compensated load in Figure 13.18, assume that the load impedance is composed of a 10 Ω resistance in series with a 0.1 H inductance. The transmission line inductance is 0.01 H and the compensation capacitor is 100 μF. The voltage at the source side has a third harmonic component of 100 V at 180 Hz. Compute the third harmonic voltage across the load.

Solution

The total impedance on the load side is

$$\overline{Z}_{\text{load}} = \frac{(R + j\omega L)\left(\frac{1}{j\omega C}\right)}{(R + j\omega L) + \left(\frac{1}{j\omega C}\right)} = \frac{(10 + j0.1\omega)\left(\frac{10^4}{j\omega}\right)}{(10 + j0.1\omega) + \left(\frac{10^4}{j\omega}\right)}$$

where

$$\omega = 2\pi f = 2\pi(3 \times 60) = 1131 \text{ rad/s}$$

Hence, the total impedance at the load side is

$$\overline{Z}_{\text{load}} = 0.0713 - j9.58 \ \Omega$$

The source current of the system is

$$\overline{I} = \frac{\overline{V}_s}{\overline{Z}_{\text{load}} + \overline{Z}_{\text{cable}}} = \frac{100}{0.0713 - j9.58 + j11.31} = 2.39 - j57.9 \text{ A}$$

Hence, the voltage across the load is

$$\overline{V}_{\text{load}} = \overline{Z}_{\text{load}} \, \overline{I} = (0.0713 - j9.58)(2.39 - j57.9) = 555.2\angle{-177.2}° \text{ V}$$

Notice that the load voltage is more than five times the third harmonic source voltage. This high voltage may cause damage to equipment.

13.2.3 Effect of Harmonics on Transmission Lines and Cables

There are two main transmission line problems associated with system harmonics: losses and excessive voltage. The harmonic currents produce line loss P_{line} in the form

$$P_{\text{line}} = \sum_{k} RI_k^2 \tag{13.48}$$

where
   k is the harmonic number
   R is the resistance of the line

These losses increase when the harmonic is of high magnitudes.

   The second problem, the high voltage across lines and cables, is much more damaging than line losses. If the harmonic is close to the resonant frequency, the voltage across the cable or line is likely to be excessive and could damage cable insulations or even cause arcing between lines and nearby grounded objects. Examine Example 13.13.

Example 13.13

For the system in Example 13.12, compute the voltage across the feeder impedance.

Solution

The voltage across the line impedance V_{line} is

$$\overline{V}_{\text{line}} = \overline{Z}_{\text{cable}} \, \overline{I} = j11.31(2.39 - j57.9) = 655.2\angle{2.4}° \text{ V}$$

The load across the line impedance is more than six times the source voltage at 180 Hz. This high voltage could damage the insulations of the cable.

13.2.4   Effect of Harmonics on Capacitor Banks

The most important problem for capacitors operating in a harmonic environment is the overvoltage across them. As seen in Example 13.12, the voltage across the capacitor, which is the same as the load voltage, could be excessive. This high voltage causes high current in the capacitor since

$$I_{capacitor} = \omega C V_{load} \tag{13.49}$$

When the harmonic frequency in Equation 13.49 is high and the load voltage for this harmonic component is also high, the capacitor current could be excessive and could damage the often expensive capacitor banks.

Example 13.14

For the system in Example 13.12, compute the capacitor current at 60 Hz and at the third harmonic voltage. Assume that the fundamental voltage is 240 V and the third harmonic voltage is 100 V.

Solution

At 60 Hz, the load impedance is

$$\overline{Z}_{load} = \frac{(R + j\omega L)\left(\frac{1}{j\omega C}\right)}{(R + j\omega L) + \left(\frac{1}{j\omega C}\right)} = \frac{(10 + j0.1 \times 377)\left(\frac{10^4}{j377}\right)}{(10 + j0.1 \times 377) + \left(\frac{10^4}{j377}\right)} = 31.3 + j61.5 \ \Omega$$

The 60 Hz current of the source is

$$\overline{I} = \frac{\overline{V}_s}{\overline{Z}_{load} + \overline{Z}_{cable}} = \frac{240}{10.0 + j37.7 + j3.77} = 1.74 + j3.21 \ \text{A}$$

Hence, the voltage across the load is

$$\overline{V}_{load} = \overline{Z}_{load} \ \overline{I} = 252.2\angle{-1.5°} \ \text{V}$$

and the current of the capacitor is

$$I_{capacitor} = \omega C V_{load} = 377 \times 10^{-4} \times 252.2 = 9.5 \ \text{A}$$

Now repeat the solution for the third harmonic. In Example 13.12, the load voltage is 555.1 V. Hence, the current of the capacitor at 180 Hz is

$$I_{capacitor} = \omega C V_{load} = (3 \times 377) \times 10^{-4} \times 555.1 = 62.8 \ \text{A}$$

This is almost 6.6 times the current at 60 Hz. Keep in mind that most capacitors are rated for the harmonic-free environment. In this case, it is expected that this capacitor would fail due to the excessive voltage across it and the excessive current through it. The excessive voltage damages the insulations inside the capacitor and the excessive current causes damaging internal heat.

13.2.5   Effect of Harmonics on Electric Machines

Harmonics can cause a variety of problems to electrical machines including harmonic losses, excessive voltage, and drag torque (braking torque). The losses inside the machine are due to the losses in its windings and core. As shown in Chapter 11, the magnetic core loss is a function of how many times the hysteresis loop is completed; one loop is completed in one cycle. Hence, the higher the frequency of the harmonic, the higher is the core loss. In addition, the various resistances of the wires of the windings increase with the frequency due to the skin effect phenomenon, thus increasing the losses inside the machine.

If the machine is compensated by a capacitor bank, the frequency of the harmonic could cause the voltage of the machine to increase to damaging levels, as explained in Section 13.2.4.

Drag torque is related to the fact that the rotating field inside the machine is a function of the sequence of the three-phase waveform as explained in Chapter 12. At 60th cycle, the rotation of the magnetic field in the airgap follows the abc sequence where

$$v_a = V_{max} \sin(\omega t)$$
$$v_b = V_{max} \sin(\omega t - 120)$$
$$v_c = V_{max} \sin(\omega t + 120)$$

(13.50)

In Equation 13.50, phase b lags phase a by 120°, and phase c leads phase a by 120°. Therefore, the magnetic field rotates in the abc sequence.

Now let us look at the fifth harmonic waveforms. The frequency of the fifth harmonic modifies Equation 13.50 to

$$v_a = V_{max} \sin 5(\omega t)$$
$$v_b = V_{max} \sin 5(\omega t - 120)$$
$$v_c = V_{max} \sin 5(\omega t + 120)$$

(13.51)

Hence,

$$v_a = V_{max} \sin(5\omega t)$$
$$v_b = V_{max} \sin(5\omega t + 120)$$
$$v_c = V_{max} \sin(5\omega t - 120)$$

(13.52)

In Equation 13.52, phase b leads phase a by 120°, and phase c lags phase a by 120°. So the magnetic field rotates in the acb sequence, which is opposite to the abc sequence of the fundamental frequency voltage. Therefore, the magnetic field rotation due to the fifth harmonic opposes the main magnetic field of the fundamental frequency. The torque produced by the fifth harmonic in this case is in opposite direction to the torque produced by the fundamental frequency. This fifth harmonic torque is therefore a dragging torque that causes the machine to run sluggishly, much like two people pushing a cart from opposite directions.

13.2.6 Effect of Harmonics on Electric Power

The instantaneous power is defined as

$$p(t) = v(t)\, i(t)$$

(13.53)

where
  v is the instantaneous voltage
  i is the instantaneous current
  p is the instantaneous power

The active power P that produces work is the average value of the instantaneous power.

$$P = \frac{1}{\tau} \int_0^\tau p(t)\, dt = \frac{1}{2\pi} \int_0^{2\pi} p(t)\, d\omega t \tag{13.54}$$

where τ is the time period. Assume that the voltage and current have different frequencies.

$$\begin{aligned} v &= V_{max} \sin(\omega t) \\ i &= I_{max} \sin(n\omega t + \theta) \end{aligned} \tag{13.55}$$

where

n is a positive integer

θ is the phase shift of the current with respect to the voltage

The power P can be computed by

$$P = \frac{1}{2\pi} \int_0^{2\pi} V_{max} \sin(\omega t) I_{max} \sin(n\omega t + \theta)\, d\omega t$$

$$P = \frac{V_{max} I_{max}}{2\pi} \left[\int_0^{2\pi} \cos((n+1)\omega t + \theta)\, d\omega t + \int_0^{2\pi} \cos((n-1)\omega t + \theta)\, d\omega t \right] \tag{13.56}$$

$$P = \frac{V_{max} I_{max}}{2\pi} \left[\int_0^{2\pi} A\, d\omega t + \int_0^{2\pi} B\, d\omega t \right]$$

The integral of A is always zero regardless of the value of n, while the integral of B is nonzero only if $n = 1$. This leads us to an important lemma for power: only voltages and currents of same frequency produce power.

To apply this lemma, let us assume that the voltage and current have harmonic components.

$$\begin{aligned} v &= V_{dc} + V_{1\,max} \sin \omega t + V_{1\,max} \sin 2\omega t + V_{3\,max} \sin 3\omega t + \cdots \\ i &= I_{dc} + I_{1\,max} \sin(\omega t - \theta_1) + I_{2\,max} \sin(2\omega t - \theta_2) + I_{3\,max} \sin(3\omega t - \theta_3) + \cdots \end{aligned} \tag{13.57}$$

According to the lemma, the power produced by these waveforms is

$$\begin{aligned} P &= V_{dc} I_{dc} + \frac{V_{1\,max} I_{1\,max}}{2} \cos\theta_1 + \frac{V_{2\,max} I_{2\,max}}{2} \cos\theta_2 + \frac{V_{3\,max} I_{3\,max}}{2} \cos\theta_3 + \cdots \\ P &= V_{dc} I_{dc} + V_1 I_1 \cos\theta_1 + V_2 I_2 \cos\theta_2 + V_3 I_3 \cos\theta_3 + \cdots \\ P &= P_{dc} + P_1 + P_2 + P_3 + \cdots \end{aligned} \tag{13.58}$$

where

P_{dc} is the power of the dc component

P_1 is the power of the fundamental frequency

P_3 is the power of the third harmonic, and so on

An important term often used with power is pf. It is a good indicator of how much of the apparent power is converted into work. For purely sinusoidal waveforms, the maximum value of pf is 1 when the load is purely resistive.

In circuits with harmonics, the pf is defined in a similar way.

$$pf = \frac{P}{VI} = \frac{P_{dc} + P_1 + P_2 + P_3 + \cdots}{\sqrt{\left(V_{dc}^2 + V_1^2 + V_2^2 + V_3^2 + \cdots\right)\left(I_{dc}^2 + I_1^2 + I_2^2 + I_3^2 + \cdots\right)}} \qquad (13.59)$$

Using the definition of the THD in Equation 13.31, we can rewrite Equation 13.59 as

$$pf = \frac{P_{dc} + P_1 + P_2 + P_3 + \cdots}{V_1 I_1 \sqrt{\left(1 + THD_V^2\right)\left(1 + THD_I^2\right)}} = \frac{P_{dc} + P_1 + P_2 + P_3 + \cdots}{P_1 \sqrt{\left(1 + THD_V^2\right)\left(1 + THD_I^2\right)}} \qquad (13.60)$$

where
THD_V is the THD of the voltage waveform
THD_I is the THD of the current

Regular power measurement devices and pf meters are designed to produce accurate results in harmonic-free environments. The readings of these devices are often inaccurate when the system is polluted with severe harmonics. These devices are extensively installed in distribution systems all over the world. Replacing these meters with more sophisticated devices that account for the harmonics is extremely expensive. Most utilities opt for inaccurate readings instead of the prohibitive costs of upgrading the millions of meters in their system.

EXAMPLE 13.15

An industrial load has the following rms harmonic components:

	V	I	θ
dc component	10	1	
First harmonic	460	10	30°
Third harmonic	90	5	0°
Fifth harmonic	30	2	10°

(a) Assume that simple devices are used to measure the power and the pf. Assume that the devices monitor only the fundamental components. What are their readings?
(b) What are the actual values of power and pf?
(c) If the price of electricity is \$0.2/kW h, compute the losses for the utility in 1 year assuming that the load is active for 10 h daily.

Solution

(a) The power meter measures the fundamental frequency component only

$$P_1 = V_1 I_1 \cos\theta_1 = 460 \times 10 \times \cos 30 = 4 \text{ kW}$$

$$pf_1 = \frac{P_1}{V_1 I_1} = \cos 30 = 0.866$$

(b) The actual value of power is

$$P = V_{dc} I_{dc} + V_1 I_1 \cos\theta_1 + V_3 I_3 \cos\theta_3 + V_5 I_5 \cos\theta_5$$

$$P = (10 \times 1) + (460 \times 10 \times \cos 30) + (90 \times 5) + (30 \times 2 \times \cos 10) = 4.52 \text{ kW}$$

The actual value of pf is

$$pf = \frac{P}{VI} = \frac{P_{dc} + P_1 + P_2 + P_3 + \cdots}{\sqrt{(V_{dc}^2 + V_1^2 + V_2^2 + V_3^2 + \cdots)(I_{dc}^2 + I_1^2 + I_2^2 + I_3^2 + \cdots)}}$$

$$pf = \frac{4520}{\sqrt{(10^2 + 460^2 + 90^2 + 30^2)(1 + 10^2 + 5^2 + 2^2)}} = 0.844$$

(c) The loss in measurements due to harmonics

$$P_{harmonics} = P - P_1 = 4520 - 4000 = 520 \text{ W}$$

The annual energy unmonitored due to the harmonics is

$$E_{harmonics} = P_{harmonics}\, t = 520 \times 10 \times 365 = 1898 \text{ kW h}$$

The annual lost revenue of the utility is

$$\text{Lost revenue} = E_{harmonics} \times \text{cost of kW h} = 1825 \times \$0.2 = \$379.6$$

The lost revenue for this one load is probably high enough to justify the installation of more accurate meters.

13.2.7 Effect of Harmonics on Communications

Any wire carrying electrical current produces a magnetic field surrounding the wire. When the magnetic fields reach other wires or electronic circuits, they induce voltages that could interfere with the operation of circuits. Devices such as cellular phones or landline phones are designed to filter out the effect of power frequency. However, high-frequency harmonic components may not be filtered out and could interfere with communication signals.

EXERCISES

1. For the system in Figure 13.2, assume $V_s = 120$ V, $Z_{cable} = 2\angle 90°$ Ω, $Z_1 = 10 + j5$ Ω, and $Z_2 = 2 + j1$ Ω. Compute the following:
   (a) Load voltage when $load_2$ is switched off
   (b) Load voltage when $load_2$ is switched on
   (c) Voltage reduction on $load_1$ due to the presence of $load_2$
2. Repeat the previous problem but assume the line is fully compensated by a series capacitor.
3. A load voltage with flicker can be represented by the following equation:

$$v_{load} = 170[1 + 2\cos(0.2t)]\cos(377t)$$

   Compute the following:
   (a) Flicker factor
   (b) VF
   (c) Frequency of the fluctuation
   (d) Human tolerance to the flicker
4. A transformer is energized through a cable whose inductive reactance is 3 Ω. The phase voltage at the source side of the cable is 8 kV. When the transformer is switched on (energized), the crest of the inrush current of the transformer can be modeled by an equivalent inductive reactance of 7 Ω. Estimate the voltage flicker at the transformer side.

5. For the system in Figure 13.12, the source voltage is 100 kV (phase value), the feeder is 20 km in length and its inductive reactance is 1.2 Ω/km. The impedance at the load side is $2\angle 20°$ kΩ. A falling tree causes a temporary fault at the middle of the feeder. The fault impedance is $80\angle 75°$ Ω. Assume the equivalent internal impedance of the source is 50 Ω. Compute the voltage sag at the load side.

6. For the system in Figure 13.13, the source voltage is 50 kV (phase value), the feeder reactive reactance X_1 is 10 Ω, and X_2 is 5 Ω. The load impedance is $3\angle 20°$ kΩ. A temporary fault occurs at the location in the figure. The fault impedance is $200\angle 80°$ Ω. Compute the voltage sag at the load side. Assume the end of the feeder was unloaded before the fault.

7. Consider the current waveform:

$$i = 2 + 100 \sin(377t) + 20 \sin(1131t) + 10 \sin(1885t) + 5 \sin(2639t) \text{ A}$$

Compute the following:
   (a) Fundamental frequency
   (b) Harmonic frequencies
   (c) rms harmonic current
   (d) Total rms current
   (e) IHD of all harmonics
   (f) THD of the current

8. A harmonic voltage source is represented by

$$v_s = 340 \sin(377t) + 100 \sin(1131t) + 30 \sin(1885t) \text{ V}$$

The voltage source is connected to a load impedance of $Z_1 = (5 + j0.2\omega)$ Ω through a cable whose inductive reactance is $Z_{cable} = j0.01\omega$ Ω. The load is compensated by a 200 μF capacitor bank connected in parallel to the load. Compute the voltage across the load and the voltage across the cable impedance.

9. Compute the capacitor current in the previous problem.

10. For the compensated load in Figure 13.18, assume the load impedance is composed of a 10 Ω resistance in series with a 0.1 H inductance. The transmission line inductance is 0.01 H and the compensation capacitor is 100 μF. The source voltage contains several harmonics. Use any simulation software and find the frequency at which the voltage across the load is at its maximum value.

11. An industrial load is powered by a purely sinusoidal voltage source. The load current has a THD of 20%, and the pf of the fundamental waveforms is 0.8. Compute the pf of the load.

12. An industrial load is powered by a purely sinusoidal voltage source of 277 V (rms). The load current has a THD of 20%, and the fundamental component of the current is $20\angle 30°$ A. Compute the load power.

13. An industrial load is powered by a voltage source whose fundamental component is 277 V (rms) and its THD is 10%. The load current has a fundamental component of 30 A and a THD of 20%. Assume no phase shift between the current components and their corresponding voltages. Compute the load power.

14 Power Grid and Blackouts

Electric power systems are probably the most complex systems ever built by man. In the United States, the power system contains several thousands of major generating units, tens of thousands of transmission lines, millions of transformers, and hundreds of millions of protection and control devices. Figure 14.1 shows the electric energy produced in the United States and worldwide. The power systems in 2006 produced over 19 PW h (19×10^{15} W h) of electrical energy worldwide and over 4.25 PW h in the United States. The worldwide consumption, although already overwhelming, is expected to increase even more primarily because of the rapid industrial development in Asia. To match the ever-growing demand for electrical energy, the capacity of power systems must continuously increase. In 2006, the United States had a generation capacity of almost 1 TW as shown in Figure 14.2. If continuously operating at their limits, all generators in the United States could have produced about 7 PW h in 2006.

Managing this staggering amount of energy is a daunting task indeed for engineers. It is remarkable that power engineers maintain the stable operation of these immense and complex power systems. Given the size and dynamic nature of the various components in the systems, it is incredible that only a few outages occur every year.

Power outages are often called blackouts where loads in a given area are left without power. Blackouts, in general, occur due to the lack of power generation or the lack of transmission lines. The questions most people ask are whether we can predict the blackouts ahead of time, and whether we can prevent them from happening? These difficult questions have no general answers for all blackout scenarios. Although it is technically feasible to build a very robust power system by constructing a large number of distributed generations and multiple redundant transmission lines, the technical solution may not be cost-effective and is often unacceptable to the public. The public views toward building new generations and new transmission lines are generally negative. Utilities often face a great deal of difficulties when expansions are proposed to meet new growth. The public, in general, prefers using renewable energy and conservation instead of building new fossil fuel or nuclear power plants near their towns. Furthermore, the utilities often face tremendous resistance when proposing new transmission lines, especially when the right of way is located near residential areas.

Renewable energy and conservation are often stated as ways to reduce the frequency of blackouts. Given the current status of the technology, these ideas cannot really save power systems from blackouts as explained in the following discussion:

- Besides hydrogenation, renewable energy is not yet a viable option to replace the large power plants (fossil, hydroelectric, and nuclear) we use today. However, in 10–20 years, the technology may provide us with high energy density renewable systems that can replace them.
- Conservation will help tremendously if it is implemented on large scales and conservation devices are imbedded in the designs of electrical equipment. Relying on individuals to reduce their energy consumptions has been proven less effective because most people buy comfort rather than electricity. Most of us do not think in terms of how many kilowatt hour we consume to make the home cool or warm. Rather, we set the temperature controllers to our comfort levels, and we let electricity flow to reach these levels.

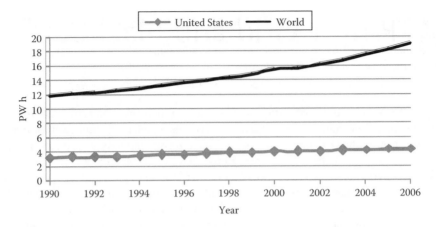

FIGURE 14.1   Net generation of electrical energy. (From the British Petroleum Annual Report, 2007.)

- Besides renewable energy and conservation, blackouts can have lesser impact if distributed generations are used. Small generating units can be installed to provide electricity to an individual building or a group of buildings either continuously or as backup systems. These generating units have different design configurations ranging from a small natural gas power plant—the size of a household water heater (power plant in a box)—to large gas or renewable systems. The main obstacle facing these systems is their high costs, which can be justified only for sensitive loads such as hospitals.

Another question that is often asked after every blackout is why power system cannot be controlled as we control phone and Internet networks. After all, they are similarly large and complex. Actually, there are several reasons that make the control of power systems unique and different from all other large and complex networks. Among them are the following:

- Power grid is not a programmable network like phone or Internet systems. Electricity cannot be sent from point-to-point in packages. The power system is rather a giant reservoir where all generators deliver their energy to the reservoir and any load can tap

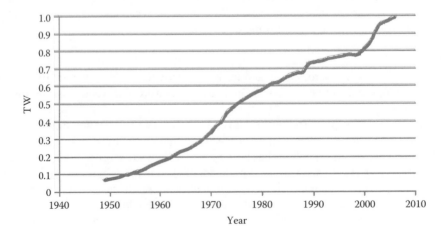

FIGURE 14.2   Generation capacity in the United States. (From the U.S. Energy Information Administration.)

from that reservoir. It is impossible to know which generator produces the energy you consume in your home at any given time.

- Since there is no effective way for storing large amounts of electrical energy, the generated power must be consumed immediately by all loads as well as the various losses in the system. A deficiency or surplus in power may lead to a blackout if not corrected within seconds (milliseconds in some cases). For the Internet network, when the packets sent are at a higher rate than what is being processed by the user's node, the extra packets are often stored temporarily until the node is available. Even if the packets are dropped out at a busy or damaged node, the network can still function normally and future packets are rerouted through the rest of the network.
- Amount of energy that must be controlled at all times in the power system is immense. The Internet and phone networks process an infinitesimal amount of energy compared to the power grid. The equipment for the power network is therefore heavy, bulky, and slow acting. The luxury of switching links and devices at a high rate in the phone and Internet networks cannot be used in power systems.
- Immense masses of generators, turbines, and motors in the power systems create relatively large delays in the control actions. Hence, instant correction is impossible for power systems. This problem is virtually nonexistent in phone and Internet networks.
- Overloaded equipment in power systems is often tripped to protect the equipment from being damaged. This could initiate outages. In phone or Internet networks, overloading the equipment often leads to just a delay in transmission.

Based on the above discussion and the status of the current technologies, the best option to reduce blackouts is to enhance the reliability of the existing power systems by building more generating plants, more transmission lines, and improving the monitoring and control systems. These options, which are discussed in the following sections, are costly solutions and must always be justified socially, environmentally, and economically. The installation of a new generating plant or a new transmission line is often resisted by nearby communities because of several reasons.

- Transmission lines have unattractive sights.
- Perception that the electromagnetic field of the power lines and substations may have negative biological effects.
- Fossil fuel power plants produce pollution, hydroelectric power plants affect fish migration, and nuclear power plants are perceived to be a radiation hazard.
- Power plants occupy large sizes of real estate.
- Substations are perceived to be incompatible with city views.

Some of these problems can be reduced by attractive new designs of transmission towers and substations and by reducing the negative environmental impacts of power plants. If and when renewable energy becomes reliable and viable, the number of existing power plants or transmission lines can be decreased.

14.1 TOPOLOGY OF POWER SYSTEMS

Early power systems were simple in design but highly unreliable. They consisted mainly of generating units connected to load centers in radial fashion as shown in Figure 14.3. The generator is connected to a transformer (xfm_g) to raise the voltage to the transmission voltage level. The transmission lines connect the generating stations to the load centers; two of them are shown in the figure. At the load centers, transformers (xfm_1 and xfm_2) are used to step down the voltage to the users' level. Another form of the radial configuration is shown in Figure 14.4.

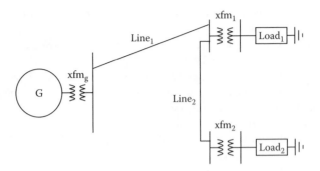

FIGURE 14.3   Radial power system.

Both systems have reliability problems. In the first system, if line$_1$ is tripped (opened) for any reason, both loads are left without power; hence the system is in a complete blackout state. If instead line$_2$ is tripped, only load$_2$ is disconnected and the system is in a partial blackout state. The second system is better designed because when any of the two lines is tripped, only the load being fed by that line is left without power. Hence, a single line tripping causes partial but not complete blackout.

14.1.1 ENHANCING POWER SYSTEM RELIABILITY BY ADDING TRANSMISSION LINES

For the radial systems in Figures 14.3 and 14.4, a loss of a single line can cause at least one load to be left without power. To correct this problem, a third line can be added as shown in Figure 14.5. In this new system, tripping any one line will not result in a power outage. The transmission system in this case is connected in a network configuration, where each load is fed by multiple lines. Thus, the network connection is more reliable than the radial connection.

14.1.2 ENHANCING POWER SYSTEM RELIABILITY BY ADDING GENERATION

The system in Figure 14.5 is unreliable in terms of generation—if the generator is tripped due to any reason, all loads will be left without power. The obvious solution to this problem is to add more generators to the system at different locations. An example is shown in Figure 14.6 for a two-generator, five-transmission lines system. Tripping out any one line, or one generator, may not result in loss of service to any load. However, there is no guarantee that the system is secure under any tripping scenario, as discussed in Section 14.2. You may find scenarios where the system is insecure especially when the system is heavily loaded and the line with most current is tripped.

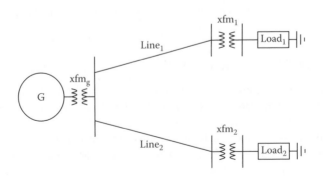

FIGURE 14.4   Another radial power system.

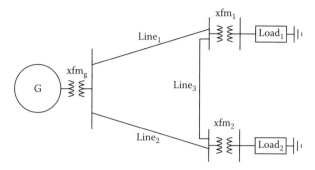

FIGURE 14.5   Network power system.

14.2 ANALYSIS OF POWER NETWORKS

To detect situations that may lead to blackouts (e.g., overcurrent, overvoltage, or undervoltage), a method known as load flow analysis is used to study the power network. The load flow is a circuit analysis tool that relates the structure of the power network and the load demands to the currents and voltages throughout the network. After performing the load flow analysis, the status of the power system can be determined. The power system is in a secure state if all currents and voltages of the major equipment (transmission lines, transformers, etc.) are within their design ranges. The system is insecure if any major equipment (transmission lines, transformers, etc.) is operating outside its normal voltage range or operating at current higher than its rating. If the load flow analysis identifies overloaded lines, the system is insecure and the operator must shift some of the loads to other lines, thus saving the system from a potential outage. For example, if the voltage at any load center is lower than its normal range, the load is probably served by insufficient number of lines and the operator must shift the flow of power through additional lines.

Because the power system is not fully monitored, the load flow analysis uses existing measurements of powers and various voltages to solve for the currents and voltages everywhere else in the network. The method, which is often numerical, uses a set of nonlinear equations relating the

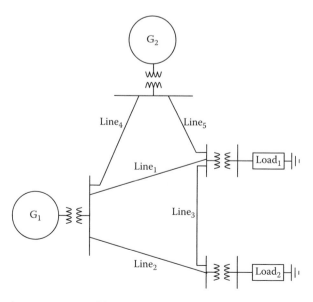

FIGURE 14.6   Network power system with multiple generators.

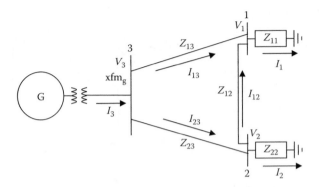

FIGURE 14.7 Current distribution of the system in Figure 14.5.

measurements to the unknowns. However, if we assume that all system impedances as well as the voltages of all generators are known, the load flow analysis turns into a solution of a set of linear equations. The procedure starts by developing the impedance diagram of the power system; the diagram for the system in Figure 14.5 is shown in Figure 14.7. Z_{11} is the impedance of load$_1$ plus its own transformer (xfm$_1$), Z_{22} is the impedance of load$_2$ including its transformer, and Z_{12} is the impedance of the transmission line between bus 1 and 2, and so on. The voltage at each bus is labeled according to the bus number, and the direction of the current can be chosen arbitrarily. The final solution of the current determines its direction; if the current is positive, the chosen direction is correct, and if it is negative the actual direction is opposite to the chosen one.

A convenient choice for the initial directions of the transmission line currents is to assume the flow is from the larger numbered bus to the smaller numbered bus. In this case, the node equations at buses 1, 2, and 3 are

$$\begin{aligned}
\bar{I}_1 &= \bar{I}_{12} + \bar{I}_{13} \\
\bar{I}_2 &= \bar{I}_{23} - \bar{I}_{12} \\
\bar{I}_3 &= \bar{I}_{13} + \bar{I}_{23}
\end{aligned} \tag{14.1}$$

The load current can be written as a function of the bus voltage and the load admittance as follows:

$$\begin{aligned}
\bar{I}_1 &= \bar{V}_1 \bar{Y}_{11} \\
\bar{I}_2 &= \bar{V}_2 \bar{Y}_{22}
\end{aligned} \tag{14.2}$$

where $\bar{Y}_{ij} = \dfrac{1}{\bar{Z}_{ij}}$.

The currents of the transmission lines can also be written as

$$\begin{aligned}
\bar{I}_{12} &= (\bar{V}_2 - \bar{V}_1)\bar{Y}_{12} \\
\bar{I}_{13} &= (\bar{V}_3 - \bar{V}_1)\bar{Y}_{13} \\
\bar{I}_{23} &= (\bar{V}_3 - \bar{V}_2)\bar{Y}_{23}
\end{aligned} \tag{14.3}$$

Substituting the currents in Equation 14.3 into Equation 14.1 yields

$$\begin{bmatrix} \bar{I}_1 \\ \bar{I}_2 \\ -\bar{I}_3 \end{bmatrix} = \begin{bmatrix} -(\bar{Y}_{12} + \bar{Y}_{13}) & \bar{Y}_{12} & \bar{Y}_{13} \\ \bar{Y}_{12} & -(\bar{Y}_{12} + \bar{Y}_{23}) & \bar{Y}_{23} \\ \bar{Y}_{13} & \bar{Y}_{23} & -(\bar{Y}_{13} + \bar{Y}_{23}) \end{bmatrix} \begin{bmatrix} \bar{V}_1 \\ \bar{V}_2 \\ \bar{V}_3 \end{bmatrix} \tag{14.4}$$

Substituting the load currents in Equation 14.2 into Equations 14.4 yields

$$
\begin{bmatrix} \bar{I}_1 \\ \bar{I}_2 \\ -\bar{I}_3 \end{bmatrix} = \begin{bmatrix} -(\bar{Y}_{12}+\bar{Y}_{13}) & \bar{Y}_{12} & \bar{Y}_{13} \\ \bar{Y}_{12} & -(\bar{Y}_{12}+\bar{Y}_{23}) & \bar{Y}_{23} \\ \bar{Y}_{13} & \bar{Y}_{23} & -(\bar{Y}_{13}+\bar{Y}_{23}) \end{bmatrix} \begin{bmatrix} \bar{Z}_{11}\bar{I}_1 \\ \bar{Z}_{22}\bar{I}_2 \\ \bar{V}_3 \end{bmatrix} \tag{14.5}
$$

Equation 14.5 can be written in the matrices form:

$$
[I] = [Y]([V]+[Z][I])
$$
$$
[I] = ([1]-[Y][Z])^{-1}[Y][V] \tag{14.6}
$$

EXAMPLE 14.1

The power system in Figure 14.7 has the following data:
  Impedance of load_1 plus its transformer is $\bar{Z}_{11}=20+j10\ \Omega$
  Impedance of load_2 plus its transformer is $\bar{Z}_{12}=30+j5\ \Omega$
  Impedance of line_1 is $\bar{Z}_{13}=j0.5\ \Omega$
  Impedance of line_2 is $\bar{Z}_{23}=j0.6\ \Omega$
  Impedance of line_3 is $\bar{Z}_{12}=j0.4\ \Omega$
  The voltage at the high-voltage side of xfm_g is fixed at 230 kV (line to line).
  The capacity of any line (maximum current the line can carry) is 6 kA.
  The range of the voltage at any load bus is $230\pm5\%$
  All impedances are referred to the high-voltage side of the transformers. Assume that all connections are in wye.

(a) Compute the power delivered to each load and the power produced by the generator.
(b) Assume that transmission line_3 is tripped; repeat part (a).
(c) Is the system in part (b) secure?
(d) Assume that line_1 is tripped while lines 2 and 3 are in service; repeat part (a).
(e) Is the system in part (d) secure?

Solution

(a) To compute the powers, we need to compute the currents of each load. We have five unknowns, I_1, I_2, I_3, V_1, and V_2. Equation 14.5 can be used to solve for the currents.

$$
\begin{bmatrix} \bar{I}_1 \\ \bar{I}_2 \\ -\bar{I}_3 \end{bmatrix} = \begin{bmatrix} -(\bar{Y}_{12}+\bar{Y}_{13}) & \bar{Y}_{12} & \bar{Y}_{13} \\ \bar{Y}_{12} & -(\bar{Y}_{12}+\bar{Y}_{23}) & \bar{Y}_{23} \\ \bar{Y}_{13} & \bar{Y}_{23} & -(\bar{Y}_{13}+\bar{Y}_{23}) \end{bmatrix} \begin{bmatrix} \bar{Z}_{11}\bar{I}_1 \\ \bar{Z}_{22}\bar{I}_2 \\ \bar{V}_3 \end{bmatrix} \tag{14.7}
$$

Equation 14.7 can be further rearranged as follows:

$$
\begin{bmatrix} 1+(\bar{Y}_{12}+\bar{Y}_{13})\bar{Z}_{11} & -\bar{Y}_{12}\bar{Z}_{22} & 0 \\ -\bar{Y}_{12}\bar{Z}_{11} & 1+(\bar{Y}_{12}+\bar{Y}_{23})\bar{Z}_{22} & 0 \\ -\bar{Y}_{13}\bar{Z}_{11} & -\bar{Y}_{23}\bar{Z}_{22} & 1 \end{bmatrix} \begin{bmatrix} \bar{I}_1 \\ \bar{I}_2 \\ -\bar{I}_3 \end{bmatrix} = \begin{bmatrix} \bar{Y}_{13} \\ \bar{Y}_{23} \\ -(\bar{Y}_{13}+\bar{Y}_{23}) \end{bmatrix} \bar{V}_3 \tag{14.8}
$$

By directly substituting the variables in this example into Equation 14.8, the currents of the system can be computed. Keep in mind that the phase voltage V_3 is $230/\sqrt{3}$.

$$
\begin{bmatrix} \bar{I}_1 \\ \bar{I}_2 \\ \bar{I}_3 \end{bmatrix} = \begin{bmatrix} 5.9\angle-27.7^\circ \\ 4.34\angle-10.6^\circ \\ 10.12\angle-20.44^\circ \end{bmatrix} \text{kA}
$$

The power consumed by $load_1$ is

$$P_1 = 3I_1^2 \, R_{11} = 3 \times 5.9_2 \times 20 = 2.08 \text{ GW}$$

The power consumed by $load_2$ is

$$P_2 = 3I_2^2 \, R_{22} = 3 \times 4.34^2 \times 30 = 1.7 \text{ GW}$$

The power produced by the generator is the sum of P_1 and P_2 since the lines have no resistance.

$$P_3 = P_1 + P_2 = 3.78 \text{ GW}$$

(b) If $line_3$ is tripped, $Y_{12} = 0$, and the current can be computed as in part (a).

$$\begin{bmatrix} \bar{I}_1 \\ \bar{I}_2 \\ \bar{I}_3 \end{bmatrix} = \begin{bmatrix} 5.87\angle-27.7° \\ 4.35\angle-0.57° \\ 10.12\angle-20.42° \end{bmatrix} \text{ kA}$$

The power consumed by $load_1$ is

$$P_1 = 3\,I_1^2 \, R_{11} = 3 \times 5.87^2 \times 20 = 2.07 \text{ GW}$$

The power consumed by $load_2$ is

$$P_2 = 3\,I_2^2 \, R_{22} = 3 \times 4.35^2 \times 30 = 1.7 \text{ GW}$$

The power produced by the generator is the sum of P_1 and P_2.

$$P_3 = P_1 + P_2 = 3.77 \text{ GW}$$

(c) To check the security of the system, we need to compute the currents in the transmission lines and compare them with their capacities. But first, we need to compute the bus voltages.

$$\bar{V}_1 = \bar{Z}_{11} \, \bar{I}_1 = 131.42 - j2.6 = 131.44\angle-1.13° \text{ kV}$$
$$\bar{V}_2 = \bar{Z}_{22} \, \bar{I}_2 = 132.31 - 2.56 = 132.33\angle-1.11° \text{ kV}$$

The upper limit of the voltage is 5% above or below the rated voltage. This makes the acceptable voltage range as 126.15–139.43 kV per phase. All the above voltages are within this voltage range. Hence, the system is secured from the voltage point of view.

The line currents are

$$\begin{bmatrix} \bar{I}_{12} \\ \bar{I}_{13} \\ \bar{I}_{23} \end{bmatrix} = \begin{bmatrix} 0 \\ 5.87\angle-27.7° \\ 4.35\angle-10.57° \end{bmatrix} \text{ kA}$$

All the currents are lower than the line capacities, which is 6 kA. The system is also secured from the current point of view. The system can continue to provide power to all customers.

(d) If $line_1$ is tripped, $Y_{13} = 0$, and the currents are

$$\begin{bmatrix} \bar{I}_1 \\ \bar{I}_2 \\ \bar{I}_3 \end{bmatrix} = \begin{bmatrix} 5.8\angle-29.9° \\ 4.3\angle-11.9° \\ 9.97\angle-22.25° \end{bmatrix} \text{ kA}$$

The power consumed by $load_1$ is

$$P_1 = 3\,I_1^2 \, R_{11} = 3 \times 5.8^2 \times 20 = 2.02 \text{ GW}$$

The power consumed by load$_2$ is

$$P_2 = 3I_2^2\, R_{22} = 3 \times 4.3^2 \times 30 = 1.66 \text{ GW}$$

The power produced by the generator is the sum of P_1 and P_2.

$$P_3 = P_1 + P_2 = 3.68 \text{ GW}$$

(e) To check the security of the system, we need to compute the currents in the remaining lines.

$$\overline{V}_1 = \overline{Z}_{11}\, \overline{I}_1 = 129.37 - j7.55 = 129.6\angle{-3.34°} \text{ kV}$$
$$\overline{V}_2 = \overline{Z}_{22}\, \overline{I}_2 = 130.53 - 5.54 = 130.65\angle{-2.43°} \text{ kV}$$

All the above voltages are within the acceptable voltage range. Hence, the system is secured from the voltage point of view. Now let us check the line currents.

$$\begin{bmatrix} \overline{I}_{12} \\ \overline{I}_{13} \\ \overline{I}_{23} \end{bmatrix} = \begin{bmatrix} 5.8\angle{-29.9°} \\ 0 \\ 9.97\angle{-22.25°} \end{bmatrix} \text{ kA}$$

Notice that line$_2$ carries 9.97 kA, which is higher than the capacity of the line (6 kA). In this case, the system is insecure and line$_2$ will eventually trip to prevent it from overheating and thus sagging too deeply and touching trees or structures. When this line is tripped, total blackout occurs and no load will be served.

14.3   ELECTRIC ENERGY DEMAND

The decision to construct new generation or transmission facilities is mainly based on reliability concerns as well as existing and predicted future load demands. However, the demand varies on hourly basis, and the maximum demand occurs probably a few times a year during heat waves or cold spills. Therefore, it is hard to justify the cost of new facilities based on the maximum demand alone. Instead, utilities regularly depend on neighboring utilities to provide support during the high demand periods.

During a typical day, the energy demand fluctuates widely. For most utilities, the peak energy consumptions occur twice in a typical day as shown in Figure 14.8: the first is around 9:00 AM when factories, shops, and offices are at their peak consumption; the second is around 6:00 PM when people are at home preparing meals, turning on lights, and watching televisions.

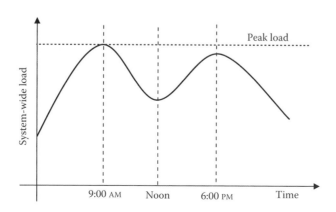

FIGURE 14.8   Daily system load.

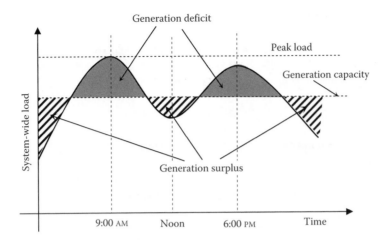

FIGURE 14.9   Setting the generation capacity.

   To provide reliable service to its customers, every utility must be able to meet the daily peak demands. Since the high energy demand occurs for a few hours every day, it is uneconomical to build generating plants to be used only during the peak loads. Instead, it would be more profitable for the utility to build its generating capacity to meet the average daily demand. But what should be done when the load is higher than the available generation? Consider, for example, the load profile in Figure 14.9. The dark areas represent the periods when the generating capacity is below the demand, and the dashed areas represent the periods when the generating capacity is higher than the demand. The obvious economical solution to the deficit and surplus powers is to trade electricity with neighboring utilities. When a utility has surplus, it sells the excess power to another utility that needs it. When it has a deficit, the utility buys the extra power from another utility with surplus. This arrangement makes sense economically, but requires all utilities to be interconnected through a mesh of transmission lines.

EXAMPLE 14.2

The load demand of a power system for a given day can be approximated by

$$P = 2 - 0.3 \times \cos(0.5417t - 1)\, e^{-0.01t}\ \text{GW}$$

where t is the time of the day in hours using the 24 h clock. Compute the following:

(a) Times of the peak demands
(b) Peak demands
(c) Average demand
(d) If the generation capacity of the utility is 2 GW, compute the power to be imported during the first peak

Solution

(a) To compute the times of the peaks, we need to set the derivative of the power equation with respect to time to zero.

$$\frac{dP}{dt} = 0$$

$$0.01 \times \cos(0.5417t - 1) + 0.5417 \sin(0.5417t - 1) = 0$$

Hence,

$$\tan(0.5417t - 1) = -0.01846$$

$$t = \frac{1 + \tan^{-1}(-0.01846)}{0.5417}$$

Solving the above equation using the four quadrants of the tangent function, the first peak is at

$$t_1 = 7.6115 = 7{:}37 \text{ AM}$$

and the second peak is at

$$t_2 = 19.2 = 7{:}12 \text{ PM}$$

(b) First and second peaks are

$$P_{\text{peak 1}} = 2 - 0.3 \times \cos(0.5417t_1 - 1)\, e^{-0.01t_1} = 2.278 \text{ GW}$$

$$P_{\text{peak 2}} = 2 - 0.3 \times \cos(0.5417t_2 - 1)\, e^{-0.01t_2} = 2.2475 \text{ GW}$$

(c) Average power during the 24 h period is

$$P_{\text{ave}} = \frac{1}{24} \int\limits_0^{24} P\, dt$$

The power equation can be rewritten in the following form:

$$P = 2 - 0.162\, e^{-0.01t} \times \cos(0.5417t) - 0.2525\, e^{-0.01t} \times \sin(0.5417t)$$

where

$$\int e^{ax}[c\cos(bx) + k\sin(bx)]\, dx = \frac{e^{ax}}{a^2 + b^2}[(ka + cb)\sin(bx) + (ca - kb)\cos(bx)]$$

Hence,

$$P_{\text{ave}} = \frac{1}{24} \int\limits_0^{24} [2 - 0.162\, e^{-0.01t} \times \cos(0.5417t) - 0.2525\, e^{-0.01t} \times \sin(0.5417t)]\, dt$$

$$P_{\text{ave}} = 2 - \left[\frac{e^{-0.01t}}{0.01^2 + 0.5417^2}[(0.08523)\sin(0.5417t) - (0.1384)\cos(0.5417t)] \right]_0^{24}$$

$$P_{\text{ave}} = 1.7688 \text{ GW}$$

(d) The power to be imported during the first peak is

$$P_{\text{import}} = P_{\text{peak 1}} - P_{\text{ave}} = 2.278 - 1.7688 = 509.2 \text{ MW}$$

14.4 TRADING ELECTRIC ENERGY

As mentioned earlier, the trading of energy between utilities makes great economic sense. For example, the two utilities in Figure 14.10 are connected by a transmission line often called tie-line. Assume that the two utilities are located in two different longitudes (two time zones). The time zone

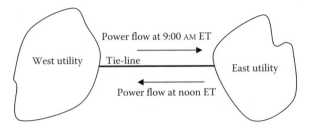

FIGURE 14.10   Two utilities at different time zones.

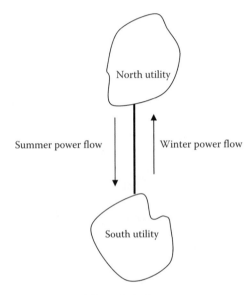

FIGURE 14.11   Trade between utilities at different latitudes.

of the western utility lags that of the eastern utility by 3 h. At 9:00 AM eastern time (ET), the eastern utility is at its peak demand while the western utility has surplus generation since it is 6:00 AM in the west. Hence, the extra power can be delivered to the eastern utility by the western utility. At noon ET, the western utility is at its peak demand while the eastern utility has surplus. Therefore, the power can flow from east to west.

Another example of the power exchange occurs between utilities at different latitudes. Consider Figure 14.11, where the northern area is assumed to be generally cooler than the southern area, for example, the northwest region and California in the United States. Heating loads are dominant in the Northwest while cooling loads are prevailing in California. In the summer, the temperature in the Northwest region is mild and the demand for electric energy is low. In California, however, it is generally hot and the energy demand is high because of the use of air-conditioning equipment. Hence, the Northwest has surplus energy while California has a deficit. As a result, the power moves from north to south. In the winter, the scenario is reversed.

EXAMPLE 14.3

The load demand for a given utility can be approximated by

$$P = 4 - \cos(0.5t - 1.2) \text{ GW}$$

where t is the time of the day in hours using the 24 h clock. The generating capacity of the utility is 4 GW. Compute the following:

(a) Energy available for export
(b) Imported energy during the time of energy deficit
(c) Net energy trade

Solution

(a) The power available for trading is the difference between the actual demand and the generation capacity.

$$\Delta P = P_{\text{ave}} - P = \cos\left(0.5t - 1.2\right)$$

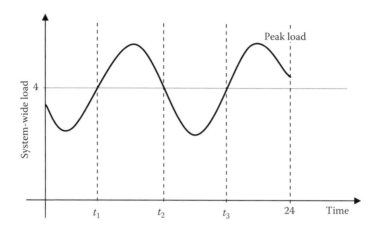

Before we compute the traded energy, we need to identify the times in the figure when the demand is equal to the capacity.

$$\Delta P = \cos\left(0.5t - 1.2\right) = 0$$

Hence,

$$t_1 = 5.542$$
$$t_2 = 11.825$$
$$t_3 = 18.108$$

The energy available for export is

$$E_{\text{export}} = \int_0^{t_1} \Delta P \, dt + \int_{t_2}^{t_3} \Delta P \, dt$$

$$E_{\text{export}} = 2\sin\left(0.5t - 1.2\right)\Big|_0^{5.542} + 2\sin\left(0.5t - 1.2\right)\Big|_{11.825}^{18.108} = 7.864 \text{ GW h}$$

(b) Imported energy is

$$E_{\text{import}} = \int_{t_1}^{t_2} \Delta P \, dt + \int_{t_3}^{24} \Delta P \, dt$$

$$E_{\text{import}} = 2\sin\left(0.5t - 1.2\right)\Big|_{5.542}^{11.825} + 2\sin\left(0.5t - 1.2\right)\Big|_{18.108}^{24} = -7.96 \text{ GW h}$$

(c) Net trade is

$$E_{\text{net}} = E_{\text{export}} - E_{\text{import}} = 7.864 - 7.96 = -96 \text{ MW h}$$

14.5   WORLD WIDE WEB OF POWER

In addition to the economic benefits, interconnecting utilities makes great sense with reference to reliability. If some generators in certain areas are out of service, the generation deficit can be compensated by importing extra power from neighboring utilities. Also, if a key transmission line is tripped, other transmission lines in the grid can reroute the power to the customers. Because of these economic and reliability benefits, power systems are continuously merging with little regard to political or geographical borders.

Although interconnected, the operation, trade, and control functions for the power systems are left to local pools, which are composed of several utilities in a geographical region. In the United States, Canada, and a small part of Mexico, the power grid is divided into the 10 power pools shown in Figure 14.12; the pools are called *reliability councils*. An example of a reliability council is the Mid-Atlantic Area Council (MAAC), which encompasses an area of nearly 130,000 km^2 and provides electricity to more than 23 million people in the northeastern region of the United States and eastern Canada. The capacity of the MAAC is approximately 60 GW and it has over 13,000 km of bulk power transmission lines.

Another example of a pool is the Western Electricity Coordinating Council (WECC), which provides electricity to 71 million people in 14 western states, 2 Canadian provinces, and a portion of Baja California state in Mexico. The WECC system consists of over one thousand generating units and tens of thousands of transmission lines. It has a capacity of 130 GW. The full WECC system is very complex and is hard to display in a single page. A small part of the WECC is shown in Figure 14.13.

All 10 coordinating councils form a nonprofit corporation called North American Electric Reliability Council (NERC). The NERC's mission is to ensure that the electric power system in North America is reliable, adequate, and secure by setting, monitoring, and enforcing standards for the reliable operation and planning of the electric grid.

The power systems are also fully interconnected in western Europe, and their equivalent to NERC is called the Union for the Co-ordination of Transmission of Electricity (UCTE). UCTE

FIGURE 14.12 (See color insert following page 300.)   U.S. regional reliability councils.

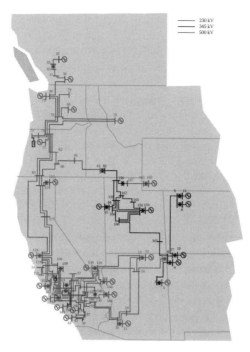

230 kV
345 kV
500 kV

FIGURE 14.13 Major components of the WECC.

covered about 20 European countries while more countries from eastern Europe are expected to be connected to the UCTE grid. Their service area includes 400 million people and its total annual consumption is approximately 2100 TW h. North Africa is also being connected with Europe through submarine cable systems under the Mediterranean Sea. In other regions of the world, the power systems are linked together whenever possible. Indeed, many consider that power engineers have succeeded in creating a borderless power grid that is truly unifying the world.

Although the interconnection of the power networks has many great advantages, it has two major drawbacks:

1. Power grid becomes incredibly complex creating enormous challenges for monitoring, operation, protection, and control of the system.
2. Major failures in one area could affect other areas, thus creating wider blackouts. This was evident in the major blackouts in the United States in 1965, 1976, and 2003.

14.6 ANATOMY OF BLACKOUTS

Blackouts can be caused by a number of natural or man-made events as well as major component failure. The list is long, but the typical triggers for blackouts include the following:

- Faults in transmission lines could lead to excessive currents in the system. The faulted section of the network must be isolated quickly to prevent any thermal damage to the transmission lines, transformers, and other major power equipment. This is done by tripping (opening) the circuit breakers on both ends of the faulted line. The loss of the transmission line could result in outages.
- Natural calamities such as lightning, earthquake, strong wind, and heavy frost can damage major power system equipment. When lightning hits a power line, the line insulators can be

damaged leading to short circuits (faults). Major earthquakes could damage substations, thus interrupting power to the areas served by the substations. Heavy winds may cause trees to fall on power lines creating faults that trip transmission lines.

- When major power system equipments such as generators and transformers fail, they may lead to outages.
- Protection and control devices may not operate properly to isolate faulty components. This is known as hidden failure and can cause the fault to affect a wider region in the system.
- Breaks in communication links between control centers in the power system may lead to the wrong information being processed, causing control centers to operate asynchronously and probably negating each other.
- Human errors can lead to tripping of important equipment.

14.6.1 BALANCE OF ELECTRIC POWER

The power balance of the system must always be maintained to ensure its stable operation, that is, the sum of power from all generators must be equal to the sum of power of all loads plus all losses. This simple relationship must be preserved at the system-wide level, and at the power pool level. For example, the power pool in Figure 14.14 represents a geographical area or a service territory of a utility or a group of utilities. The power pool has a total power generation P_g from all its power plants, and a total system load P_l representing customers demand plus the losses in the system. The pool imports power P_{import} from neighboring utilities and exports power P_{export} to other neighboring utilities. The balanced power equation of the pool is

$$P_g + P_{import} = P_l + P_{export} \tag{14.9}$$

Now, let us assume that the tie-line transmitting P_{import} is lost for some reason. Then, the power balance is not maintained, and

$$P_g < \left(P_l + P_{export}\right) \tag{14.10}$$

The fundamental question is how to balance the power in the pool? In this simple example, we have four options:

1. Reduce P_{export}, which may result in a power deficit in the neighboring utility.
2. If the generated power is below the capacity of the pool, increase the generation P_g.

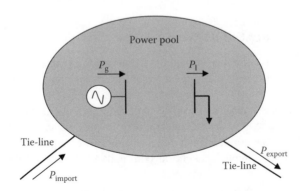

FIGURE 14.14 Power balance of a pool.

3. Reduce the demand by disconnecting some loads. This is called *rolling blackouts*. If some loads are disconnected for some time, then other loads are disconnected afterward, and so on.
4. Find another utility that can transmit the needed power through other transmission routes.

All these solutions must be implemented within a short time (milliseconds in some cases), otherwise the system may collapse. Why? The answer is in the following sections.

14.6.2 BALANCE OF ELECTRICAL AND MECHANICAL POWERS

The synchronous generator is an electromechanical converter as depicted in Figure 14.15. Its input power is mechanical P_m and its output is electrical power P_g. If you ignore the internal losses, the balanced operation of the generator is achieved when

$$P_m = P_g \tag{14.11}$$

Keep in mind that the magnitude of the electrical power P_g is determined by customers' actions in addition to any import or export transaction, as given in Equation 14.9. Although the utilities do not often know in advance their customers' intentions, the utility must continually adjust the mechanical power into its generators so that Equation 14.11 is satisfied at all times. As long as the balance is maintained, the system is stable and all loads are served.

Among the problems that lead to blackouts is the uncompensated imbalance between the mechanical and electrical powers of the generators. When the input power to the generator P_m is greater than the output power P_g, the excess power is stored in the rotating mass of the generator–turbine unit in the form of extra kinetic energy. If $P_m < P_g$, the electric power shortage is recovered from the kinetic energy stored in the rotating mass. The dynamic equation of the electromechanical powers of the generator can be written as

$$M_{eq} \frac{d\omega}{dt} = P_m - P_g \tag{14.12}$$

where
M_{eq} is the equivalent inertia of the rotating mass of the generator–turbine unit
ω is the angular speed of the rotor of the generator
$d\omega/dt$ is the angular acceleration of the rotor
$M_{eq}(d\omega/dt)$ is the kinetic power

As explained in Chapter 12, the speed of the synchronous machine ω_s is constant during the steady-state operation (stable operation) and is equal to

$$\omega_s = 2\pi \left(\frac{n_s}{60}\right) \text{ rad/s}$$

$$n_s = 120 \left(\frac{f}{p}\right) \text{ rpm} \tag{14.13}$$

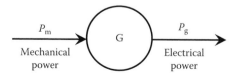

FIGURE 14.15 Power balance of a synchronous generator.

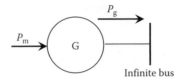

FIGURE 14.16   Synchronous generator connected to infinite bus.

where

f is the frequency of the output voltage of the generator
p is the number of poles inside the machine
n_s is the synchronous speed of the magnetic field, which is the same as the rotor speed in the steady state

Now, let us assume that the synchronous machine is connected to an infinite bus as shown in Figure 14.16. As discussed in Chapter 12, the infinite bus is a term used to describe a large system that has constant frequency and constant voltage. The equivalent circuit of the synchronous generator (also see Figure 12.39) is depicted in Figure 14.17a. E_f is the equivalent field voltage, V_t is the terminal voltage of the generator (the infinite bus voltage in our case), and X_s is the synchronous reactance of the generator.

As explained in Chapter 7, any phasor has three attributes: magnitude, angle, and frequency. Although we draw the phasor as a stationary vector, it is actually rotating at its own frequency. In our example, and as shown in Figure 14.17b, V_t rotates at the frequency of the infinite bus voltage f, which is constant (60 Hz in the United States or 50 Hz in Europe). E_f rotates at frequency f_n corresponding to the actual rotor speed n. As given in Chapter 12, f_n can be computed by using Equation 14.13.

$$f_n = \left(\frac{p}{120}\right) n \qquad (14.14)$$

When the rotor of the generator rotates at the synchronous speed ($n = n_s$), the frequency of E_f is equal to the system frequency, that is, $f_n = f$, and the angle δ is constant.

Now let us use Equation 14.12 to discuss the three possible scenarios in Table 14.1. In scenario 1, the mechanical and electrical powers are equal. Hence, the acceleration $d\omega/dt = 0$, and the rotor speed ω is equal to the synchronous speed ω_s. Thus, $f_n = f$. Consequently, as shown in Figure 14.17, the power angle δ is constant. The difference between $\overline{E}_f$ and $\overline{V}_t$, which is $\overline{I}_a \overline{X}_s$, is also constant. Since X_s is fairly constant, the current in the machine is constant. This is a balanced (steady-state) operation and the generator is said to be in synchronism with the rest of the power system.

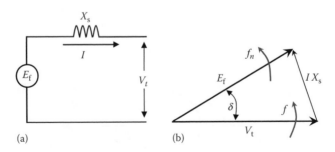

FIGURE 14.17   Model and phasor diagram of a synchronous generator. (a) Equivalent circuit and (b) phasor diagram.

TABLE 14.1

Effects of Power Imbalance

Scenario	Power Equation	Rotor Acceleration $\frac{d\omega}{dt}$	Rotor Speed ω	Kinetic Energy of Rotating Mass
1	$P_m = P_g$	$\frac{d\omega}{dt} = 0$	$\omega = \omega_s$	Unchanged—machine is in synchronism. This is the steady-state operation
2	$P_m > P_g$	$\frac{d\omega}{dt} > 0$	$\omega > \omega_s$	Increases—machine is out of synchronism
3	$P_m < P_g$	$\frac{d\omega}{dt} < 0$	$\omega < \omega_s$	Decreases—machine is out of synchronism

For scenario 2 where $P_m > P_g$, the generator accelerates and the rotor speed becomes greater than the synchronous speed. Hence, $f_n > f$, and the machine in this case is said to be out of synchronism with the rest of the power system. Because of the difference in frequencies, δ keeps increasing; thus, the current inside the machine is continuously increasing. When the current reaches the thermal limit of the machine, the generator is disconnected from the grid to prevent any further damage to the machine. This can happen in just a few milliseconds.

In scenario 3 when $P_m < P_g$, the rotor decelerates and its speed goes below the synchronous speed. Hence, $f_n < f$, and the machine is out of synchronism with the rest of the power system. Unless corrected, E_f will eventually lag V_t making the generator operate as a motor and reversing the flow of power. During this period, the current inside the machine will increase to a high level that shuts down the generator.

Based on the above discussion, we can conclude that every generator in the power network must operate under scenario 1, which satisfies Equation 14.11. Thus, the power plant controller must continuously adjust the mechanical power of the turbine to match the changes in the load demand to keep the generators of the plant in synchronism with the rest of the power network.

14.6.2.1 Control Actions for Decreased Demand

Substituting the value of P_g in Equation 14.9 into Equation 14.11 yields

$$P_m = P_1 + P_{export} - P_{import} \tag{14.15}$$

Assume that P_{import} and P_{export} are constants. When the load demand P_1 is reduced, the mechanical power into the generator must be reduced to match the new demand. This is a relatively simple process if the change in demands is slow (minutes or hours). For thermal power plants, the reduction of P_m is done by gradually reducing the steam entering the turbines. For hydroelectric power plants, the governor reduces the amount of water entering the turbine.

However, if the reduction in demand is sudden and large, the control action is stern and very fast. For thermal power plants, the steam is allowed to escape in the air, thus rapidly reducing the amount of steam entering the turbine. However, for hydroelectric power plants, changing the water flow is a slow process because rapid closure of the water valve can damage the penstock due to the unexpected excessive water hammers. Instead, resistive loads, known as braking resistances, are connected to the system to consume the excess electric energy until the water flow is reduced.

14.6.2.2 Control Actions for Increased Demand

When the electrical demand P_1 is greater than the output mechanical power of the turbine P_m, the steam of the thermal power plant or the water of the hydroelectric power plant must increase to match the new demand, as explained in Equation 14.15. This is achievable if the change in demand is slow. However, matching a rapid increase in demand is a much difficult process as the increase in

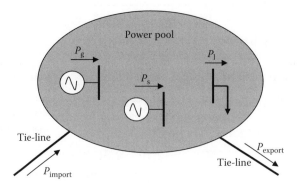

FIGURE 14.18 Power balance of a pool with a spinning reserve.

mechanical power is a slow process; for hydroelectric plants, the water time constant is about 7–10 s. Until the mechanical power increases, the hope is that enough kinetic energy is stored in the system to match the new demand for a short period. If not, one solution would be to increase the import power or decrease the export power. This must be coordinated with the other utilities to ensure that they are able to cope with the new changes. If none of these solutions is possible, the utility may have to resort to rolling blackouts to maintain the power balance in Equation 14.15.

To reduce the reliance on the rolling blackouts to balance the system powers, utilities install enough spinning reserves in each pool. The spinning reserves are rotating generators in power plants that are on standby mode to generate electricity very quickly. These spinning reserves are normally steam power plants with their steam temperature and pressure kept at values necessary for generating electricity on short notice. When the demand is rapidly increasing, the utility can use its own spinning reserve before it reaches out to neighboring utilities. This way, the reliability of the system is greatly enhanced. Figure 14.18 shows the modified power pool with the spinning reserve P_s.

With the spinning reserve, the stable operation of the power pool is when

$$(P_g + P_{import} + P_s) > (P_l + P_{export}) \tag{14.16}$$

Hence,

$$(P_g + P_{import} + P_s) = \lambda(P_l + P_{export}) \tag{14.17}$$

where λ is defined as the *pool power margin*. For a robust system, the margin must be greater than 1.

EXAMPLE 14.4

A power pool has a generation capacity of 1.5 GW. The trade commitments and forecasted load at four different times are given in the following table:

Time	P_l	P_{export} (MW)	P_{import} (MW)
T_1	800 MW	200	0
T_2	1 GW	200	0
T_3	1.2 GW	400	100
T_4	1.8 GW	600	200

Compute the spinning reserve at each of the four times to maintain a power pool margin of at least 110%.

Solution

Use Equation 14.17 to compute the spinning reserve.

At t_1, $P_s = \lambda(P_1 + P_{\text{export}}) - (P_g + P_{\text{import}}) = 1.1(800 + 200) - (1500 + 0) = -400$ MW

There is no need for a spinning reserve at t_1.

At t_2, $P_s = \lambda(P_1 + P_{\text{export}}) - (P_g + P_{\text{import}}) = 1.1(1000 + 200) - (1500 + 0) = -180$ MW

There is no need for a spinning reserve at t_2.

At t_3, $P_s = \lambda(P_1 + P_{\text{export}}) - (P_g + P_{\text{import}}) = 1.1(1200 + 400) - (1500 + 100) = 160$ MW

At t_4, $P_s = \lambda(P_1 + P_{\text{export}}) - (P_g + P_{\text{import}}) = 1.1(1800 + 600) - (1500 + 200) = 940$ MW

14.7 BLACKOUT SCENARIOS

Most major blackouts occur during heavy loading conditions. The worst scenarios occur when the power plants are generating electricity close to their capacities, the transmission lines are heavily loaded, or the spinning reserves are low.

For example, consider the power pools shown in Figure 14.19. The pool in the middle is our system, which is connected to two neighboring pools (external pool 1 and external pool 2). Two tie-lines connect our power pool to the external pools. Assume that our power pool is heavily loaded, and $P_1 > P_g$. To compensate for the deficit in power, our power pool imports P_1 from external pool 1, and P_2 from external pool 2. Thus, our power pool is balanced. Hence,

$$P_g + P_1 + P_2 = P_1 \tag{14.18}$$

If we add the spinning reserve P_s of our pool to Equation 14.18, we get

$$(P_g + P_1 + P_2 + P_s) > P_1 \tag{14.19}$$

So far, the spinning reserve is not needed. Assume that all tie-lines are heavily loaded (close to their capacities). Now, let us assume that tie-line 2 is lost for some reason. In this case, our power pool has a deficit in power since

$$(P_g + P_1) < P_1 \tag{14.20}$$

To compensate for the deficit, our power pool uses its own spinning reserve. However, assume that the spinning reserve is not enough to compensate for the lost power P_2; that is,

$$(P_g + P_1 + P_s) < P_1 \tag{14.21}$$

To correct the problem, tie-line 1 needs to deliver more power to our pool. This requires two conditions: (1) external pool 1 can provide more power than what is scheduled and (2) tie-line 1 can

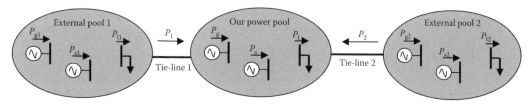

FIGURE 14.19 Blackout scenario.

handle the extra currents. If these two conditions cannot be met, our power pool will be left with a deficit in generation and must quickly trip some of its loads (rolling blackout). Otherwise, its own generating plants will shut down to prevent them from being damaged and our power pool may experience a total blackout. Needless to say that unless they reduce their generation quickly, the other two external pools will also face blackouts after the tie-lines are opened.

14.7.1 Great Northeast Blackout of 1965

The Great Northeast Blackout of 1965 left 30 million people without electricity in New England, New York, and Ontario. Before the blackout, Ontario Hydro was exporting about 1.7 GW to the United States through five tie-lines, including the Massena tie-line. The blackout started in the following sequence:

1. Relay protecting one of the main transmission lines in the Ontario Hydro system was set too low. The line was heavily loaded and the relay tripped the transmission line. The line was directly connected to a hydroelectric plant on the Niagara River.
2. Because of the line tripping, the flow of power was shifted to the remaining four lines. These lines were heavily loaded before the power shift and were overloaded after the shift.
3. Overloading the lines resulted in the tripping of the remaining lines successively.
4. Tripping of all lines resulted in a shortage of about 1.5 GW in the Canadian system, and the power reversed its flow to Canada from the United States.
5. Reversal of power overloaded the Massena tie-line even further and it was tripped. The Canadian system was then completely isolated and became deficient in generation and collapsed.
6. U.S. system was also generation deficient since it was dependent on about 1.7 GW from Canada. The New England and New York systems collapsed, and a total blackout occurred in their systems in a matter of seconds. It took 24 h to restore the system.

14.7.2 Great Blackout of 1977

This blackout affected 8 million people mainly in New York City. The blackout was the second one for New York City in 12 years and lasted for 25 h.

On the hot evening of July 13, 1977, the electrical demand of New York City was very high at about 6 GW. About 3 GW of the power was imported from neighboring utilities. The area was also experiencing a lightning storm. The blackout sequence was as follows:

1. One of the main tie-lines serving Manhattan, the Bronx, Brooklyn, Queens, and Staten Island was stricken by lightning, which damaged the insulators of the lines causing the lines to trip.
2. Power was shifted to the other remaining lines feeding New York City, which were heavily loaded.
3. Spinning reserve of New York City was used to alleviate the stress on the remaining tie-lines. The solution was adequate and the system survived the loss of the line.
4. However, the storm was severe and after 20 min another lightning strike hit another major tie-line and it was tripped.
5. All other lines carried currents beyond their thermal limits and sagged. One of the tie-lines sagged too deeply and touched a tree causing a short circuit, and the line was tripped.
6. New York had a severe deficit in generation, and the system loads could not be served with the existing generations and transmission systems. The system collapsed in seconds.

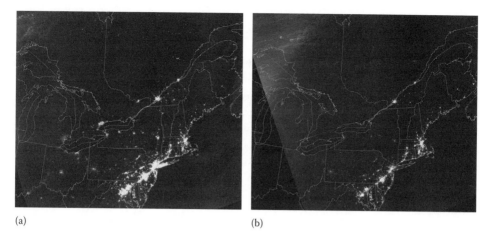

(a)   (b)

FIGURE 14.20 (See color insert following page 300.) Satellite pictures of the Northeast United States on August 14, 2003, (a) just before the blackout and (b) just after the blackout. (Images courtesy of NASA.)

14.7.3 GREAT BLACKOUT OF 2003

This blackout is the worst in history, so far. About 50 million people were affected by the blackout, and the power was out for several days. NASA published interesting satellite photos of Northeast United States before and after the blackout. These pictures are shown in Figure 14.20. Notice how dark Manhattan Island and the Midwest region are during the blackout.

On the hot summer day of August 14, 2003, the systems in the eastern and midwestern United States as well as Ontario, Canada were heavily loaded and the blackout started in the following sequence:

1. Six hundred and eighty megawatt coal generation plant in Eastlake, Ohio, went off-line amid high energy demand.
2. An hour later, a transmission line in northeastern Ohio tripped because it was overloaded.
3. Outage put an extra strain on other transmission lines, and some lines sagged and came in contact with trees causing short circuits that tripped the lines.
4. Utilities in Canada and the eastern United States experienced wild power swings causing stress on their systems. Power plants and high-voltage electric transmission lines in Ohio, Michigan, New York, New Jersey, and Ontario shut down.
5. In less than 3 min, 21 power plants were shut down including 7 nuclear power plants. A total of 200 power plants eventually went off-line.

EXERCISES

1. What is a spinning reserve?
2. State two advantages of interconnecting power pools.
3. State some differences between the power grid and the Internet.
4. What is the advantage of a network grid over a radial grid?
5. What is the advantage of connecting pools at different time zones?
6. Search the Web and the literature to find the capacity of the ERCOT power pool in the United States. Find the area covered and the size of the population served.
7. Search the Web and the literature and find the largest power pool in the United States.
8. Power system in Figure 14.5 has the following data:
   Impedance of load$_1$ plus its transformer is $\overline{Z}_{11} = 50 + j5\ \Omega$
   Impedance of load$_2$ plus its transformer is $\overline{Z}_{22} = 40 + j2\ \Omega$

Impedance of line$_1$ is $\overline{Z}_{13} = 0 + j4 \ \Omega$

Impedance of line$_2$ is $\overline{Z}_{23} = 0 + j5 \ \Omega$

Impedance of line$_3$ is $\overline{Z}_{12} = 0 + j3 \ \Omega$

All impedances are referred to the high-voltage side of the transformers. The voltage at the high-voltage side of xfm$_g$ is fixed at 500 kV (line to line). The capacity of any line (maximum current the line can carry) is 15 kA. Assume that all connections are in wye. Compute the power delivered to each load and the power produced by the generator

9. For the system in the previous problem assume that transmission line$_3$ is tripped; repeat the solution. Is the system secure?

10. Load demand for a given day can be approximated by

$$P = 2 + 2e^{\frac{-(t-9)^2}{8}} \ \text{GW}$$

where t is the time of the day in hours using the 24 h clock. Compute the following:

(a) Peak demand

(b) Time of the peak demands

(c) Average daily demand

11. For the system in the previous problem, assume that the generation capacity is 1.9 GW. Compute the imported energy to compensate for all demands higher than 1.9 GW.

12. Power pool has a generation capacity of 1.5 GW and a demand of 1 GW. The trade commitments of the pool are $P_{\text{import}} = 500$ MW and $P_{\text{export}} = 800$ MW. Compute the spinning reserve that maintains the pool margin at 130%.

15 Future Power Systems

Thomas Edison and Nikola Tesla invented the power system. Their system, however, has gone through tremendous developments and improvements over the past century. Today's power system is probably the most complex system known to man. It is immense in capacity, huge in size, complex in operation, and fast in response. Despite the various blackouts we experience every few years, the reliability of the power system is extremely high considering the complexity and enormous size of the system.

The question often asked is what the future power system would look like. Is it going to be the same system we know today, but more expanded and more complex? Will we use the same energy resources? Will the energy resources last for future generations? Will we keep locating our generating plants near cities? Will we continue polluting the environment? Will we be able to harvest electricity from the natural dynamics of earth? Will we be able to provide electricity to remote places on earth and space?

These are tough questions, which may require a crystal ball to answer. Nevertheless, researchers are working hard to develop what they perceive as the power system of the future. The following are some of the ideas and concepts that are being researched. Some of them are in the conceptual phase and others are currently being implemented.

15.1 LESS-POLLUTING POWER PLANTS

Since Thomas Edison built the first power plant by the end of the nineteenth century, fossil fuel has been the main resource of generating electricity. As we have seen in this book, burning fossil fuels to produce the steam needed to generate electricity also introduces polluting agents into the environment. The health effects and environmental damages due to industrial wastes are well studied, monitored, and documented in most of the world. By the middle of the twentieth century, these studies created strong awareness among the public, which led to the creation of several environmental advocacy groups and agents. The public no longer accepts industrial processes that harm the environment.

One of the most polluting fossil fuels is coal, which accounts for over 25% of the electricity generated worldwide and over 50% in the United States and China. The wide use of coal is due to its abundance and low cost; coal is probably the main reason for the low price of electricity in most parts of the world. However, because of the environmental concern regarding coal burning, the public are demanding that scientists and engineers improve the burning processes of fossil fuels to meet the various goals set by several health and environmental organizations as well as international treaties.

The research in the area of clean burning of fossil fuels is rapidly progressing. Scientists believe that it is possible in the future to make fossil fuels such as coal burn at near-zero emissions. To achieve this goal, several technologies are being developed to make future coal burning systems dramatically more efficient and cleaner than they are today. Besides the technical challenges, the new processes must keep electricity affordable to ensure compliance in countries with growing economies. Among the new technologies are coal gasification, advanced combustion systems, and carbon sequestration.

Coal gasification is a process by which coal is converted into combustible gases known as synthesis gas (syngas). During the gasification process, coal reacts with pure oxygen and steam at an elevated pressure to produce gas with varying amounts of carbon monoxide, carbon dioxide, hydrogen, methane, and nitrogen. The various components of the syngas can be used as fuels or raw materials for chemical processes and fertilizers.

Carbon sequestration refers to the long-term storage of carbon dioxide (CO_2). Since CO_2 is one of the main greenhouse gases, it would seem reasonable to capture it and trap it in storage facilities. Scientists are recommending storing the CO_2 underground or in oceans. Some researchers are even proposing the terrestrial biosphere as a storage place for CO_2.

Hydroelectric power plants, which are renewable energy systems, have some negative environmental impacts as discussed in Chapter 5. In most countries, current hydroelectric systems and future developments must comply with several environmental laws. The main concern with hydroelectric systems is their blockage of water flow, which negatively affects the fish population and the biota of the river. To alleviate this problem, several techniques must be implemented. These techniques, though some of them are still under research, include building fish ladders to facilitate fish migration, redesigning turbine blades to allow small fish to pass through without injuries, optimizing the operation of the power plant to maintain the water flow and reservoir height to the levels needed to protect spawning and incubation of fish, and compensating for fish habitat losses through restoration or replacement.

Besides solving the fish problem, research is needed to reduce the effects of coastline erosions stemming from the lack of silt on seashores. Silt management is also important to farmland on the banks of the rivers as silt revitalize the soil. Oxygen depletion that kills deepwater fish behind the dam and the increase of nitrogen downstream from the dam are two other areas of research.

It has been wrongly perceived that implementing rigid environmental enhancement measures may lead to higher costs and fewer jobs. In fact, studies have shown that the extra cost added to the products due to the compliance with environmental laws is offset by the reduction in medical expenses. Indeed, numerous studies show that health has been improved in areas where pollution was reduced. Furthermore, reducing the pollution is an industry by itself. It requires the skills of engineers, chemists, physicists, mathematicians, biologists, meteorologists, hydrologists, geologists, medical doctors, several government agencies, and many more. This field is wide open to new ideas and methods.

15.2 ALTERNATIVE RESOURCES

Because of the environmental problems associated with the existing forms of energy resources (fossil, hydroelectric, and nuclear), researchers are looking for other resources that are less polluting and have less impact on the environment. Among the alternative resources are the energy of the sun, wind, tide, geothermal, biomass, hydrogen, and others. Some of these technologies are in the early stages of development, and much work is needed to make them viable for serious consideration. Others are further along in their development, but more work is needed to increase their efficiency and reliability and make them cost-effective.

Among the highly researched areas in energy resources is hydrogen, which is used to generate electricity using fuel cells. However, before hydrogen becomes competitive with other fossil fuels, serious technical challenges must be addressed to transform the oil-based economy to a hydrogen-based economy. Engineers need to develop systems that produce abundant amounts of hydrogen at affordable costs. They also need to develop extensive infrastructure to store and distribute hydrogen, which is safe and economical.

15.3 DISTRIBUTED GENERATION

The power plants we know today are located in areas where energy resources exist or can be easily accessed. This necessitates the construction of large regional generating plants, which is probably the most cost-effective arrangement. To transmit power to users, power lines hundreds of miles long are needed. The reliability of this power system is highly dependent on transmission systems; if the is tripped off, the service is interrupted.

An alternative method is to install distributed generation systems where the power plants are located in neighborhoods or even at homes. For example, since natural gas can reach a large number of homes, it can be used to generate electricity at homes by using small size thermal power plants—the size of regular water heaters.

Another idea is to use fuel cells to generate electricity for home use. These devices use hydrogen or natural gas to produce electricity without introducing pollution to the immediate environment.

The concept of distributed generation should make electricity more reliable and more available. With this technology, blackouts that last for days will no longer affect large areas.

15.4 POWER ELECTRONICS

Electrical apparatus and equipment are often designed to provide optimum performance at a narrow range of loading conditions. When these devices operate at different conditions, their efficiency is normally reduced. To address this problem, the voltage and frequency must be adjusted continuously using semiconductor power electronic devices. Indeed, most of the appliances we use today are much more efficient than their predecessors. Every year, newer power electronic devices and circuits are introduced, which enhance the operation of various equipment and improve their overall efficiency.

In addition to the enhancement of efficiency, power electronic devices make possible the development of technologies such as fuel cells, photovoltaic, superconducting magnetic energy storage systems, and magnetic-levitated systems. Power electronic devices allow us to use motors and actuators in applications that were never envisioned before such as angio-surgeries, micro-technology, and mechatronics.

In high-power applications, power electronic devices are used to control the bulk power flow in transmission lines through devices such as the flexible alternating current (ac) transmission system (FACTS) and voltage compensators. In addition, power electronics allow us to monitor systems, process information, and take continuous control actions.

15.5 ENHANCED RELIABILITY

Although power systems are highly reliable, we tend to disagree with this notion when a blackout occurs. This is because we have become so dependent on electric power that blackouts are often costly and highly annoying. To totally eliminate blackouts is probably an unrealistic task. However, reducing the frequency and impact of blackouts could be possibly achieved.

At any given moment, the power system is moving enormous amounts of power through its various components. For stable operation, a delicate balance must always be maintained between demand and generation. If this balance is altered, blackouts could occur unless the balance is rapidly restored. The speed with which the energy balance is restored depends on the availability of fast and synchronized sensing, fast-acting relays and breakers, quick topology restoration, and effective system control. Although the research in these areas is continuously advancing, predicting all possible scenarios for blackouts is a very challenging task. Furthermore, to be able to analyze and control the system online requires much more powerful computers and better knowledge of complex system techniques than we have attained so far.

The reliability of the power system also depends on the redundancy of the transmission systems. However, due to public concern, fewer and fewer transmission systems are allowed to be constructed. Public resistance to the construction of new lines is based on ecstatic considerations and concern regarding magnetic fields. Newer designs and technologies are needed to address these two issues.

15.6   MONITORING AND CONTROL OF POWER SYSTEMS

One of the most critical aspects of controlling our massive power system is to sense and synchronize various signals measured at locations thousands of miles apart. Regular terrestrial communication networks have shortcomings with the exact synchronization of high-frequency signals. In ongoing research, engineers are considering using geosynchronous (e.g., GPS) and low-earth orbit (LEO) satellites for better monitoring and control of the power systems. This is one of the intriguing ideas that will eventually lead to more effective and more reliable power system operation.

15.7   INTELLIGENT OPERATION, MAINTENANCE, AND TRAINING

The human presence in hazardous environments, such as in nuclear power plants, will eventually be replaced by intelligent autonomous robots. A swarm of these robots could provide the maintenance and operation inside the reactors where humans cannot be located. Robots can also be developed to crawl inside the millions of tubes to check the health status of power cables and equipment and even perform maintenance work.

The power system will eventually benefit from the virtual reality technology as it will make it possible for experts located miles away to perform maintenance on high-voltage equipment while it is energized, much like the remote surgery approach that is being developed in the medical field.

Virtual reality could also provide an excellent training system that could lead to more experienced operators. This is because various scenarios can be imposed on the system and the operators could learn how to steer the system to a secure and stable region. This is similar to the flight training simulators for aircraft pilots.

15.8   SMART GRID

Distribution networks in today's utilities play passive roles as they just receive the power from the transmission networks and deliver it to the loads (customers). Thus, the power flow in the distribution network is mostly unidirectional. The infrastructure of the current distribution networks, their protection and monitoring devices, as well as their control systems are all designed to operate in this passive environment. However, because renewable energy and distributed generation systems are often installed into the distribution networks, the networks must be redesigned to allow for more vibrant bidirectional operations and control environments. One of the concepts for the future distribution network infrastructure is called *smart grid*.

The smart grid is the vision for future distribution network, which combines all the technologies mentioned in this chapter and more to allow for the integration of renewable energy systems and distributed generation systems into the distribution network. The concept of the smart grid was conceived at the end of the twentieth century by European and American researchers. It is an ambitious structure where conventional and renewable energy generators, energy storage systems, demand-side management, wide-area monitoring, and intelligent control are all embedded into the distribution network.

Figure 15.1 shows an artistic depiction of a smart grid. The renewable energy systems are connected directly into the distribution network. Renewable resources such as hydrogen have their own infrastructure to power loads and fuel cell vehicles. This requires safe and efficient storage systems for the hydrogen as well as underground pipe systems to transmit the gas. When the electrical energy demand is low, the hydrogen can be generated by the renewable energy resources through, for example, electrolysis processes. Surplus electric energy can be stored in super capacitors (at the bottom left of the figure) for long periods with minimum losses. In addition, thermal storage can be used to directly heat buildings. This could be a simple system where rocks are heated when electricity is cheap or abundant. The heat is then extracted when electricity is expensive. The customers can also have their equipment controlled by the utility where at peak demands, the unessential loads are disconnected. This way, customers receive lower energy bills and the

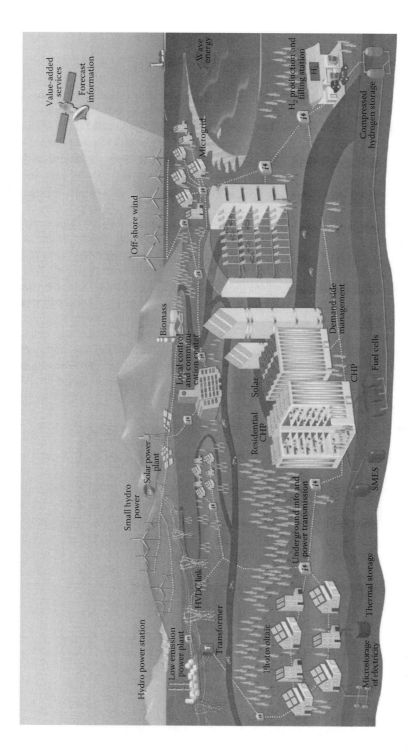

FIGURE 15.1 (See color insert following page 300.) Concept of smart grid. (Image courtesy of European Commission on Smart Grid, EUR22040 report.)

utility manages its resources much better. To optimize the operation of the smart grid, local and network information is collected and transmitted to a central control system where sophisticated algorithms operate the network efficiently and reliably at all times.

15.9   SPACE POWER PLANTS

Solar energy density in the outer space is much higher than that at the surface of the earth. The U.S. government is pioneering the conceptual idea of having massive solar power plants orbiting the earth. The envisioned system consists of football-field sized solar cells that could generate large amounts of power comparable to large terrestrial power plants. The generated electric power can be used to support spacecraft and long space missions as well as beam the energy down to the earth. The concept of a large-scale space solar power plant was considered improbable and silly in the 1970s. However, NASA and the National Science Foundation revived the idea by the end of the twentieth century and several research programs are working on developing the needed technologies.

A space power plant is compatible with the environment and it could lead to less dependency on the conventional energy resources. The plant could potentially provide electric energy to remote and inaccessible areas all over the world. However, the idea of having satellites beaming their energy toward earth has remarkable technological challenges. Beaming the energy to earth may require the use of microwaves and highly accurate positioning systems. The health risks associated with the exposure to electric and magnetic fields generated by beaming the power to earth is a significant public concern.

To make it possible to place football-field sized solar arrays in space, the solar cell technologies must improve substantially. The solar cells must be lightweight, thin, flexible, highly efficient, and able to withstand the harsh environment of space. Today's solar cells are not adequate because they are heavy, bulky, brittle, and have low efficiency. However, nanotechnology offers a potential solution to this problem.

FIGURE 15.2 (See color insert following page 300.)   International Space Station. (Image courtesy of NASA.)

Furthermore, space power stations located miles apart in orbits pose maintenance challenges. Among the ideas proposed is to use swarms of autonomous robots as maintenance crew. The research in this area is fascinating and highly challenging.

An example of a smaller-scale space power station can be found in the International Space Station (ISS) shown in Figure 15.2. The electrical energy in the ISS is used to operate the life-support systems as well as allowing the crew to operate the station and perform scientific experiments. The power system is integrated out of photovoltaic technology, which is much higher in efficiency and energy density than the commercial devices. The system includes storing devices to provide continuous power during the eclipse part of the orbit. The system is highly efficient as the heat generated during the conversion of sunlight into electricity is used in various other processes on the ISS.

EXERCISES

All questions in this exercise are research papers. Students are free to use external resources to address the issues:

1. What are the most important changes that need to be implemented to reduce the pollution associated with power plants?
2. Can coal be burned without polluting the environment? What are the latest technologies in reducing coal pollution?
3. If we replace all coal power plants in the world by wind energy systems, estimate how much reduction in CO_2 is achieved.
4. What are the problems associated with the cyclic and stochastic nature of some renewable energy systems?
5. Find a diagram of a simple distribution network and compare it with the system in Figure 15.1. Identify the key differences.
6. Can you think of ways to store electrical energy besides batteries?
7. What are the major challenges associated with building infrastructure for hydrogen?
8. What is intelligent control?
9. What are the advantages and disadvantages of demand-side management?
10. Why is extensive monitoring needed for the smart grid?
11. Identify the major technical obstacles facing the development of the space power system.
12. Assume that the space power system is a reality and identify its benefits to the society.
13. Can you direct microwaves and reduce their scattering?
14. Can you identify technologies that we need to develop for future power systems besides the ones discussed in this chapter?

Appendix A: Units and Symbols

Variable	Unit	Variable	Unit
Time t	Second (s)	Length l	Meter (m)
Electric current I	Ampere (A)	Electric potential E or V	Volt (V)
Electric resistance R	Ohm (Ω)	Resistivity ρ	Ω m
Capacitance C	Farad (F) or s/Ω	Inductance L	Henry (H) or Ω s
Charge q	Coulomb (C) or A s	Frequency f	Hertz (Hz) or cycles/s
Magnetic flux ϕ	Weber (Wb) or V s	Magnetic field density B	Tesla (T) or V s/m^2
Magnetic field intensity H	A/m	Reluctance $\Re$	A turn/Wb
Magnetomotive force $\Im$	A turn	Permeability μ	H/m or newton per square ampere (N/A^2)
		Electric real power P	Watt (W)
Electric reactive power Q	Volt-ampere reactive (V Ar)	Electric apparent power S	V A
Mechanical power P	Horsepower (hp)	Energy or work	W s or joule (J)
Linear speed v	m/s	Angular speed ω	Revolutions per second (r/s)
Linear acceleration	m/s^2	Angular acceleration	r/s^2
Force	Newton (N)	Torque	N m
Weight w	Gram (g)	Volume	m^3
Temperature	Degree Celsius ($^\circ$C) or kelvin (K)		

Appendix B: Conversions

PREFIX

Prefix	Value	Symbol	Prefix	Value	Symbol
yotta	10^{24}	Y	deci	10^{-1}	d
zetta	10^{21}	Z	centi	10^{-2}	c
exa	10^{18}	E	milli	10^{-3}	m
peta	10^{15}	P	micro	10^{-6}	μ
tera	10^{12}	T	nano	10^{-9}	n
giga	10^{9}	G	pico	10^{-12}	p
mega	10^{6}	M	femto	10^{-15}	f
kilo	10^{3}	k	atto	10^{-18}	a
hecto	10^{2}	h	zepto	10^{-21}	z
deca	10^{1}	da	yocto	10^{-24}	y

DISTANCE

1 mi = 1.609 km
1 yd = 0.91440 m
1 ft = 0.30480 m
1 in. = 0.02540 m

AREA

1 sq mi = 2.59 km^2
1 ha = 0.01 km^2
1 acre = 4046.9 m^2

VOLUME

1 cu ft = 0.02832 m^3
1 m^3 = 1.308 cu yd
1 m^3 = 1000 L
1 fl oz = 0.02957 L
1 U.S. gal = 3.78540 L
1 L of water = 1 kg
1 U.S. gal of water = 3.7854 kg

MASS

1 lb = 0.45360 kg
1 oz = 28.35 g
1 ton = 1000 kg

FORCE

1 lb/sq in. $= 7.03 \times 10^2$ kg/m^2
1 lb/sq in. $= 6.89$ kPa

SPEED

Speed of light $= 299{,}792{,}458$ m/s
1 kn $= 1.151$ mi/h
1 kn $= 1.852$ km/h

ENERGY

British thermal unit (BTU) $= 252$ cal
British thermal unit (BTU) $= 1.0544$ kJ (kW s)
1 N m $= 1$ W s

POWER

1 kW $= 1.34$ hp

TEMPERATURE

Kelvin $= °$Celsius $+ 273.15$
$°$Celsius $= (°$Fahrenheit $- 32) \times 5/9$

MAGNETIC FIELDS

1 T $= 10^4$ G

Appendix C: Key Parameters

Elementary charge constant $(q) = 1.602 \times 10^{-19}$ C

Boltzmann's constant $k = 1.380 \times 10^{-23}$ J/K

One mole of a substance contains 6.002×10^{23} entities (molecules, ions, electrons, or particles)

Avogadro number $N_A = 6.002 \times 10^{23}$/mol

Extraterrestrial solar power density $= 1.353$ kW/m^2

Standard atmospheric pressure at sea level $= 101,325$ Pa or N/m^2

Specific gas constant for air $= 287$ W s/(kg K)

Air permittivity $\varepsilon_o = 8.85 \times 10^{-12}$ F/m

Table of Resistivity

Aluminum	4.0×10^{-8} Ω m
Copper	1.673×10^{-8} Ω m
Nickel	10^{-7} Ω m
Steel	7.2×10^{-7} Ω m

Appendix D: Inductors

The inductor described in Chapter 10 can be designed using a ring core (toroid). The toroid is made of a magnetic material and a copper wire wrapped around the core as shown in the figure. The inductance of the toroid is directly related to the number of turns in the toroid, the geometry of the core, and the material used to build the core.

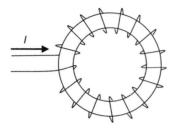

The inductance L is defined as the flux linkage λ produced by the current I

$$L = \frac{\lambda}{I} = \frac{N\phi}{I} = \frac{N^2\phi}{NI}$$

where
  N is the number of turns of the coil
  I is the current of the coil
  ϕ is the magnetic flux
  λ is the flux linkage

Since the reluctance $\Re$ of the core is defined as

$$\Re \equiv \frac{NI}{\phi}$$

hence,

$$L = \frac{N^2}{\Re}$$

The reluctance can be calculated using the geometry of the core and its permeability:

$$\Re = \frac{l}{\mu_o \mu_r A}$$

where
  A is the cross section of the core
  l is the circumference of the core
  μ_o is the absolute permeability ($\mu_o = 4\pi \times 10^{-7}$ H/m)
  μ_r is the relative permeability of the core material

Hence, the inductance of the toroid is

$$L = \mu_o \mu_r \left(\frac{A}{l}\right) N^2$$

Appendix E: Key Integrals

$$V_{\text{average}} = \frac{1}{2\pi} \int_0^{2\pi} v \, d\omega t$$

$$V_{\text{rms}} = \sqrt{\frac{1}{2\pi} \int_0^{2\pi} v^2 \, d\omega t}$$

$$\int e^{ax} \cos(bx) = \frac{e^{ax}}{a^2 + b^2} \left[a \cos(bx) + b \sin(bx) \right]$$

$$\int_{t=0}^{24} e^{\frac{-(t-t_0)^2}{2\sigma^2}} \, dt \approx \sqrt{2\pi} \sigma$$

$$\int e^{ax} \left[c \cos(bx) + k \sin(bx) \right] dx = \frac{e^{ax}}{a^2 + b^2} \left[(ka + cb) \sin(bx) + (ca - kb) \cos(bx) \right]$$

Index